Ultracold Atomic Physics

The field of ultracold atomic physics has developed rapidly during the last two decades. It currently encompasses a broad range of topics in physics, with a variety of important applications in topics ranging from quantum computing and simulation to quantum metrology, and can be used to probe fundamental many-body effects, such as superconductivity and superfluidity. Beginning with the underlying and including the most cutting-edge experimental developments, this textbook covers essential topics, such as Bose–Einstein condensation of alkali atoms, studies of BEC-BCS crossover in degenerate Fermi gas, synthetic gauge fields and Hubbard models, and many-body localization and dynamical gauge fields. Key physical concepts, such as symmetry and universality, highlight the connections between different systems, and theory is developed with plain derivations supported by experimental results. This self-contained and modern text will be invaluable for researchers, graduate students, and advanced undergraduates studying cold atom physics, from both a theoretical and experimental perspective.

Hui Zhai is Changjiang Chair Professor of Physics at the Institute for Advanced Study, Tsinghua University, and his research is focused on ultracold atomic physics, condensed matter physics, and machine learning. He was awarded a National Natural Science Foundation of China (NSFC) award for distinguished young scholars and the Rao Yutai Prize by the Chinese Physical Society.

Ultracold Atomic Physics

HUI ZHAI
Tsinghua University

CAMBRIDGE
UNIVERSITY PRESS

University Printing House, Cambridge CB2 8BS, United Kingdom

One Liberty Plaza, 20th Floor, New York, NY 10006, USA

477 Williamstown Road, Port Melbourne, VIC 3207, Australia

314–321, 3rd Floor, Plot 3, Splendor Forum, Jasola District Centre, New Delhi – 110025, India

79 Anson Road, #06–04/06, Singapore 079906

Cambridge University Press is part of the University of Cambridge.

It furthers the University's mission by disseminating knowledge in the pursuit of
education, learning, and research at the highest international levels of excellence.

www.cambridge.org
Information on this title: www.cambridge.org/9781108498685
DOI: 10.1017/9781108595216

© Hui Zhai 2021

First published 2021

Printed in the United Kingdom by TJ Books Limited, Padstow Cornwall

A catalogue record for this publication is available from the British Library.

Library of Congress Cataloging-in-Publication Data
Names: Zhai, Hui, author.
Title: Ultracold atomic physics / Hui Zhai.
Description: New York: Cambridge University Press, 2021. | Includes index.
Identifiers: LCCN 2020039636 (print) | LCCN 2020039637 (ebook) | ISBN
9781108498685 (hardback) | ISBN 9781108595216 (ebook)
Subjects: LCSH: Nuclear physics. | Electrons.
Classification: LCC QC770 .Z522 2020 (print) | LCC QC770 (ebook) | DDC 539.7–dc23
LC record available at https://lccn.loc.gov/2020039636
LC ebook record available at https://lccn.loc.gov/2020039637

ISBN 978-1-108-49868-5 Hardback

Additional resources for this publication at www.cambridge.org/zhai.

Cambridge University Press has no responsibility for the persistence or accuracy of
URLs for external or third-party internet websites referred to in this publication
and does not guarantee that any content on such websites is, or will remain,
accurate or appropriate.

Contents

Preface

Since 1995, the field of ultracold atomic physics has been developing very quickly, and it is still expanding. The physics of ultracold atomic systems has covered a broad range of topics and has had an impact on several other fields, such as condensed matter physics, quantum information and computation, nuclear physics, and high-energy physics. This textbook tries to cover most of the major achievements of ultracold atomic physics in the past 25 years, although it is not possible to cover all of them. These achievements range from the early-stage developments, such as Bose–Einstein condensation of alkali atoms, to the studies of BEC-BCS crossover in degenerate Fermi gas, synthetic gauge fields, and the Hubbard models, and recent progress, such as many-body localization and dynamical gauge fields. To cover these topics, the book consists of four parts. Full-color versions of certain figures can be found in the resources tab for this book at cambridge.org.

- Part I introduces basic atomic physics relevant to ultracold atomic systems, in order to be self-contained. The part consists of two chapters, one on single-atom physics, such as atomic structures and atom–light interaction, and the other on two-body collision physics. This part provides basics for readers to understand, for instance, how to trap and manipulate ultracold atoms with light and how to tune the interaction by magnetic field. It is precisely these control tools that make the ultracold atomic physics possible. When discussing atomic structure and two-body collision, we not only cover the widely used alkali-metal atoms but also introduce the alkaline-earth-metal atoms, which have been used by more laboratories in recent years. When discussing atom–light interaction, we not only the scalar light shift, which is the basics mechanism for optical trapping and optical lattices, but also vector light shift and STIRAP, which are essential for generating a synthetic gauge field and creating ultracold molecules. We also extend the discussion of the two-body problem to the three-body problem, where the famous Efimov effect has been extensively studied in ultracold atomic systems.

- Part II is about interacting Bose gas. This part consists of two chapters, one focusing on the interaction effect and the other focusing on topology and spin effects. The interaction effect is mainly about Bose condensate and superfluidity. One exception is the one-dimensional system, where we highlight that the interaction effect is so strong that it destroys condensate. The topology effect mainly concentrates on topological defects in a Bose–Einstein condensate, both spinless and spinful. Finally, we also discuss the spin-orbit coupling effect, arising from the synthetic gauge field, in a Bose–Einstein condensate. The spin-orbit coupling effect has been studied extensively in electronic systems in condensed matter physics, but in ultracold atomic systems, it is the first time

that the spin-orbit coupling effects are studied in the Bose system, which has been a
major topic in ultracold atomic physics in the past decade.

- Part III is about Fermi gas. This part also consists of two chapters, one on Fermi liquids
and the other on Fermi superfluids. For Fermi liquids, we use polaron as an example to
discuss a number of basic qualities and their universal relations in a Fermi liquid, and
we use quantum point contact as an example to discuss the transport property of Fermi
liquids. Both polaron and quantum point contacts have been focused experimental topics of ultracold atomic physics over the past 10 years or so. For Fermi superfluid, we
first introduce the basics of the BCS theory, also for the purpose of being self-contained.
Then, we generalize the BCS theory to discuss the BEC-BCS crossover across a Feshbach resonance. We both introduce the theoretical concepts and describe the crossover,
and we also review the representative experimental results for the crossover.

- Part IV is about lattice physics. Part II and Part III consider uniform systems, and this
part considers lattice effects by applying optical lattices to ultracold atoms. This part
also consists of two chapters, one on the noninteracting band effect and the other on
the interacting effect. The noninteracting band effect mainly focuses on various kinds of
topological bands, including how to realize such topological bands and how to reveal the
unique physical effects of topological bands in ultracold atomic systems. The interaction effect mainly focuses on Bose and Fermi Hubbard models, and we also discuss the
interplay between interaction and disorder potential, which has led to the new developments in many-body localization seen in the past 10 years. Being an isolated system, an
ultracold atomic gas is an ideal platform for experimental studies of many-body localization, and so far, most experiments about many-body localization have been carried
out in ultracold atomic systems.

When I selected and organized the topics for this book, I paid special attention to the
following considerations. I hope that, with these considerations, this book is accessible
for most readers, especially for experimentalists; for junior researchers, including senior
undergraduate students; and for readers outside the field of ultracold atomic physics.

- A few key physics concepts are emphasized throughout the book for example, symmetry and universality. Many studies in ultracold atomic physics have illustrated the
importance and power of these concepts. I hope that by introducing these examples, the
book can also benefit readers outside the field of ultracold atomic physics.
Symmetry plays a crucial role in many physics discussions, and it is one of the key concepts that we continually highlight in this book. For example, first of all, the concepts
of the symmetry of Hamiltonian and the symmetry of the state are discussed in Sections 1.1, 3.5, and 4.5, which lead to the relation between symmetry and degeneracy,
as well as the concept of symmetry breaking. In Section 4.5, we have also emphasized
how these concepts can help us understand the orders of phase transition. In Section 8.1,
these concepts are revisited by introducing the concept of emergent symmetry. Second,
the relation between symmetry and topology, especially the symmetry-protected topological phenomenon, is introduced in Sections 7.2 and 7.3. Third, in Section 8.2, we
discuss another use of symmetry, that is, two different systems are related by a symmetry,
and how this can help us understand one system with the knowledge of the other system.

Here the example is the Fermi Hubbard model, where the repulsive Fermi Hubbard is related to the attractive Fermi Hubbard model by the particle-hole symmetry. Finally, a special symmetry, known as the scaling symmetry, is encountered several times in the discussion of the Efimov effect in Section 2.6, of the Tonks–Girardeau gas in Section 3.4, and of the unitary Fermi gas in Section 6.2.

Universality is another important concept in physics, which states that many microscopically different systems can share the same low-energy physics described by very few parameters. We have discussed several such examples in this book. In Section 2.2, the low-energy scattering of different interatomic potentials can be universally described by the s-wave scattering length. In Section 5.1, different Fermi liquids can be described by a few parameters known as the Fermi liquid parameters. In Section 8.1, the quantum critical regime at different microscopic models can be described by a very few critical exponents.

- Connections between different physics contents are highlighted. Many seemingly different physics can have connections, sometimes because of a common mathematical structure behind them. For example, the synthetic gauge field has become a major topic of study since about 2010 in ultracold atomic physics, created by the atom–light interaction; however, this effect actually already existed in the magnetic trapping of atoms, even prior to the birth of the field. This connection is explicitly discussed when I introduce magnetic trapping in Section 2.1. Other examples include discussion of topology and mean-field theory. For topology, in Sections 4.2 and 4.4, we discuss various kinds of topological defects in a Bose–Einstein condensate, and in Sections 7.2 and 7.3, we discuss various kinds of topological band structures for noninteracting fermions. These two are different physics, but they share the same mathematical descriptions. For mean-field theory, we discuss the BCS mean-field theory for fermions in Section 6.1 and the mean-field theory for the Bose–Hubbard model in Section 8.1. The physics of these two systems are also very different, and these two mean-field theories also look quite different. However, there are common physical insights behind these two, which are highlighted in Section 8.1. In this book, we use many boxes to discuss the concepts and connections across different chapters. By building up connections between different physics phenomena, we hope this can help readers to understand the physics more deeply.

- Discussions of theories are always supported by experimental results in ultracold atomic systems. If one looks at the literature on ultracold atoms, there are a lot more theory papers than experimental works. When I choose topics for this book, aside from discussing some open issues, I only selected those theories that have been confirmed by experiments. All the discussions of theories are supported by experimental results. Though I do not go into the experimental details, I hope this can provide readers with direct physical pictures and intuitions. In addition, for the discussion of theories, I try to use the back-of-envelope calculations, though sometimes being less rigorous. I try to avoid using advanced theoretical tools, such as Green's functions and field theory approaches. Students with quantum mechanics and statistical mechanics backgrounds should be able to understand most parts of the results. I hope this can make this book friendly to students, including senior undergraduate students and those mostly focusing on experiments.

I have taught a course on Ultracold Atomic Physics at Tsinghua University, Beijing, since 2012. For the past eight years, every year, about 80–100 students attend this course. More than half of them are not from Tsinghua University but rather from other universities and institutions in the Beijing area, or even from other cities. Although they do not get grades from the course, they attend all the lectures because of their interest in the physics. They ask many excellent questions in and after class. These questions help me improve my lecture notes, which have led to this book as it stands now. Here I give special thanks to all of them. This book would not have been possible without their enthusiasm in the course.

In the past 10 years, I have enjoyed fruitful collaborations with my previous students and postdocs, and many things of which I have written in the book I learned from them during these collaborations. The book would also not have been possible without their contributions. I thank them for many valuable suggestions and for help in finalizing the manuscript: Chao Gao, Zheyu Shi, Pengfei Zhang, Yanting Cheng, Ran Qi, Zeng-Qiang Yu, Wei Zheng, Yu Chen, Ren Zhang, Boyang Liu, Mingyuan Sun, Zhigang Wu, and Juan Yao. Most of them are currently already faculty in different universities and institutions in China. I wish them great success in their careers.

I sincerely thank my thesis advisor, Professor Chen-Ning Yang, who brought me into the field of ultracold atomic physics nearly 20 years ago. As Professor Yang always said, it is good luck for someone to grow up together with a young field. His taste in physics and style of doing research, his guidance and encouragement, have had an important impact on my scientific career. This book is a special gift to Professor Yang's coming one-hundredth birthday. I thank Professor Tin-Lun Ho and many other senior scientists for their support over all these years. The Institute for Advanced Study of Tsinghua University (IASTU) provides a very special academic environment, and I am very glad that I can carry out my research there.

Last, but not least, I am grateful to all my family members for their support all these years. Together with my wife, it was great fun to watch two kids grow up during the eight years over which I wrote this book.

Part I

Atomic and Few-Body Physics

1   A Single Atom

Learning Objectives

- Discuss the Coulomb interaction, the spin-orbit coupling, and the hyperfine coupling as the three effects that determine atomic structure.
- Highlight the importance of the separation of energy scales of these three effects.
- Introduce different atomic structures of alkali-metal, alkaline-earth-metal, and magnetic atoms.
- Introduce the long-lived excited states in alkaline-earth atoms, and their applications, such as to atomic optical clocks.
- Discuss the Zeeman structure of atoms in a magnetic field.
- Discuss the idea of magnetic trapping, which can naturally lead to the emergence of a synthetic gauge field.
- Introduce the scalar light shift and its applications, such as laser trapping, optical lattices, and laser cooling.
- Introduce the vector light shift and its applications, such as the light-induced Zeeman field and synthetic spin-orbit coupling.
- Discuss the synthetic spin-orbit coupling and various kinds of gauge fields generated by the vector light shift.
- Introduce the basic idea of the stimulated Raman adiabatic passage.

1.1 Electronic Structure

Let us first consider a general Hamiltonian of Z electrons moving around a nucleus that contains the Coulomb interaction, the spin-orbit coupling, and the hyperfine coupling. These are the three effects that determine the electronic structure of an atom. Here we should emphasize the important role of the separation of energy scales; that is to say, the typical energy scales of these three terms are quite different. Thanks to the separation of energy scales, we can analyze them one by one, which enables us to obtain a clear picture of the electron structure.

Coulomb Interaction between Electron and Nucleus. Each electron moves around the nucleus with an attractive Coulomb interaction between the electron and the nucleus, which is described by

$$\hat{H}_0 = \sum_{i=1}^{Z} \left(-\frac{\hbar^2 \nabla_i^2}{2m^*} + V_{\text{ei}}(\mathbf{r}_i) \right),\qquad(1.1)$$

where $i = 1, \ldots, Z$ labels the electrons; $\mathbf{r}_i$ labels the coordinate of electron centering at the nucleus; $m^* = mM/(m+M)$ is the reduced mass, where m is the electron mass and M is the nucleus mass; $V_{\text{ei}}(\mathbf{r}) = -Z\kappa/r$ is the Coulomb potentials between the electron and the nucleus, where $\kappa = e^2/(4\pi\epsilon_0)$; e is the electron charge; and ϵ_0 is the vacuum permittivity. The eigenstates are characterized by three quantum numbers (n, l, m). Usually for the spherical symmetric potential, because of the $SO(3)$ rotational symmetry, the energy spectrum only depends on n and l and does not depend on m. However, for the $1/r$ Coulomb potential, such as with a hydrogen atom, the eigenspectrum is

$$E = -\frac{m^* Z^2 \kappa^2}{2\hbar^2 n^2},\qquad(1.2)$$

which only depends on the principal quantum number n and is independent of angular momentum quantum number l; l can take integer values from 0 to $n - 1$. This extra degeneracy is a consequence of $1/r$ potential, which leads to an $SO(4)$ symmetry larger than the three-dimensional rotational symmetry [102]. The separation of these energy levels is of the order of electron volts ($\sim 10^{14}$Hz) because it originates from the Coulomb interaction. The energy levels, usually named as the term-diagrams, are schematized in Figure 1.1, where $1, 2, 3, \ldots$ label the principal quantum number n and $s, p, d, \ldots$ represent the angular momentum quantum number l. The term-diagram for a hydrogen atom is shown in Figure 1.1(a).

Coulomb Interaction between Electrons. The repulsive Coulomb interaction between electrons is given by

$$\hat{V}_{\text{c}} = \sum_{i<j} V_{\text{ee}}(\mathbf{r}_i - \mathbf{r}_j),\qquad(1.3)$$

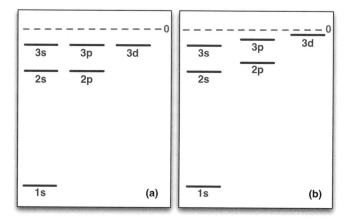

Figure 1.1 Schematic of the term-diagram: (a) the hydrogen atom without screening effect and (b) an alkali-metal atom with screening effect. A color version of this figure can be found in the resources tab for this book at cambridge.org/zhai.

where $V_{ee}(\mathbf{r}) = \kappa/r$. Here we discuss a couple of physical consequences of this term. First, the inner electrons of the fully filled levels screen the positive Ze charge of the nucleus, and thus, the valence electron experiences a reduced Coulomb potential. When the electronic orbit of the outermost electron is far from the nucleus, approximately, it experiences a fully screened field of all the rest of the $Z - 1$ electrons. That is to say, for large enough r, the effective attraction between the electron and the nucleus becomes a Coulomb potential with effective charge unity, that is, $-\kappa/r$. The closer this electron approaches toward the nucleus, the more it experiences the unscreened nuclear potential with charge Ze. The attraction between the electron and the nucleus recovers $-Z\kappa/r$ for sufficiently small r. Therefore the effective potential seen by the valance electron is no longer proportional to $1/r$, and hence, the enlarged $SO(4)$ symmetry no longer exists. Consequently, the eigenstates with the same n but different l are no longer degenerate, and the energy level becomes

$$E = -\frac{m^* Z^2 \kappa^2}{2\hbar^2 (n - \delta(n,l))^2}, \tag{1.4}$$

where $\delta(n,l)$ is a function depending on n and l and is also called the "quantum defect" [157]. The term-diagram with the screening effect is schematized in Figure 1.1(b). Normally, the energy level with larger l becomes higher. This energy splitting is also of the order of electron volts, because it also originates from the Coulomb interaction.

Second, let us consider two electrons in two orbits, say, $\psi_1(\mathbf{r})$ and $\psi_2(\mathbf{r})$. Because the total wave function of two electrons has to be antisymmetric, and if these two electrons form a spin singlet, the wave function in the spin space is antisymmetric, and their spatial wave function has to be symmetric, that is, $\psi_1(\mathbf{r}_1)\psi_2(\mathbf{r}_2) + \psi_1(\mathbf{r}_2)\psi_2(\mathbf{r}_1)$. If these two electrons form a spin triplet, the wave function in the spin space is symmetric, and their spatial wave function should be antisymmetric, that is, $\psi_1(\mathbf{r}_1)\psi_2(\mathbf{r}_2) - \psi_1(\mathbf{r}_2)\psi_2(\mathbf{r}_1)$. In the latter case, the wave function vanishes when two electrons come close enough, which reduces the repulsive interaction energy. Thus, the energies of the triplet states are lower than the energy of the singlet state. In other words, the Coulomb repulsion favors the total spin S of electrons to be maximized. This argument can be generalized to cases with more than two electrons and to cases with more than two quantum states, which gives the early day explanation of the first Hund's rule.[1] Also, for a given S, the short-range repulsion is minimized when the total angular momentum L is maximized, which gives the second Hund's rule. The characteristic energy scale of the Hund's rules is also of the order of electron volts.

The Spin-Orbit and Hyperfine Couplings. The Hamiltonian for the spin-orbit coupling is given by

$$\hat{H}_{so} = \sum_i \alpha_f^i \hat{\mathbf{S}}_i \cdot \hat{\mathbf{L}}_i, \tag{1.5}$$

---

[1] There are more advanced discussions of the origin of the first Hund's rule in later quantum chemistry calculations that we will not discuss in detail here.

and it describes the coupling between the electronic spin $\hat{\mathbf{S}}_i$ and its orbital angular momentum $\hat{\mathbf{L}}_i$ with strength $\alpha_{\mathrm{f}}^i{}^2$, giving rise to the fine structure. The origin of the spin-orbit coupling can be intuitively understood as follows. Sitting in the rest frame of an electron, the nucleus moves around the electron. Because the nucleus is charged, the circulating motion of the nucleus gives rise to an electric current, and the strength of the current is proportional to the angular momentum of the relative motion between the electron and the nucleus. The circulating current further induces a magnetic field, which acts on the spin of electrons. This leads to the spin-orbit coupling given by Eq. 1.5.

As one can see from this picture, because this process involves the magnetic effect induced by the electric current, it is naturally weaker than the Coulomb interaction, because the latter is purely electronic. In fact, the characteristic energy scale of the spin-orbit coupling is typically of the order of 10^{-3}eV ($\sim 10^{11}$Hz), and in many cases it is much weaker than the Hund's rule coupling originating from the Coulomb interaction. Originally, this spin-orbit coupling is between the spin and orbital angular momentum of each individual electron; however, because the Hund's rule coupling locks the electron spins of all valance electrons to an eigenstate of the total electron spin $\hat{\mathbf{S}}$, and locks the angular momentum of all valance electrons to an eigenstate of the total angular momentum $\hat{\mathbf{L}}$, it is more convenient to express the leading order effect of the spin-orbit coupling in terms of $\hat{\mathbf{S}}$ and $\hat{\mathbf{L}}$ as $\alpha_{\mathrm{f}}\hat{\mathbf{S}} \cdot \hat{\mathbf{L}} + \cdots$. Here the first term represents the coupling between $\hat{\mathbf{S}}$ and $\hat{\mathbf{L}}$ with strength α_{f}, which is called the LS coupling. The residual terms represented by $\cdots$ denote the difference between the actual coupling (Eq. 1.5) and the LS coupling term. Because S and L are not really good quantum numbers for Eq. 1.5, these residual terms compete with the Hund's rule and can change the quantum number S and L. Nevertheless, $\hat{\mathbf{J}} = \hat{\mathbf{S}} + \hat{\mathbf{L}} = \sum_i \hat{\mathbf{J}}_i$ still commutes with this coupling.

The hyperfine interaction couples the electronic degrees of freedom $\hat{\mathbf{S}}$ and $\hat{\mathbf{L}}$ to the nucleus spin $\hat{\mathbf{I}}$. In general, $\hat{\mathbf{S}}$ and $\hat{\mathbf{L}}$ couple to $\hat{\mathbf{I}}$ differently. Nevertheless, the characteristic energy scale for the hyperfine coupling is of the order of 10^{-6}eV ($\sim 10^8$–10^9Hz), which is much smaller compared with the spin-orbit coupling. This is because the nuclear magneton is much smaller than the Bohr magneton. Since the LS coupling already locks $\hat{\mathbf{S}}$ and $\hat{\mathbf{L}}$ to an eigenstate of $\hat{\mathbf{J}}$, we express the leading order effect of the hyperfine coupling in terms of $\hat{\mathbf{J}}$ and $\hat{\mathbf{I}}$ as $\alpha_{\mathrm{hf}}\hat{\mathbf{J}} \cdot \hat{\mathbf{I}} + \cdots$, where α_{hf} is the strength of this coupling. This gives rise to the hyperfine structure. Only with the first term, J is still a good quantum number, but the residual term represented by $\cdots$ can change the quantum number J, which is due to $\hat{\mathbf{S}}$ and $\hat{\mathbf{L}}$ coupled to $\hat{\mathbf{I}}$ differently.

Zoo of Ultracold Atoms. So far, three classes of atoms have been cooled to quantum degeneracy in cold atom experiments. They are (1) alkali-metal atoms, including hydrogen (H), lithium (Li), sodium (Na), potassium (K), rubidium (Rb), and cesium (Cs); (2) alkaline-earth-metal (-like) atoms, including strontium (Sr), calcium (Ca), and ytterbium (Yb). In the periodic table, ytterbium does not belong to the alkaline-earth-metals, but its outer electronic structure is the same as alkaline-earth-metal atoms; and (3) atoms with large electronic magnetic moments, which are called "magnetic atoms" here. These include chromium (Cr), dysprosium (Dy), and erbium (Er). We also anticipate that more atomic

---

^{2} In general, α_{f}^i should also depend on spatial position. Here we ignore this dependence for simplicity.

Table 1.1 The electronic structure and the nuclear spin of the alkali-metal atoms used in current experiments

Atom	Valance electron	Label $^{2S+1}L_J$	Nuclear spin I
Li	$2s^1$	$^2S_{\frac{1}{2}}$	^{7}Li ($I = 3/2$, B); ^{6}Li ($I = 1$, F)
Na	$3s^1$	$^2S_{\frac{1}{2}}$	^{23}Na ($I = 3/2$, B)
K	$4s^1$	$^2S_{\frac{1}{2}}$	^{40}K ($I = 4$, F); ^{39}K ($I = 3/2$, B); ^{41}K ($I = 3/2$, B)
Rb	$5s^1$	$^2S_{\frac{1}{2}}$	^{85}Rb ($I = 5/2$, B); ^{87}Rb ($I = 3/2$, B)
Cs	$6s^1$	$^2S_{\frac{1}{2}}$	^{133}Cs ($I = 7/2$, B)

Note: F denotes fermion, and B denotes boson. Here the symbol $^{2S+1}L_J$ labels the electronic structure of each atom.

species can be cooled to quantum degeneracy in the future. Here we will discuss the electronic structure and the spin structure at zero magnetic field of these three classes based on the aforementioned terms.

Alkali-Metal Atoms. So far, all atomic species in Table 1.1 have been cooled to quantum degeneracy, among which ^{87}Rb and ^{23}Na are the most-studied ultracold bosonic isotopes and ^{40}K and ^{6}Li are the most-studied fermionic isotopes. Following are a few key points about this class of atoms:

- In the ground state, since there is only one electron in the s-orbital, $S = 1/2, L = 0$, and $J = 1/2$. The ground state is always labeled by $^2S_{1/2}$. There is no spin-orbit coupling in the ground state, and the atomic spin structure is determined by the hyperfine coupling. The hyperfine spin is defined as $\hat{\mathbf{F}} = \hat{\mathbf{I}} + \hat{\mathbf{J}}$, and F is a good quantum number for an alkali-metal atom at the zero magnetic field. For instance, for ^{87}Rb, $I = 3/2$, so the total F can be either 1 or 2. At zero magnetic field, the energy splitting between $F = 1$ states and $F = 2$ states is a few times 10^9Hz.
- For the first excited state, the valence electron is the p-orbital, and thus $L = 1$. Due to the LS coupling, the total J can be either $1/2$ or $3/2$. Thus the excited states are split into $^2P_{\frac{3}{2}}$ and $^2P_{\frac{1}{2}}$, as shown in Figure 1.2(a). Historically, this splitting was discovered in the absorption spectra of lights due to sodium atoms, and they are named as D_1 and D_2 lines. Using sodium as an example, this fine splitting is 2.1×10^{-3}eV ($\simeq 5 \times 10^{11}$Hz), and the splitting between the ground state $^2S_{\frac{1}{2}}$ and these two states is about 2.1eV ($\simeq 5 \times 10^{14}$Hz). Because the fine-structure splitting is much smaller compared with the excitation energy, normally both $^2P_{\frac{3}{2}}$ and $^2P_{\frac{1}{2}}$ should participate in the optical transition, which are key processes for trapping and manipulating alkali-metal atoms, as we shall discuss in Section 1.3 in detail. In addition, $^2P_{\frac{3}{2}}$ and $^2P_{\frac{1}{2}}$ are further split by the hyperfine coupling, and the hyperfine splitting is even smaller compared with the fine-structure splitting.

Alkaline-Earth-Metal (-Like) Atoms. Table 1.2 contains the alkaline-earth-metal atoms (Ca and Sr) and alkaline-earth-metal-like atom (Yb) that have been cooled to quantum

Table 1.2 The electronic structure and the nuclear spin of alkaline-earth-metal (-like) atoms used in current experiments

Atom	Valance electron	Label $^{2S+1}L_J$	Nuclear spin I
Yb	$4f^{14}6s^2$	$^1S_0$	^{174}Yb ($I = 0$, B); ^{171}Yb ($I = 1/2$, F); ^{173}Yb ($I = 5/2$, F)
Ca	$4s^2$	$^1S_0$	^{40}Ca ($I = 0$, B)
Sr	$5s^2$	$^1S_0$	^{84}Sr ($I = 0$,B); ^{87}Sr ($I = 9/2$, F)

Note: F denotes fermion, and B denotes boson.

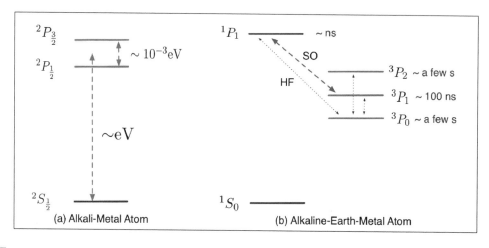

(a) Alkali-Metal Atom   (b) Alkaline-Earth-Metal Atom

Figure 1.2 Schematic of the electronic structure: (a) an alkali-metal atom and (b) an alkaline-earth-metal atom. The dashed line in (a) denotes the excitation energy, and the dashed lines in (b) denote that these states are coupled either by the spin-orbit coupling (SO) or by the hyperfine coupling (HF) process. A color version of this figure can be found in the resources tab for this book at cambridge.org/zhai.

degeneracy. Alkaline-earth-metal atoms have several unique properties compared with the alkali-metal atoms:

- For the ground state, two electrons occupy the s-orbital, and therefore, the total electron spin $S = 0$ and the angular momentum $L = 0$. All bosonic isotopes of alkaline-earth-metal atoms have zero nuclear spin, and the fermionic isotopes have nonzero nuclear spin I, which can be very large. However, because of $J = 0$, and consequently, the absence of the hyperfine coupling, the nuclear spin is decoupled from the electronic spin degree of freedom. Therefore, the nuclear spin nearly does not participate in two-body interactions, and the interaction possesses an $SU(N)$ symmetry with large-N [193]. We will discuss this in Section 2.3.

- For the first excited states, one electron still occupies the s-orbital, but the other electron is excited to the p-orbital. Thus, these states have $L = 1$. The electronic structure of these excited states is shown in Figure 1.2(b). First of all, because of the Hund's rule, the Coulomb energies for the $S = 1$ states (^3P_J) are lower than that for the $S = 0$ state

($^1P_1$). Second, due to the LS coupling, all states within the $S = 1$ manifold split into $J = 0$, 1, and 2, denoted by $^3P_0$, $^3P_1$, $^3P_2$, respectively.

- As we will see in Section 1.3, because the optical transition is dominated by the dipole transition, and the dipole transition does not change the electronic spin quantum number S, the direct coupling between these excited states with $S = 1$ (^3P_J) and the ground state with $S = 0$ is forbidden because of different quantum number S. That is to say, at the leading order, the dipole transition can only couple the ground state to the $S = 0$ excited state ($^1P_1$) and cannot couple the ground state to the $S = 1$ manifold (^3P_J).

- As we have mentioned above, the spin-orbit coupling term does not conserve the quantum number S and L, and it can mix two states as long as they have the same J. Thus, among the three states with $S = 1$, $^3P_1$ states can be coupled to $^1P_1$ states by the spin-orbit coupling. Through this coupling, there exists a small but finite dipole transition matrix element between the $^3P_1$ states and the ground state. This gives rise to a lifetime for $^3P_1$ states of about a few hundred nanoseconds. And for $^3P_0$ and $^3P_2$, because their quantum numbers J are different from that of $^1P_1$, they cannot be coupled to $^1P_1$ by the spin-orbit coupling term.

- For fermionic isotopes, the coupling between $^3P_0$ or $^3P_2$ and $^1P_1$ can be induced by the hyperfine coupling. As we have discussed above, after including the hyperfine coupling, J is also not a good quantum number. However, the coupling mediated by the hyperfine coupling is much weaker, and hence, the lifetimes of $^3P_0$ and $^3P_2$ states are much longer than for $^3P_1$ states, and the lifetime can be many seconds. These long-lived electronic excited states can be used as an important tool for precision measurement. On one hand, the spontaneous emission rates of these states are so small, and on the other hand, the coupling is not completely forbidden because of these residual couplings. Taking advantage of these properties, the transition between $^3P_0$ and $^1S_0$ induced by laser coupling can be used for the purpose of realizing the atomic optical clock. Therefore, these states are also called the "clock state." The atomic optical clock has reached an accuracy of 10^{-19}s nowadays, and it is the most accurate clock we have now [25]. If one were to start to run such a clock from the beginning of the universe until now, this clock would be expected neither to gain nor to lose even one second. Such a clock can now be used to test fundamental physics [92].

- For bosonic isotopes, due to the absence of the nuclear spin, there is absolutely no one-photon dipole transition for $^3P_0$ and $^3P_2$. In this case, the coupling to ground state has to be induced by higher-order processes. The lifetime of these two states can be many years long, and for all practical purposes, these states can be viewed as not decayed.

Magnetic Atoms. Table 1.3 contains three atoms whose total angular momentum of electron J is very large. For chromium, five d-orbitals and one s-orbital are all half-filled, and thus all six electrons are spin polarized because of the first Hund's rule, which gives $S = 3$ and $L = 0$. Dysprosium and erbium are open-shell lanthanide atoms. For dysprosium, 7 f-orbitals are filled by 10 electrons, and thus there are 4 unpaired electrons. Because of the first Hund's rule, these four unpaired electrons are spin polarized, which gives rise to a total electronic spin $S = 2$. And because of the second Hund's rule, these four unpaired electrons give maximized angular momentum $L = 6$. Similarly, for erbium, 7 f-orbitals

Table 1.3 The atomic structure of high-spin magnetic atoms like Cr and lanthanide Dy and Er

Atom	Valance electron	Label $^{2S+1}L_J$	Nuclear spin I
Cr	$3d^5 4s^1$	$^7 S_3$	^{52}Cr $(I = 0, B)$; ^{53}Cr $(I = 3/2, F)$
Dy	$4f^{10} 6s^2$	$^5 I_8$	^{162}Dy $(I = 0, B)$; ^{163}Dy $(I = 5/2, F)$
Er	$4f^{12} 6s^2$	$^3 H_6$	^{168}Er $(I = 0, B)$

Note: F denotes fermion, and B denotes boson.

are filled by 12 electrons, and thus there are 2 unpaired electrons, which gives $S = 1$ and a maximized angular momentum $L = 5$. Furthermore, it turns out that for both Dy and Er, the spin-orbit coupling favors a maximum J, that is, $J = 8$ for dysprosium and $J = 6$ for erbium. The atomic structures of these atoms also have strong effects on the interaction between these atoms:

- In the presence of a finite magnetic field, **J** can be easily polarized, which results in a magnetic moment $d = 6\mu_B$ for chromium, $d = 10\mu_B$ for dysprosium, and $d = 7\mu_B$ for erbium. Therefore, the magnetic moment is about one order of magnitude larger than that of the alkali-metal atoms, and hence the magnetic dipole interaction between two atoms is two orders of magnitude larger.
- In the presence of a finite magnetic field, because the angular momentum L is nonzero for dysprosium and erbium, the electron cloud is anisotropic, so that the short-range Van der Waals potential is also anisotropic. This effect does not exist in chromium, whose angular momentum is zero.
- In the limit of a vanishing magnetic field, **J** becomes depolarized, and the spin rotational symmetry is restored. These atoms exhibit the aspects of high-spin particles, and their interactions depend on spin, as we will discuss in Section 4.3.

1.2 Magnetic Structure

Now we consider the effect of a static magnetic field on the atomic structure. Because electrons are charged, in principle, the electron motion inside an atom can also be affected by the presence of magnetic field. However, this effect is too small compared with the Coulomb interaction, such that we can safely ignore the change of electron orbital due to the magnetic field. We only focus on the Zeeman effect acting on the electron spin **S**, orbital angular momentum **L**, and nuclear spin **I**. The energy scale of the Zeeman splitting is comparable with the hyperfine splitting for a typical magnetic field of hundreds of Gauss in the laboratory.

Hence, here we consider an atom as a point neural particle carrying **S**, **L**, and **I**. Now let us focus on the ground state of alkali-metal atoms. For example, for ^{87}Rb atoms with $S = 1/2, L = 0$, and $I = 3/2$, the ground state spin structure is determined by

$$\hat{H}_s = B(\mu_B g_S \hat{S}_z + \mu_N g_I \hat{I}_z) + \alpha_{hf} \, \hat{\mathbf{J}} \cdot \hat{\mathbf{I}}, \tag{1.6}$$

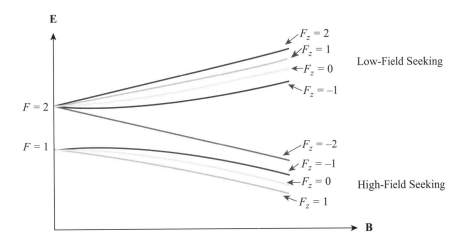

Figure 1.3 Schematic of the Zeeman energy structure. Here we consider the electronic ground state of a ^{87}Rb atom with $J = 1/2$ and $I = 3/2$. F_z labels the good quantum number of each state. A color version of this figure can be found in the resources tab for this book at cambridge.org/zhai.

where μ_B and μ_N are the Bohr magneton and the nuclear magneton, respectively, and $\mu_N \ll \mu_B$. g_S and g_I are the Landé g-factors. Here we first consider the situation that the magnetic field is spatially uniform and its direction is chosen as the $\hat{z}$ direction. For this Hamiltonian, F_z is a good quantum number, and its spectrum can be solved exactly. Here we analyze the behavior in the small B and large B regimes, respectively. By smoothly connecting the small B and large B regimes, one naturally obtains the qualitative feature for the energy diagram, as shown in Figure 1.3

- In the small B-field regime, when $B\mu_B g_s \ll \alpha_{hf}$, the hyperfine coupling dominates. The hyperfine coupling splits the energy between states with $F = 1$ and the states with $F = 2$. Within the three $F = 1$ states, or the five $F = 2$ states, the Zeeman field simply creates a linear Zeeman energy and a quadratic Zeeman energy proportional to F_z and F_z^2, respectively. The reason that there exists a quadratic Zeeman effect is precisely because the hyperfine spin contains both electronic spin and nuclear spin components, and they couple to the external magnetic moment differently.
- In the large B-field regime, when $B\mu_B g_s \gg \alpha_{hf}$, the Zeeman energy of electron spin dominates. The energies of four states with $S_z \approx -1/2$ decrease as B increases, and the energies of the other four states with $S_z \approx 1/2$ increase as B increases.

Magnetic Trap. In the presence of a magnetic field, the energies of some spin states increase with an increasing magnetic field strength. That is to say, if an atom is prepared in such a state, it can be trapped in the regime where the magnetic field strength has a local minimum. Atoms in these states are called the "low-field seeking" atoms. The energies of some other states decrease with an increasing magnetic field strength. These states can be trapped in the regime where the magnetic field strength has a local maximum. Atoms in these states are called the "high-field seeking" atoms. This is the basic idea of the magnetic trapping. It is not difficult to show that, due to the constraint from the Maxwell equations, the magnetic field cannot have a local maximum if there is no electronic current inside a vacuum chamber. It can also be shown that the magnetic field cannot point to the same direction in order to create a local minimum of magnetic field strength in space. Thus, the natural idea is to trap the "low-field seeking" atoms by a minimum of the magnetic field strength, using a spatially varying magnetic field texture. For instance, a so-called quadrupole trap has a magnetic field configuration $\mathbf{B} = B_0(x, y, -2z)$, where $\mathbf{r} = (x, y, z)$ is the spatial coordinate and the magnetic field strength has a minimum at $\mathbf{r} = 0$.

Emergent Synthetic Gauge Field. Above we have studied the Zeeman energy level of an atom in a uniform magnetic field, which results in the idea of trapping atoms near a minimum of the magnetic field strength. On the other hand, we have also noticed that the magnetic field cannot point to the same direction in order to have a local minimum of its strength. Thus, to fill the gap, we have to consider the motion of an atom in a spatially varying magnetic field configuration $\mathbf{B}(\mathbf{r})$. Let us consider the Schrödinger equation for the motion of an atom as

$$i\hbar\frac{\partial\psi}{\partial t} = \left(-\frac{\hbar^2}{2m}\nabla^2 + \hat{H}_{\mathrm{s}}(\mathbf{r})\right)\psi, \tag{1.7}$$

where

$$\hat{H}_{\mathrm{s}}(\mathbf{r}) = \mu_{\mathrm{B}}g_S\mathbf{B}(\mathbf{r})\cdot\hat{\mathbf{S}} + \mu_{\mathrm{N}}g_I\mathbf{B}(\mathbf{r})\cdot\hat{\mathbf{I}} + \alpha_{\mathrm{hf}}\hat{\mathbf{J}}\cdot\hat{\mathbf{I}}. \tag{1.8}$$

We introduce a unitary matrix $\mathcal{U}(\mathbf{r})$ to diagonalize $\hat{H}_{\mathrm{s}}(\mathbf{r})$ as $\Lambda(\mathbf{r}) = \mathcal{U}^\dagger(\mathbf{r})\hat{H}_{\mathrm{s}}(\mathbf{r})\mathcal{U}(\mathbf{r})$ for every $\mathbf{r}$. For each $\mathbf{r}$, we can always choose a local coordinate such that the magnetic field direction is taken as local $\hat{z}$ direction, therefore, $\Lambda(\mathbf{r})$ has the same energy level as shown in Figure 1.3 for the Hamiltonian Eq. 1.6 that only depends on the strength $|\mathbf{B}(\mathbf{r})|$.

Denoting $\tilde{\psi} = \mathcal{U}^\dagger(\mathbf{r})\psi$, the Schrödinger equation for $\tilde{\psi}$ can be written into the adiabatic spin bases as

$$i\hbar\frac{\partial\tilde{\psi}}{\partial t} = \left(\frac{1}{2m}(-i\hbar\nabla - \mathbf{A})^2 + \Lambda(\mathbf{r})\right)\tilde{\psi}, \tag{1.9}$$

where $\mathbf{A}(\mathbf{r}) = i\hbar\mathcal{U}^\dagger(\mathbf{r})(\nabla\mathcal{U}(\mathbf{r}))$. Here we have used

$$\mathcal{U}^\dagger(-i\hbar\nabla)\mathcal{U} = -i\hbar\nabla - i\hbar\mathcal{U}^\dagger(\mathbf{r})(\nabla\mathcal{U}(\mathbf{r})). \tag{1.10}$$

Notice that $\mathcal{U}^\dagger(\mathbf{r})\mathcal{U}(\mathbf{r}) = 1$, which means

$$\mathcal{U}^\dagger(\mathbf{r})(\nabla\mathcal{U}(\mathbf{r})) + (\nabla\mathcal{U}^\dagger(\mathbf{r}))\mathcal{U}(\mathbf{r}) = 0, \tag{1.11}$$

and $\mathcal{U}^\dagger(\mathbf{r})(\nabla\mathcal{U}(\mathbf{r}))$ is purely imaginary. Thus $\mathbf{A}$ is a real field. Eq. 1.9 takes the same form as the Schrödinger equation for the motion of a particle in an external gauge field [73]. In cold atom literatures, this emergent gauge field $\mathbf{A}$ is called the "synthetic gauge field."

Gauge fields are classified as abelian and non-abelian. For abelian ones, different components of the gauge field commute with each other, for instance, when $\mathbf{A}$ is a number. For non-abelian ones, different components of the gauge field do not commute with each other. Here, in general, $\mathbf{A}$ is a matrix, and different spatial components of $\mathbf{A}$ do not commute with each other, which represents a non-abelian gauge field. The off-diagonal component of **A** gives rise to the transition between different spin eigenstates in local coordinates. If these off-diagonal components are too small compared with the difference between the diagonal components of $\Lambda(\mathbf{r})$, the transition effect becomes negligible, and one can safely assume that the atoms always stay in the same adiabatic spin eigenstate. It is usually the case when the magnetic field strength is large and the spatial variation is small. Using the adiabatic approximation, an atom in this adiabatic spin state can effectively be viewed as a spinless particle, which experiences a potential only depending on the strength $|\mathbf{B}(\mathbf{r})|$. This manifests the magnetic trapping as discussed above. The only modification is the presence of the diagonal component of the gauge field **A**, say, denoted by A_{ii}, corresponding to the adiabatic spin eigenstate labeled by i. This gives rise to an abelian gauge field. This abelian gauge field has a physical effect when it leads to a nonzero synthetic magnetic field $\mathbf{B}_{\text{syn}}$, given by $\mathbf{B}_{\text{syn}} = \nabla \times \mathbf{A}_{ii}(\mathbf{r})$.

The discussion above says that the motion of a neural atom with spin in a spatially varying magnetic field $\mathbf{B}(\mathbf{r})$ is equivalent to the motion of a spinless charged particle in a synthetic magnetic field $\mathbf{B}_{\text{syn}}$. This synthetic magnetic field $\mathbf{B}_{\text{syn}}$ should not be confused with the real magnetic field **B**. The real magnetic field **B** only acts on the spin degree of freedom of an atom and does not couple to its motion because atoms are neutral. However, the synthetic gauge field only acts on the motion of the particle. This equivalence is emergent from the Berry phase effect, and the basic idea is illustrated in Figure 1.4. Let us consider a neutral atom with spin moving in a spatially varying magnetic field. Following the adiabatic approximation, when an atom moves, its spin direction is always aligned with the local magnetic field direction. Therefore, when the atom follows a closed trajectory in space, its wave function acquires an extra phase that is proportional to the solid angle expanded by the spin direction along the trajectory. On the other hand, for a charged particle moving in a magnetic field, due to the Aharonov–Bohm effect, it also requires a phase for any closed trajectory which is proportional to the total flux enclosed by the trajectory. If these two

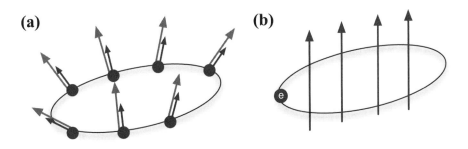

Figure 1.4 Emergence of synthetic gauge field. (a) A neutral and spinful atom (represented by the filled circles) moves in a magnetic field with spatially varying magnetic field directions (represented by longer arrows), and its spin direction (represented by arrows attached to the filled circles) always follows the local magnetic field direction. (b) A charged spinless particle moves in a magnetic field (represented by the vertical arrows). These two motions are equivalent when (i) the adiabatic approximation for case (a) is valid and (ii) the phases accumulated along any trajectory are always equal between cases (a) and (b). A color version of this figure can be found in the resources tab for this book at cambridge.org/zhai.

Box 1.1   Different Kinds of "Magnetic Fields"

It is really important to distinguish these three concepts of the real magnetic field, the light-induced magnetic field, and the synthetic magnetic field. Here we summarize the difference and the relations between these three "magnetic fields." First of all, the former two act on the spin degree of atoms, and the third one acts on the spatial motion of atoms. Second, both the real and the light-induced magnetic fields can polarize spin of atoms in a spatially dependent way, which leads to the synthetic magnetic field. Third, the synthetic magnetic field generated by the light-induced magnetic field can be much stronger than that generated by the real magnetic field. The synthetic gauge field is an active research topic in cold atom physics, and more discussion can be found in Box 7.3.

phases are always equal for any trajectory, then the spatial motion of a neutral spinful atom in a spatially varying magnetic field can be effectively described by a "charged" spinless particle in a synthetic magnetic field. This also tells us that the synthetic magnetic flux generated in this way depends on how fast the magnetic field direction varies in space. However, it is hard for a real magnetic field to vary very rapidly in space. In Section 1.3 we will introduce the vector light shift, which can generate a light-induced Zeeman field for atoms that can vary in the scale of the order of interparticle spacing. By combining these two effects, we can generate a strong synthetic magnetic field. To further clarify these concepts, we summarize different terminology of "magnetic fields" in Table 1.1.

Finally, we should also mention that this adiabatic approximation can break down either when the off-diagonal components of $\mathbf{A}$ become very large due to fast variation of the magnetic field direction or when the energy differences of the diagonal components of $\Lambda(\mathbf{r})$ become quite small. For instance, the latter happens at $\mathbf{r} = 0$ of the quadruple trap, where a few spin states become degenerate. Around the $\mathbf{r} = 0$ regime, the off-diagonal components of $\mathbf{A}$ are always important, which drives transitions between different adiabatic spin states. Therefore, there is significant probability that a "low-field seeking" state can flip into a "high-field seeking" state, which is also known as the "Majorana transition." When the transition takes place, atoms cannot be trapped by the magnetic trap. This is one of the major challenges in achieving a Bose–Einstein condensate in a magnetic trap before 1995. This is actually an effect of the non-abelian gauge field. It is interesting to note that though the synthetic gauge field became a major research subject in cold atom physics after around 2010, its effect already existed even prior to the birth of this field. The JILA group and the MIT group came up with different methods to solve this problem. The JILA group applies an oscillating offset magnetic field to deal with this problem [140]. Treating a time-periodic system requires the Floquet theory, which will also be discussed in Section 7.4.

1.3 Light Shift

In the previous sections, we discussed the structure of a single atom. In this section, we will discuss how a single atom interacts with a laser light.

General Framework. Atoms experience an effective potential in the presence of the laser field. Here we will discuss such potential using the ground state alkali-metal atoms as an example. One can see that the first two leading order contributions to this light-induced potential are the scalar potential and the vector potential, which are known as the scalar light shift and the vector light shift, respectively.

As discussed in Section 1.1, let us consider all states in the ground state $^2S_{\frac{1}{2}}$ manifold with the electron angular momentum $L = 0$, and all states in the electronic excited states $^2P_{\frac{1}{2}}$ and $^2P_{\frac{3}{2}}$ manifolds with $L = 1$. Here, by "manifold," we mean that both electronic and nuclear spin degrees of freedom are included, and each manifold contains multiple spin states. The energy splitting E_{ex} between the ground and the excited manifolds is of the order of the electron volt, which usually lies in the energy window of a visible light field, as shown in Figure 1.5. We denote the detuning $\Delta_e = E_{ex} - \hbar\omega$, where ω is the laser frequency. Usually, Δ_e is comparable to or larger than the fine-structure splitting and is much larger than the hyperfine splitting. Δ_e is also much smaller than the detuning of the other electronic excited states, that is to say, all the other electronic excited states are far detuned. With these energy scale considerations, our model is established as follows:

- The contributions from both $^2P_{\frac{1}{2}}$ and $^2P_{\frac{3}{2}}$ are important and should be treated on equal footing.
- The hyperfine coupling and the Zeeman energy are safely ignored.
- Except for $^2S_{\frac{1}{2}}$, $^2P_{\frac{1}{2}}$ and $^2P_{\frac{3}{2}}$ manifolds, all the other electronic states are not included.

Hence, we write down our model as follows [61]:

$$\hat{H} = \hat{H}_{at} + \hat{H}_{d}. \tag{1.12}$$

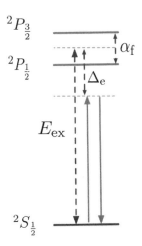

Figure 1.5   Schematic of the level diagram for atom-light interaction. Here we use the alkali-metal atoms as an example. Three relevant energy scales, the excitation energy E_{ex}, the detuning Δ_e, and the fine structure splitting α_f, are marked on the diagram. A color version of this figure can be found in the resources tab for this book at cambridge.org/zhai.

$\hat{H}_{at}$ denotes the Hamiltonian for the atom part and it is given by

$$\hat{H}_{at} = E_{ex}\mathcal{P}_e + \alpha_f \hat{\mathbf{S}} \cdot \hat{\mathbf{L}}. \tag{1.13}$$

Here we define a projection operator $\mathcal{P}_g$ to the ground states manifold with $L = 0$ and a projection operator $\mathcal{P}_e$ to the excited states manifold with $L = 1$. In our model, $\mathcal{P}_g + \mathcal{P}_e = 1$. We have also set the ground state energy as zero energy, and the $\hat{\mathbf{S}} \cdot \hat{\mathbf{L}}$ vanishes when acting on the ground state manifold. It is also important to note that $\hat{H}_{at}$ itself does not provide any coupling between the ground and the excited states.

Here the atom–light interaction refers to the interaction between electrons inside the atom and the electromagnetic field of the laser. Considering a time-dependent electric field applied to electrons, the Hamiltonian can be written as

$$\hat{H} = \sum_{i=1}^{Z} \frac{1}{2m^*}(\hat{\mathbf{p}}_i + e\mathbf{A}(t))^2 + \ldots, \tag{1.14}$$

where $\mathbf{r}_i$ denotes the coordinate of a valance electron in the rest frame of the nucleus, and $\ldots$ represents the Coulomb interactions and other terms discussed in Section 1.1. Typically, the wave length of a laser is much longer than the size of an atom, therefore, on the scale of electronic wave function inside an atom, we can ignore the spatial dependence of $\mathbf{A}$ and only consider its temporal dependence. As we will see below, this leads to the dipolar coupling.

Here, by performing a gauge transformation, we consider $\hat{\mathcal{U}}^\dagger \hat{H} \hat{\mathcal{U}} \to \hat{H}$, and $U^\dagger \psi \to \psi$, and here

$$\hat{\mathcal{U}} = e^{-\sum_i ie\mathbf{A}\cdot\mathbf{r}_i/\hbar}. \tag{1.15}$$

Under this gauge transformation, $\hat{\mathbf{p}}_i + e\mathbf{A}$ changes back to $\hat{\mathbf{p}}_i$ and the kinetic energy returns to the normal $\hat{\mathbf{p}}_i^2/(2m^*)$. However, because this gauge transformation is time dependent, there will be an extra term $i\hbar(\partial_t\hat{\mathcal{U}})^\dagger\hat{\mathcal{U}}$ added into the Hamiltonian, which is given by

$$i\hbar(\partial_t\hat{\mathcal{U}}^\dagger)\hat{\mathcal{U}} = -e\mathbf{r}_i \cdot \frac{\partial\mathbf{A}}{\partial t} = e\mathbf{r}_i \cdot \mathbf{E}, \tag{1.16}$$

where $\mathbf{E}$ is the electric field of the laser given by $\mathbf{E} = -\partial\mathbf{A}/\partial t$. Hence, it results in the dipole coupling $\hat{H}_d = \mathbf{d} \cdot \mathbf{E}$, where $\mathbf{d} = \sum_i e\mathbf{r}_i$ is the electron dipole operator, and it gives rise to the transition between the ground and the excited states.

First, we consider a single laser field, and

$$\hat{H}_d = \mathbf{d} \cdot \mathbf{E} = \sum_{j=x,y,z} d_j E_j^0 \cos(\phi_j - \omega t), \tag{1.17}$$

where $E_j^0 \cos(\phi_j - \omega t)$ is the jth component of the electric field and $j = x, y, z$ denotes three spatial components of the laser field. For example, in this notation, for a light linearly polarized along $\hat{x}$, $E_x^0 \neq 0$ and $E_y^0 = E_z^0 = 0$; for a light linearly polarized along the $(\hat{x} + \hat{y})/\sqrt{2}$ direction, $E_x^0 = E_y^0 \neq 0$, $\phi_x = \phi_y$ and $E_z^0 = 0$; and for a light circularly polarized in the xy plane, $E_x^0 = E_y^0 \neq 0$, $\phi_x = \phi_y \pm \pi/2$ and $E_z^0 = 0$.

Here we first simply the Hamiltonian by employing the rotating wave approximation. We first define a unitary transformation as

$$\hat{\mathcal{U}}(t) = e^{-i\omega t \mathcal{P}_e} = (1 - \mathcal{P}_e) + \mathcal{P}_e e^{-i\omega t} = \mathcal{P}_g + \mathcal{P}_e e^{-i\omega t}. \tag{1.18}$$

and apply this unitary transformation to the Hamiltonian. The rotating wave approximation ignores $e^{\pm i2\omega t}$ terms because they are high-frequency oscillating terms. With this approximation, it is straightforward to show that the Hamiltonian is reduced to

$$\hat{H}_d = \hat{\mathcal{U}}^\dagger(t)\hat{H}_d\hat{\mathcal{U}}(t) \approx \frac{1}{2} \sum_{j=x,y,z} \left(\mathcal{E}_j^* \mathcal{P}_g d_j \mathcal{P}_e + \mathcal{E}_j \mathcal{P}_e d_j \mathcal{P}_g \right), \tag{1.19}$$

where we have also used $\mathcal{P}_g d_j \mathcal{P}_g = \mathcal{P}_e d_j \mathcal{P}_e = 0$ because of the rotational symmetry of the electron wave functions. Here $\mathcal{E}_j$ is defined as $E_j^0 e^{i\phi_j}$. Unlike **E**, here $\mathcal{E}$ is time-independent, which encodes the phase and amplitude information of the laser field. In this notation, for a light linearly polarized along $\hat{x}$, $\mathcal{E} = E_0 \mathbf{e_x}$; for a light linearly polarized along the $(\hat{x}+\hat{y})/\sqrt{2}$ direction, $\mathcal{E} = \frac{E_0}{\sqrt{2}}(\mathbf{e_x}+\mathbf{e_y})$; and for a light circularly polarized in the xy plane, $\mathcal{E} = \frac{E_0}{\sqrt{2}}(\mathbf{e_x}+i\mathbf{e_y})$.

Moreover, a time-dependent term $i\hbar(\partial_t\hat{\mathcal{U}}^\dagger)\hat{\mathcal{U}}$ can be absorbed in $\hat{H}_{at}$ so that $\hat{H}_{at}$ becomes

$$\hat{H}_{at} = \Delta_e \mathcal{P}_e + \alpha_f \hat{\mathbf{S}} \cdot \hat{\mathbf{L}}, \tag{1.20}$$

where $\Delta_e = E_{ex} - \hbar\omega$. $\Delta_e > 0$ is called the red detuning and $\Delta_e < 0$ is called the blue detuning. In this way, we obtain the new $\hat{H}_{at} + \hat{H}_d$ as a time-independent effective Hamiltonian. We will revisit such a time periodical problem in Section 7.4 with a general theoretical framework known as the Floquet theory.

This formalism can be straightforwardly generalized to the situations with multiple laser beams. In this case, the electric field consists of contributions from all laser beams, and we write the dipole coupling as

$$\hat{H}_d = \mathbf{d} \cdot \mathbf{E} = \sum_\kappa \sum_{j=x,y,z} d_j E_{j,\kappa}^0 \cos(\phi_{j,\kappa} - \omega_\kappa t), \tag{1.21}$$

where κ labels different laser fields. Here different lasers can have different frequencies, but their difference should be small. That is to say, suppose we denote the average of these frequencies as ω; their difference should be much smaller than ω. Therefore, we can still safely drop all terms with $e^{\pm i(\omega_\kappa + \omega)t}$ and retain all terms with $e^{\pm i(\omega_\kappa - \omega)t}$. After the rotating wave approximation, it is easy to show that Eq. 1.19 still holds, except that the definition of $\mathcal{E}_j$ should be modified as

$$\mathcal{E}_j = \sum_\kappa E_{j,\kappa}^0 e^{i\phi_{j,\kappa} - i(\omega_\kappa - \omega)t}. \tag{1.22}$$

Note that if all lasers share the same frequency, $\mathcal{E}$ is time-independent. If there are multiple frequencies, $\mathcal{E}$ is also time-dependent.

The physical meaning of the rotating wave approximation can be understood more clearly when we treat the laser as a quantum field labeled by photon number. Let us now consider an atom in the ground state (denoted by $|g\rangle$) in a laser field with N photons (denoted by $|N\rangle$). In the second quantized form, the $\mathbf{d} \cdot \mathbf{E}$ coupling can either create a photon or annihilate a photon when it couples an atom from the ground state to the excited state. Thus, we shall consider following two different second-order perturbation processes. Starting from the state $|g\rangle|N\rangle$, the atom can either be first excited to the excited state (denoted by $|e\rangle$) by absorbing a photon, such that the intermediate state is $|e\rangle|N-1\rangle$ with energy Δ_e, and then return to the ground state $|g\rangle|N\rangle$ by emitting a photon. The atom can also be first excited to the excited state by emitting a photon, such that the intermediate state is $|e\rangle|N+1\rangle$ with energy $E_{ex} + \hbar\omega$, and then return to the ground state $|g\rangle|N\rangle$ by absorbing a photon. The second process has a much larger intermediate state energy, and therefore, the probability is considerably smaller. The rotating wave approximation is basically to ignore the second process. With the first process alone, by taking the energy of photons into account, the energy detuning changes from E_{ex} in Eq. 1.13 to Δ_e in Eq. 1.20.

Hereafter we will consider how this second-order process generates an effective potential for atoms in the ground state manifold after eliminating the excited states $|e\rangle|N-1\rangle$. Following the standard perturbation theory, the effective Hamiltonian for atoms in the ground manifold is derived as

$$\hat{H}_{\text{eff}} = -\mathcal{P}_g \hat{H}_d \mathcal{P}_e \hat{H}_{\text{at}}^{-1} \mathcal{P}_e \hat{H}_d \mathcal{P}_g = -\frac{1}{4} \sum_{i,\,j=x,y,z} \mathcal{E}_i^* \hat{\mathcal{D}}_{ij} \mathcal{E}_j, \tag{1.23}$$

where $\hat{\mathcal{D}}_{ij}$ is a rank-2 Cartesian tensor operator defined as

$$\hat{\mathcal{D}}_{ij} = \mathcal{P}_g d_i \mathcal{P}_e \hat{H}_{\text{at}}^{-1} \mathcal{P}_e d_j \mathcal{P}_g. \tag{1.24}$$

$\hat{\mathcal{D}}_{ij}$ is purely a property of atoms. By using the fact that $\mathcal{P}_g + \mathcal{P}_e = 1$ and $\mathcal{P}_g d_i \mathcal{P}_g = 0$, $\mathcal{P}_e$ in Eq. 1.24 can be eliminated. Hereinafter we should focus on the properties of $\hat{\mathcal{D}}_{ij}$ in different circumstances.

The Scalar Light Shift. We first consider a special case with $\alpha_f = 0$; then $\hat{H}_{\text{at}}$ is simplified as $\Delta_e \mathcal{P}_e$, and

$$\hat{\mathcal{D}}_{ij} = \frac{1}{\Delta_e} \mathcal{P}_g d_i d_j \mathcal{P}_g. \tag{1.25}$$

Since the ground state has $L = 0$, which has the spatial reflection symmetry and rotational symmetry, it is easy to see that $\mathcal{D}_{ij} = 0$ if $i \neq j$, and all three $\mathcal{D}_{jj}$ ($j = x, y, z$) are the same. Therefore,

$$\hat{\mathcal{D}}_{ij} = -4u_s \delta_{ij}, \quad u_s = -\frac{e^2}{12\Delta_e} \langle g|r^2|g\rangle, \tag{1.26}$$

which gives rise to an effective Hamiltonian

$$\hat{H}_{\text{eff}} = u_s \mathcal{E}^2. \tag{1.27}$$

This term acts identically on different spin states of the ground manifold and cannot flip the spin, and this term also does not depend on the polarization of the lasers. This is because the dipole coupling only acts on the orbital degree of freedom of the valance electron, and therefore, in the absence of $\hat{\mathbf{S}} \cdot \hat{\mathbf{L}}$ coupling, the orbital degree of freedom is decoupled from the spin degree of freedom. Hence, this term is called the *scalar light shift*, and u_s is called the atom's scalar ac polarizability. Below we shall discuss some important applications of the scalar light shift.

Trapping Atoms with Laser. The scalar potential only depends on the intensity of the laser. For the red detuning case, $u_s < 0$, atoms will be trapped in the place where the laser intensity is a local maximum. Hence, one can trap atoms with a focused laser beam. This is the basic mechanism of the laser trapping. In this optical trap, all spin states in the ground state manifold will experience identical potentials, and one can fully utilize the spin degree of freedom of atoms, as we will discuss in Section 4.3.

However, there is one complication we should notice. So far what we have considered is the second-order process that the photon absorption is followed by the stimulated emission. In fact, there is another process, that is, the photon absorption is followed by the spontaneous emission. The spontaneous emission is what causes the finite lifetime of the excited state and can be described by adding an imaginary part Γ in the excited state energy, where Γ is the line width of the excited states. Hence, this process can be described by adding an imaginary part to the excited state energy in the expression of the ac polarizability. Consequently, u_s acquires an imaginary part as

$$u_s \propto \frac{1}{\Delta_e - i\Gamma} = \frac{\Delta_e}{\Delta_e^2 + \Gamma^2} + i\frac{\Gamma}{\Delta_e^2 + \Gamma^2}. \tag{1.28}$$

For the off-resonant case with $\Delta_e \gg \Gamma$, the real part behaves as $1/\Delta_e$ and the imaginary part behaves as Γ/Δ_e^2. Since the real and imaginary parts have difference power dependence on Δ_e, in the far-detuned regime with $\Delta_e \gg \Gamma$, the imaginary part can be strongly suppressed compared with the real part. Since the real part provides the trapping effect, one usually works in this far-detuned regime for laser trapping. The laser trapping was awarded the Nobel Prize in 2018.

Optical Lattice. Now let us consider applying two counterpropagating laser beams, say, along $\hat{x}$. These two lasers have the same frequency and the same polarization; for instance, both are linearly polarized along $\hat{y}$. Suppose both two lasers have the same strength E^0; according to Eq. 1.22, we have

$$\mathcal{E}_y = E^0 e^{ikx} + E^0 e^{-ikx} = 2E^0 \cos(kx). \tag{1.29}$$

The electric field intensity has a spatial periodical modulation due to the interference between two lasers. Hence, the scalar light shift gives rise to a periodic lattice potential $V(x) \propto \cos^2(x)$, which is now well known as the optical lattice. Optical lattices will be the main topic of the last part of this book.

Cooling Atoms with Laser. If Δ_e becomes small, the imaginary part of the scalar potential also becomes important, which means the process with spontaneous emission becomes important. There is an important difference between the process with the stimulated emission and the process with the spontaneous emission. For the former, because it is the same photon that is absorbed and emitted, the momentum transfer is canceled, and atoms do not receive momentum transfer after the entire second-order process. But for the latter, since the photon can go any direction in the spontaneous emission process, on average, the momentum transfer during the emission process is canceled out, and therefore, atoms get kicked by the photons of the laser field during absorbing the photon. Effectively, atoms feel a force proportional to Γ/Δ_e^2 and the momentum of the laser. Now considering two counterpropagating lasers with same frequency, if an atom is at rest, the force from the left laser cancels with the force from the right laser, because the momentum transfers received from the left and the right lasers are of equal strength and opposite sign. However, if an atom moves toward the right, due to the Doppler effect, Δ_e for two lasers depends on the velocities of the atoms, and they are different. Thus, the force cannot be canceled out, and it can be shown that the net force is always opposite to the velocity of the atom [138]. Thus, this effect slows down the motion of atoms. That is the basic mechanism of the laser cooling, which is one of the most important steps to cool atoms toward the quantum degeneracy. Laser cooling was recognized with the Nobel Prize in 1997.

However, due to various limitations, it is very hard to reach quantum degeneracy of atomic gases directly by laser cooling. Directly reaching quantum degeneracy by laser cooling was first achieved more than 20 years after achieving Bose–Einstein condensation in atomic gases [76, 160]. In most experiments, one needs to perform evaporative cooling after laser cooling. The basic idea of evaporative cooling is quite straightforward. Considering that the trapping potential has a finite depth U_0, atoms with kinetic energy larger than U_0 have a significant chance to escape from the trap. Now let us gradually decrease U_0, thus, more and more atoms with larger kinetic energy escape from the trap, and the average kinetic energy of the remaining atoms decreases. Therefore, the temperature of the remaining atomic gas decreases. However, we should notice that the evaporative cooling pays the price of losing atoms in order to lower the temperature. When losing atoms, the degenerate temperature also decreases. Therefore, the efficiency of the evaporative cooling becomes crucial. That is to say, lowering the temperature has to be faster than lowering the degenerate temperature; otherwise, the system can never reach quantum degeneracy. The efficiency of evaporative cooling certainly depends on how to lower the trap depth U_0 as a function of time t, and controlling this dynamical process is the key challenge to reaching quantum degeneracy through evaporative cooling.

Vector Light Shift. Now let us consider the situation with finite $\hat{\mathbf{S}} \cdot \hat{\mathbf{L}}$ coupling. As we discussed above, fine structure splitting α_f can be comparable to detuning Δ_e. Here, just to simplify the calculation, we consider the situation $\alpha_f/\Delta_e \ll 1$, and the main conclusion does not change if $\alpha_f \sim \Delta_e$. When $\alpha_f/\Delta_e \ll 1$, we can expand $\hat{H}_{at}^{-1}$ to the leading order of α_f/Δ_e, and we obtain

$$\hat{H}_{at}^{-1} = \left(\frac{1}{\Delta_e} - \frac{\alpha_f}{\Delta_e^2} \hat{\mathbf{S}} \cdot \hat{\mathbf{L}} \right) \mathcal{P}_e, \tag{1.30}$$

and therefore,

$$\hat{\mathcal{D}}_{ij} = \mathcal{P}_g d_i \frac{1}{\Delta_e} d_j \mathcal{P}_g - \frac{\alpha_f}{\Delta_e^2} \mathcal{P}_g d_i (\hat{\mathbf{S}} \cdot \hat{\mathbf{L}}) d_j \mathcal{P}_g. \tag{1.31}$$

As shown above, the first term gives the scalar part. And for the second term, one has

$$\mathcal{P}_g d_i (\hat{\mathbf{S}} \cdot \hat{\mathbf{L}}) d_j \mathcal{P}_g = \mathcal{P}_g (\hat{\mathbf{S}} \cdot \hat{\mathbf{L}}) d_i d_j \mathcal{P}_g - \mathcal{P}_g [(\hat{\mathbf{S}} \cdot \hat{\mathbf{L}}), d_i] d_j \mathcal{P}_g \tag{1.32}$$

$$= -i\hbar \epsilon_{lim} S_l \mathcal{P}_g d_m d_j \mathcal{P}_g, \tag{1.33}$$

where the first term vanishes because ground state has $\hat{\mathbf{L}}\mathcal{P}_g = 0$, and for the second term, we have used the commutative relation $[\hat{L}_l, \hat{d}_i] = i\hbar\epsilon_{lim}\hat{d}_m$, where ϵ_{jim} is the Levi–Civita symbol. Thus we reach

$$\hat{\mathcal{D}}_{ij} = -4u_s\delta_{ij} + i\frac{\hbar\alpha_f}{\Delta_e}\epsilon_{ijl}S_l(-4u_s) \simeq -4u_s\left(\delta_{ij} + i\frac{\hbar\alpha_f}{\Delta_e}\epsilon_{ijl}S_l\right), \tag{1.34}$$

where the second term will give rise to a vector light shift. In this case, the effective Hamiltonian is given by

$$\hat{H}_{\text{eff}} = -\frac{1}{4}\sum_{i,j=x,y,z}\mathcal{E}_i^*\hat{\mathcal{D}}_{ij}\mathcal{E}_j = u_s|\mathcal{E}|^2 + iu_v(\mathcal{E}^* \times \mathcal{E})\cdot\hat{\mathbf{S}}, \tag{1.35}$$

where $u_v = \hbar\alpha_f u_s/\Delta_e$ is the vector polarizability. The physical meaning of the vector light shift is a Zeeman field $\mathbf{B} = iu_v\mathcal{E}^* \times \mathcal{E}$ acting on the electronic spin degree of freedom of an alkali-metal atom.

Again, because the dipole coupling acts on the orbital degree of freedom, the vector light shift has to reply on the $\hat{\mathbf{S}}\cdot\hat{\mathbf{L}}$ coupling that hybridizes the orbital degree of freedom with the electronic spin degree of freedom. In this case, the $\hat{\mathbf{S}}\cdot\hat{\mathbf{L}}$ exists only in the excited states, and therefore, the vector light shift is smaller than the scalar term by an order of $\hbar\alpha_F/\Delta_e$. Consequently, the vector light shift scales with Δ_e also as $1/\Delta_e^2$. Thus, the vector light shift has the same scaling as the spontaneous emission process. In the discussion of the scalar light shift, we have noticed that, because the scalar light shift and the spontaneous emission scale with Δ_e differently, one can suppress the effect of spontaneous emission compared with the scalar potential by increasing the detuning. However, this no longer works for the vector light shift. Because the vector light shift and the spontaneous emission scale with Δ_e in the same way, one cannot suppress the spontaneous emission process without reducing the strength of the vector light shift. This discussion of atom–light interaction can also be generalized to lanthanide atoms like Dysprosium and Erbium [41]. There, the f-orbital is not fully filled for the ground state, and electric spin and angular momentum are not good quantum numbers for atoms in their ground state. Therefore, the dipole coupling can always act on the electronic spin degree of freedom. Thus, for these lanthanide atoms, the vector light shift scales in the same way as the scalar light shift, and they both scale as $1/\Delta_e$ [41, 24]. Indeed, the vector light shift has been realized in the lanthanide atoms with suppressed spontaneous emission [41, 24].

Light-Induced Zeeman Energy. Below we will discuss a couple of examples as the applications of the vector light shift. If an atom is illuminated in a linearly polarized light, then $\mathcal{E}^* \times \mathcal{E} = 0$, and the vector light shift vanishes. If the light is circularly polarized, for instance, $\mathcal{E} = \frac{E_0}{\sqrt{2}}(\mathbf{e_x} + i\mathbf{e_y})$, then

$$\mathcal{E}^* \times \mathcal{E} = iE_0^2\mathbf{e_z}, \tag{1.36}$$

which gives rise to a light-induced Zeeman energy proportional to $u_v E_0^2 S_z$. This difference between a linear polarized light and a circular polarized light lies between their difference in symmetry. The presence of a linear polarized light does not break the time reversal symmetry, but the presence of a circular polarized light does. Since the presence of a Zeeman field breaks the time-reversal symmetry, symmetry-wise, it is compatible with the presence of a circular polarized light, but is not compatible with the presence of a linearly polarized light.

Since u_v depends on the fine-structure splitting and detuning, if two different atoms are illuminated in the same circular polarized laser, they will experience different light-induced Zeeman fields. In general, because the spin-orbit coupling is stronger for heavier atoms, the vector light shift is also larger for heavier atoms for a given detuning. One experiment that can demonstrate this effect is the spin exchanging scattering between ^{23}Na and ^{87}Rb atoms [108]. The spin exchanging scattering will be discussed in detail in Section 2.3. Here let us first mention it briefly. Considering the $F = 1$ spin states of both ^{23}Na and ^{87}Rb atoms, each of them has three magnetic states labeled by $|F_z\rangle$. For instance, one of the spin-exchanging scattering processes can take place as

$$|0\rangle_{\text{Na}}|-1\rangle_{\text{Rb}} \leftrightarrow |-1\rangle_{\text{Na}}|0\rangle_{\text{Rb}}. \tag{1.37}$$

This spin-exchanging scattering reaches a resonance at a certain magnetic field when the Zeeman energy difference between the incoming and the outgoing states vanishes, that is,

$$\Delta E = (E_{|0\rangle_{\text{Na}}} + E_{|-1\rangle_{\text{Rb}}}) - (E_{|-1\rangle_{\text{Na}}} + E_{|0\rangle_{\text{Rb}}}) = 0, \tag{1.38}$$

at which the period of the spin-exchanging oscillation reaches a maximum. The Zeeman energy contains a contribution from both the static magnetic field and the light-induced Zeeman field. In the common practical condition [108], given a typical value of the optical trapping laser, the light-induced Zeeman energy for ^{87}Rb is equivalent to applying a magnetic field of $\sim$mG, but the light-induced Zeeman energy for ^{23}Na is only equivalent to applying a magnetic field of $\sim \mu$G, which can be safely ignored. This difference is due to the difference in their fine-structure splitting. Therefore, the magnetic field strength to reach the spin-exchanging resonance is different for different laser intensities. This has been experimentally observed, and the experimental results are shown in Figure 1.6.

Synthetic Spin-Orbit Coupling. Now we consider atoms in a real static magnetic field along $\hat{z}$ that gives rise to a Zeeman energy, assumed to be hF_z for simplicity. In addition, let us consider two counterpropagating Raman beams along $\hat{x}$, with one polarized along $\hat{y}$ and the frequency ω_1 and the other polarized along $\hat{z}$ and the frequency ω_2, as shown in Figure 1.7. The electronic field is given by

$$\mathbf{E} = E_1 e^{ik_0 x + i\omega_1 t} \mathbf{e_y} + E_2 e^{-ik_0 x + i\omega_2 t} \mathbf{e_z}. \tag{1.39}$$

This is the situation of two lasers with different frequencies. With the help of Eq. 1.22, one can obtain

$$\mathcal{E}^* \times \mathcal{E} = E_1 E_2 (e^{-i2k_0 x - i\delta\omega t} - e^{i2k_0 x + i\delta\omega t}) \mathbf{e_x}, \tag{1.40}$$

where $\delta\omega = \omega_2 - \omega_1$. Then, atoms in this laser field experience a light-induced Zeeman field as

$$\hat{H}_{\text{eff}} = iu_v E_1 E_2 (e^{-i2k_0 x - i\delta\omega t} - e^{i2k_0 x + i\delta\omega t}) \hat{S}_x. \tag{1.41}$$

In the low-field regime, by using the Wigner–Eckart theorem, one can project the Hamiltonian Eq. 1.43 into the hyperfine eigenstate bases, where the $\hat{S}_{x,y,z}$ terms become $\hat{F}_{x,y,z}$

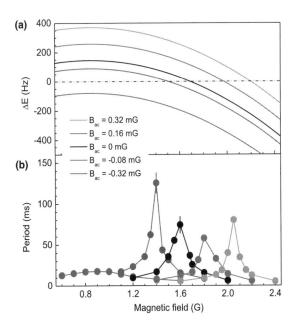

Figure 1.6 Physical effect of the light-induced Zeeman energy. (a) The energy difference between different spin states ΔE defined in Eq. 1.38. (b) The observed spin exchanging oscillation periods as a function of the real magnetic field strength, for different light-induced magnetic fields. B_{ac} in the figure denotes the light-induced Zeeman field for ^{87}Rb at different intensities of the circular polarized light. Reprinted from Figure [108]. A color version of this figure can be found in the resources tab for this book at cambridge.org/zhai.

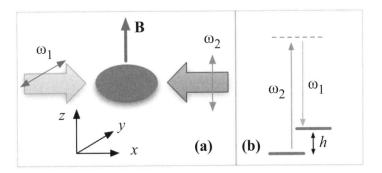

Figure 1.7 Schematic of the synthetic spin-orbit coupling for atoms. (a) The laser configuration of the Raman coupling scheme. Two counterpropagating Raman beams with different polarization directions and different frequencies are applied along $\hat{x}$, and **B** is the real magnetic field applied along the $\hat{z}$ direction. (b) Raman coupling couples two Zeeman levels split by a Zeeman energy denoted by h. A color version of this figure can be found in the resources tab for this book at cambridge.org/zhai.

terms, respectively.[3] Thus, the total Hamiltonian for the spin part contains both the real static Zeeman field and the light-induced Zeeman field as

$$\hat{H}_s = h\hat{F}_z + i\Omega(e^{-i2k_0x-i\delta\omega t} - e^{i2k_0x+i\delta\omega t})\hat{F}_x, \tag{1.42}$$

where Ω denotes the coupling constant proportional to $E_1 E_2$. Let us consider the situation that $\delta\omega$ is near-resonant with h. Following the discussion above, we apply a unitary transformation $\hat{\mathcal{U}} = e^{-i\delta\omega t\hat{F}_z/\hbar}$ and implement the rotating-wave approximation to drop the terms with $2\delta\omega$ frequency; the full Hamiltonian can be reduced to

$$\hat{H}_s = (h - \delta\omega)\hat{F}_z + \Omega(\sin(2k_0x)\hat{F}_x - \cos(2k_0x)\hat{F}_y). \tag{1.43}$$

This Hamiltonian represents a spatially dependent Zeeman field that varies the spin direction at the scale of the laser wavelength.

The discussion above mainly concerns the motion of electrons inside an atom. Now, we consider an atom as a point particle moving in such a spatially dependent Zeeman field, and the Hamiltonian is given by

$$\hat{H} = \frac{\hbar^2\hat{\mathbf{k}}^2}{2m} + \hat{H}_s, \tag{1.44}$$

where $\mathbf{k}$ now stands for the spatial motion of the atom. By applying a spatially dependent spin rotation $\hat{\mathcal{U}} = e^{-i2k_0x\hat{F}_z}$ to the Hamiltonian, one can obtain that

$$\hat{H} = \frac{\hbar^2}{2m}(\hat{k}_x - 2k_0\hat{F}_z/\hbar)^2 + \frac{\hbar^2\hat{\mathbf{k}}_\perp^2}{2m} + (h - \delta\omega)\hat{F}_z - \Omega\hat{F}_y. \tag{1.45}$$

Below we will analyze several different situations of the model Eq. 1.45.

- When $h - \delta\omega$ is much larger than Ω, we can only keep the lowest spin branch by implementing the adiabatic approximation. Without loss of generality, we consider $h - \delta\omega > 0$. Nearby the minimum of the dispersion, the dispersion can be well approximated by $\frac{1}{2m^*}(k_x - k_{min})^2$ [112]. For small Ω, $k_{min} \approx 2k_0 + o(\Omega)$. This can be viewed as realizing a constant gauge $U(1)$ gauge field $\hat{H} = \frac{1}{2m^*}(k_x - A_x)^2$ where A_x is a constant. A constant $U(1)$ gauge field has no physical effect because there is neither an electric nor a magnetic field, and it can be gauged away by a gauge transformation $e^{ik_{min}x}$.

- In the regime discussed above, if Ω depends on the y coordinate, then it means k_{min}, or equivalently, to say, A_x, depends on y. This leads to a synthetic magnetic field $B_{syn} = -\partial A_x/\partial y$ [113]. This emergent synthetic magnetic field can also be understood in terms of spatial twisting of spins by the light-induced magnetic field, as discussed in Section 1.2. If Ω depends on time, then A_x also depends on time. This realizes a synthetic electric field given by $E_{syn} = -\partial A_x/\partial t$ [115]. However, the spatial and temporal dependences of these gauge fields are completely fixed by the external classical fields, which are the profile of the laser intensity in this case. There is no quantum dynamics of these gauge fields. We will summarize the timeline of simulating various kinds of $U(1)$

---

[3] According to the Wigner–Eckart theorem, the projection from $\hat{S}_{x,y,z}$ terms to $\hat{F}_{x,y,z}$ terms acquires a constant coefficient given by $\frac{F(F+1)-I(I+1)+J(J+1)}{2F(F+1)}$. For ^{87}Rb, $I = 3/2$, $J = 1/2$, and for the $F = 1$ manifold, this constant is $-1/4$.

gauge fields in Box 7.3, where we will discuss how to realize a $U(1)$ gauge field with its own dynamics.

- In the regime when $h - \delta\omega$ is comparable to Ω, we need to keep all spin components. Here we again consider Ω a constant independent of spatial and temporal coordinates, but we need to keep both the vector gauge potential $A_x = 2k_0\hat{F}_z/\hbar$ and the scalar potential $-\Omega\hat{F}_y$ terms as matrices. These two terms do not commute with each other, and this realizes a non-abelian gauge field. Unlike the abelian case, a non-abelian gauge field has a physical effect even though they are constants. As one can see, if one wants to gauge away the vector potential, one needs to make a gauge transformation $e^{2ik_0x\hat{F}_z}$, and this gauge transformation does not commute with the scalar potential term.

  In this case, the Hamiltonian can also be written as

$$\hat{H} = \frac{\hbar^2\hat{\mathbf{k}}^2}{2m} + \mathbf{h_k} \cdot \hat{\mathbf{F}}, \tag{1.46}$$

where

$$\mathbf{h_k} = \left(0, -\Omega, \left(h - \delta\omega - \frac{\hbar^2}{m}k_0k_x\right)\right). \tag{1.47}$$

In this way, this Hamiltonian can be viewed as a momentum-dependent Zeeman field. For eigenstates, the spin direction of the atom is locked by its momentum, which gives rise to the spin-orbit coupling effect. Here we should note that one should not confuse this synthetic spin-orbit coupling with the real spin-orbit coupling discussed in atomic structure. For the synthetic spin-orbit coupling, "orbit" refers to the spatial motion of atoms, and "spin" refers to the total spin of an atom. In nature, neutral atoms do not possess the spin-orbit coupling effect, and here it is a synthetic effect generated by the atom–light interaction. The real spin-orbit coupling refers to the fine structure coupling given by nature, where "orbit" means the motion of electrons around the nucleus inside the atom and "spin" refers to the spin of electrons. Nevertheless, from the discussion above, it is interesting to note that generating the synthetic spin-orbit coupling relies on the real spin-orbit coupling.

- We have not included the quadratic Zeeman energy in the discussion above. Considering atoms like ^{87}Rb with $F = 1$, in the moderate magnetic field, there also exists sizable quadratic Zeeman energy denoted by qF_z^2, as discussed in Section 1.2. Especially, let us consider the situation that $h - \delta\omega \approx q \gg \Omega$, such that

$$\delta = h - \delta\omega - q \ll \Omega \tag{1.48}$$
$$h - \delta\omega + q \approx 2q \gg \Omega. \tag{1.49}$$

In this case, two states with $|F = 1, F_z = 0\rangle$ and $|F = 1, F_z = -1\rangle$ are nearly degenerate, and their energy separation with $|F = 1, F_z = 1\rangle$ is much larger than the coupling. Hence, we introduce a pseudo-spin-1/2 to represent $|F = 1, F_z = 0\rangle$ and $|F = 1, F_z = -1\rangle$ states, described by the Pauli matrices $\boldsymbol{\sigma}$, and ignore $|F = 1, F_z = 1\rangle$. In the spin-1/2 subspace described by the Pauli matrix, the Hamiltonian can be written as

$$\hat{H} = \frac{\hbar^2}{2m}(\hat{k}_x - k_0\sigma_z)^2 + \frac{\hbar^2 \hat{\mathbf{k}}_\perp^2}{2m} + \frac{\delta}{2}\sigma_z - \Omega\sigma_y, \tag{1.50}$$

where δ denotes $h - \delta\omega - q$. Upon a spin rotation, the Hamiltonian can also be written in the form used most often in literature:

$$\hat{H} = \frac{\hbar^2}{2m}(\hat{k}_x - k_0\sigma_z)^2 + \frac{\hbar^2 \hat{\mathbf{k}}_\perp^2}{2m} + \frac{\delta}{2}\sigma_z + \Omega\sigma_x. \tag{1.51}$$

This Hamiltonian has been realized in both Bose condensate [114] and degenerate Fermi gas [181, 32]. The effects of this spin-orbit coupling in ultracold Bose and Fermi gases have been extensively studied in cold atom physics [186]. We will discuss how this synthetic spin-orbit coupling affects the properties of a Bose condensate in Section 4.5.

1.4 Stimulated Raman Adiabatic Passage

In Section 1.3 we have discussed that atoms in their ground state can experience an effective potential due to the two-photon process via intermediate excited states. In this section, we will discuss a dynamical process that can transfer atoms from one low-energy state to another low-energy state, also through a two-photon process via an intermediate state. This is known as Stimulated Raman Adiabatic Passage (STIRAP).

Here we introduce the simplest version of the STIRAP [178]. As shown in Figure 1.8, the two low-energy states are denoted by $|1\rangle$ and $|2\rangle$ with energy E_1 and E_2, and the excited state is denoted by $|e\rangle$ with energy E_{ex}. A laser called a "pump laser" couples $|1\rangle$ to $|e\rangle$ with energy ω_p and a time-dependent coupling strength $\Omega_p(t)$, and another laser called a "stoke laser" couples $|2\rangle$ to $|e\rangle$ with energy ω_s and a time-dependent coupling strength $\Omega_s(t)$. Here it is important to note that the time dependence of $\Omega_s(t)$ and $\Omega_p(t)$ should be slow enough compared to other time scales. Here we introduce two detunings $\Delta_p = E_{ex} - E_1 - \hbar\omega_p$ and $\Delta_s = E_{ex} - E_2 - \hbar\omega_s$, and STIRAP requires $\Delta_p = \Delta_s$, which will be denoted by Δ below. Therefore, under the rotating wave approximation, the time-dependent Hamiltonian is given by

$$\hat{H}(t) = \hbar|\Psi\rangle\mathcal{H}(t)\langle\Psi|^T, \tag{1.52}$$

where $|\Psi\rangle$ denotes $(|1\rangle, |2\rangle, |e\rangle)$ and $\langle\Psi|^T$ denotes $(\langle1|, \langle2|, \langle e|)^T$, and $\mathcal{H}(t)$ is a 3×3 matrix given by

$$\mathcal{H}(t) = \begin{pmatrix} 0 & 0 & \Omega_p(t) \\ 0 & 0 & \Omega_s(t) \\ \Omega_p(t) & \Omega_s(t) & \Delta \end{pmatrix}. \tag{1.53}$$

This Hamiltonian can in fact be rewritten as

$$\hat{H}(t) = A(t)\left[\left(\frac{\Omega_p(t)}{A(t)}|1\rangle + \frac{\Omega_s(t)}{A(t)}|2\rangle\right)\langle e| + \text{h.c.}\right] + \Delta|e\rangle\langle e|, \tag{1.54}$$

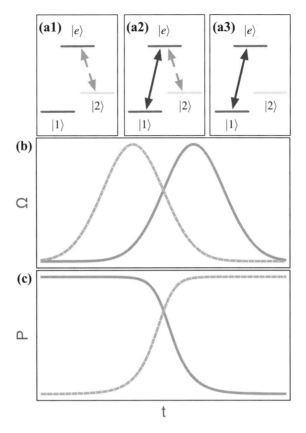

Figure 1.8 Schematic of the STIRAP scheme. (a) How two lasers couple three different states at three different stages. The dashed line and the solid line denote the stoke laser and the pump laser, respectively. (b) The temporal profiles of Ω_s (dashed line) and Ω_p (solid line) as a function of time. (c) The population in the $|1\rangle$ state (solid line) and the population in the $|2\rangle$ state (dashed line) as a function of time. A color version of this figure can be found in the resources tab for this book at cambridge.org/zhai.

where $A(t) = \sqrt{\Omega_p^2(t) + \Omega_s^2(t)}$. Now we define a "bright" state $|B\rangle$ as

$$|B\rangle = \frac{\Omega_p(t)}{A(t)}|1\rangle + \frac{\Omega_s(t)}{A(t)}|2\rangle \qquad (1.55)$$

and another "dark state" $|D\rangle$ orthogonal to $|B\rangle$ as

$$|D\rangle = -\frac{\Omega_s(t)}{A(t)}|1\rangle + \frac{\Omega_p(t)}{A(t)}|2\rangle. \qquad (1.56)$$

In the Hilbert space spanned by $\{|1\rangle, |2\rangle\}$, only $|B\rangle$ couples to $|e\rangle$, and $|D\rangle$ state does not couple to $|e\rangle$ at all. That is to say, for any given time t, $\mathcal{H}(t)$ can be diagonalized by a unitary matrix $\mathcal{U}(t)$, and there will always be an instantaneous eigenstate $|D(t)\rangle$ whose eigenenergy always remains as zero. The other two instantaneous eigenstates will be in superposition of $|B\rangle$ and $|e\rangle$, and their energies will be $\pm\sqrt{A^2(t) + \Delta^2/4}$.

The key idea of STIRAP relies on this dark state $|D(t)\rangle$. If one solves the time-dependent Schrödinger equation, there will be an extra $i\hbar(\partial_t \mathcal{U}^\dagger(t))\mathcal{U}(t)$ term when one rotates into the instantaneous eigenstate bases, as we discussed in Section 1.3. The off-diagonal matrix elements of this term can couple different instantaneous eigenstates. However, since the three instantaneous eigenstates are always separated by a finite energy difference, if $\mathcal{H}(t)$ changes sufficiently smoothly as a function of time, $\mathcal{U}(t)$ also varies sufficiently smoothly, and therefore the off-diagonal matrix elements can be made sufficiently small compared with the energy separation of the three instantaneous eigenstates. Therefore, in this adiabatic regime, the system can nearly remain in one of the instantaneous eigenstates during the time evolution. In this case of STIRAP, we would like to keep the system in the dark state $|D(t)\rangle$.

The idea of the STIRAP is to utilize the dark state $|D(t)\rangle$ to transfer atoms from $|1\rangle$ to $|2\rangle$ by properly designing the time dependence of $\Omega_p(t)$ and $\Omega_s(t)$. To fulfill this goal, at initial time, $\Omega_p/\Omega_s \to 0$ and therefore $|D\rangle \to |1\rangle$; and at the final time of the transfer, $\Omega_s/\Omega_p \to 0$ and therefore $|D\rangle \to |2\rangle$. The time sequence is very counterintuitive. It basically constitutes three steps: (1) initially, atoms are populated in state $|1\rangle$, however, one first opens up the coupling between $|2\rangle$ and $|e\rangle$, as shown in Figure 1.8(a1); (2) in the intermediate stage, both lasers' coupling is open, and atoms are transferred from $|1\rangle$ to $|2\rangle$, as shown in Figure 1.8(a2); and (3) when atoms have been gradually transferred to the $|2\rangle$ state, the coupling between $|e\rangle$ and the less occupied $|1\rangle$ should remain open, as shown in Figure 1.8(a3). We schematically show an example of the temporal profiles of $\Omega_s(t)$ and $\Omega_p(t)$ in Figure 1.8(b). Correspondingly, the populations of the $|1\rangle$ state and $|2\rangle$ state are given by $\Omega_s^2(t)/A^2(t)$ and $\Omega_p^2(t)/A^2(t)$, respectively, and they are plotted in Figure 1.8(c).

The STIRAP scheme transfers atoms from one state to another in a coherent way, and the energy difference between these two states is taken away by photons such that there is no inelastic energy transfer. The STIRAP scheme has two major advantages that make the STIRAP scheme very stable:

- Because this dark state does not involve the excited state component $|e\rangle$, it is stable against the spontaneous emission of the $|e\rangle$ state.
- This scheme is not sensitive to the details of the temporal profile of $\Omega_p(t)$ and $\Omega_s(t)$, and it works as long as the temporal profiles satisfy the aforementioned initial and long time conditions.

STIRAP has been widely used in atomic and molecular physics, as well as chemistry [178]. In cold atom physics, STIRAP has been mostly used for producing ultracold ground state molecules [127]. In this application, it transfers the two-body molecular state instead of the single-particle atomic states, but the working principle is the same as discussed above. The experiment starts with an ultracold sample of the Feshbach molecules whose energy is very close to scattering threshold, and the size is as large as the interparticle spacing. The Feshbach molecule will be discussed in Section 2.4. The goal is to transfer them into the ground state molecule, whose energy is more than 10^{14}Hz below the threshold, and the size of the molecule is a few times the Bohr radius. This process is shown in Figure 1.9. In this case, the excited state is chosen as one of the excited molecular states that has reasonably large coupling matrix elements with both the initial Feshbach molecule

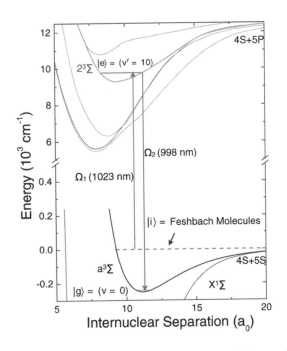

Figure 1.9 STIRAP for producing ground-state molecules. Schematic of using the STIRAP method to transfer a molecule from a Feshbach molecule to the ground state molecule through a molecular excited state. Reprinted from Ref. [127]. A color version of this figure can be found in the resources tab for this book at cambridge.org/zhai.

and the final ground state molecule. The transfer efficiency can be larger than 90%. As the STIRAP process is coherent, the final sample of the ground state molecules remains at very low temperature. Using this method, a degenerate gas of ground state molecules has been produced [45].

Exercises

1.1 Considering an electron with Coulomb interaction between electron and nucleus, the Hamiltonian is given by

$$\hat{H} = -\frac{\hbar^2 \nabla_i^2}{2m^*} - \frac{Z\kappa}{r}, \tag{1.57}$$

and considering the Laplace–Runge–Lene vector defined as

$$\hat{\mathbf{J}} = \frac{1}{2m^*}\left(\hat{\mathbf{p}} \times \hat{\mathbf{L}} - \hat{\mathbf{L}} \times \hat{\mathbf{p}}\right) - Z\kappa\frac{\mathbf{r}}{r}. \tag{1.58}$$

Show that $[\hat{\mathbf{J}}, \hat{H}] = 0$. This operator $\hat{\mathbf{J}}$ and the angular momentum operator $\hat{\mathbf{L}}$ together form the $SO(4)$ algebra. This only works for the Coulomb potential.

1.2   Considering the excited state of the alkali-metal atom with $L = 1$ and $S = 1/2$, write down the eigenstates for a Hamiltonian with spin-orbit coupling

$$\mathbf{H} = \alpha_f \hat{\mathbf{S}} \cdot \hat{\mathbf{L}}, \tag{1.59}$$

where α_f is a constant.

1.3   Considering a ^{87}Rb or ^{6}Li atom in magnetic field $B\hat{z}$, the Hamiltonian can be written as Eq. 1.6.

(1) Solve the full energy spectrum E as a function of B.

(2) Analyze how E depends on B for small B, and find both the linear and the quadratic Zeeman shifts. For the linear Zeeman field, one can also use the Wiger–Eckart theorem to determine the coefficient g_F.

(3) Analyze the spin structure at large B.

1.4   Show that in a regime without electric current, the magnetic field strength can not have a local maximum.

1.5   Estimate the regime where the Majorana transition becomes significant for ^{87}Rb atoms in a pure quadrupole trap with $\mathbf{B} = B_0(x, y, -2z)$.

1.6   Derive Eq. 1.19 following the rotating wave approximation.

1.7   Considering the Hamiltonian Eq. 1.42,

(1) show, by using the rotating wave approximation with $\hat{\mathcal{U}} = e^{-i\delta \omega t \hat{F}_z/\hbar}$, that the Hamiltonian can be reduced to the form of Eq. 1.43.

(2) show, by using a unitary transformation $\hat{\mathcal{U}} = e^{-i2k_0 x \hat{F}_z}$, that the Hamiltonian Eq. 1.43 can be mapped to Eq. 1.51.

1.8   Considering a time-dependent Hamiltonian

$$\hat{H} = \omega \hat{F}_z + B_0 \cos(\omega_0 t)\hat{F}_x, \tag{1.60}$$

show that when ω_0 is close to ω, by rotating wave approximation, the Hamiltonian can become a time-independent one as

$$\hat{H} = \Delta \hat{F}_z + \frac{B_0}{2}\hat{F}_x, \tag{1.61}$$

where $\Delta = \omega - \omega_0$.

1.9   Considering a Hamiltonian with four coupled states $|0\rangle$, $|1\rangle$, $|2\rangle$, and $|3\rangle$,

(1) construct a Hamiltonian that exhibits two dark states. Write down the wave function of these two dark states.

(2) discuss the condition when this Hamiltonian has one dark state.

Two-Body Interaction

Learning Objectives

- Highlight "short-ranged," "dilute," and "low energy" as three main features of interactions between ultracold atoms.
- Introduce the important concept of the phase shift.
- Introduce the s-wave scattering length as a universal parameter describing the low-energy interaction between ultracold atoms.
- Discuss the relation between divergent scattering length, low-energy bound state, and jump of phase shift.
- Discuss the relation between the scattering length and the scattering amplitude.
- Discuss under what conditions a positive scattering length describes repulsive interaction.
- Discuss the conditions when an algebraically decayed potential can be treated as a finite range one.
- Introduce two types of zero-range single-channel potentials to capture the universal low-energy s-wave interaction between ultracold atoms.
- Introduce the concepts of renormalization condition and renormalizable theory.
- Discuss how the spin rotational symmetry imposes constraints on interaction forms for both alkali-metal and alkaline-earth-metal atoms.
- Introduce Feshbach resonance as an important tool to tune scattering length.
- Compare the two-channel Feshbach resonance with the single-channel shape resonance, and compare wide and narrow resonances.
- Introduce a zero-range two-channel model.
- Introduce the confinement-induced resonance to tune interaction strength by an external potential.
- Summarize three key conditions for a Feshbach resonance, and unify the optical Feshbach resonance, the orbital Feshbach resonance, and the confinement induce resonance all in terms of these three conditions.
- Introduce the Efimov effect as an important three-body effect at the vicinity of the two-body scattering resonance.
- Highlight the symmetry aspect of the Efimov effect.
- Discuss various connections between few-body and many-body physics.
- Illustrate that few-body calculation can be used to determine properties of many-body systems by using high-temperature expansion as an example.

2.1 Scattering Length

Interactions between particles play the most important role in quantum many-body physics. One of the major common goals of both ultracold atomic physics and condensed matter physics is to understand interaction effects in quantum many-body systems. Ultracold atomic physics studies dilute gases of neutral atoms, and condensed matter physics mainly focuses on electronic gases in solid. Many fundamental differences between the many-body phenomena in these two systems can be traced back to the different interaction forms between particles in these two systems. The interaction between electrons are the Coulomb repulsion, and the interatomic potential contains an attractive Van der Waals potential and a strong repulsion at very short distance. This interaction between ultracold atoms possesses the following three key features that are important for our subsequent discussions.

- Short Ranged: The Van der Waals interaction $V(\mathbf{r})$ is short ranged, and to a certain extent, we can approximate $V(\mathbf{r}) \simeq 0$ when $r > r_0$, where r_0 is the range of the potential.
- Dilute: The ultracold atomic gas is very dilute, and the typical distance d between two atoms is much larger than r_0.
- Low Energy: The temperature of ultracold atomic gas is very low; that is to say, the incoming energy $E = \hbar^2 k^2/(m)$ of the scattering state is very low compared with the short-range potential energy, that is, $kr_0 \ll 1$, or equivalently, $\hbar^2 k^2/(m) \ll \hbar^2/(mr_0^2)$.

With the first two points, it seems that for most of the time, any two atoms are far separated at a distance where the interaction potential is zero. Thus, classically atoms do not experience any forces mutually and the gas looks like a noninteracting one. However, as we will show in this chapter, in the quantum regime, this system is not only an interacting one but also sometimes can become a strongly interacting one.

The Phase Shift. Let us consider a two-body Schrödinger equation in the relative coordinate

$$\left[-\frac{\hbar^2}{2\bar{m}}\nabla^2 + V(\mathbf{r})\right]\Psi = E\Psi, \tag{2.1}$$

where $\mathbf{r}$ stands for the relative coordinate between two atoms, $\bar{m}$ is the reduced mass of two particles, and $\bar{m} = m/2$ for particles with equal mass m. Here we focus on the situation that $V(\mathbf{r})$ is spherical symmetric,[1] and we can expand the wave function in terms of different angular momentum partial waves as

$$\Psi(\mathbf{r}) = \sum_{l=0}^{+\infty} \frac{\chi_{kl}(r)}{kr}\mathcal{P}_l(\cos\theta), \tag{2.2}$$

and different partial waves are decoupled. It is easy to show that

$$\frac{d^2\chi_{kl}}{dr^2} - \frac{l(l+1)}{r^2}\chi_{kl} + \frac{2\bar{m}}{\hbar^2}(E - V(r))\chi_{kl} = 0. \tag{2.3}$$

---

[1] For atoms like dysprosium and erbium with partially filled f-shells, as we discussed in Section 1.1, the Van der Waals interaction is anisotropic in the presence of an external magnetic field, and there also presents an anisotropic dipolar interaction. Here we do not discuss this situation.

We first consider the s-wave scattering channel with $l = 0$. Because of the short-ranged nature of the potential, for $r > r_0$, $V(r) = 0$, and in this regime a general solution to Eq. 2.3 is given by

$$\chi_k = A \sin(kr + \delta_k), \tag{2.4}$$

where δ_k is called the *phase shift*. The phase shift is the most important quantity for low-energy scattering in a dilute quantum gas. Because of the dilute nature of the ultracold atomic gas stated above, atoms have a negligible chance to come close enough to explore the details of the interaction potential, and therefore we are only concerned about the wave function in the regime $r \gg r_0$. It is clear that all the interaction effects are contained in the phase shift δ_k.

However, the phase shift is determined by the behavior of the wave function at short distance. To determine δ_k, we need the information in the regime with $r < r_0$. We shall match the boundary condition at $r = r_0$ to give

$$\left. \frac{\chi'(r > r_0)}{\chi(r > r_0)} \right|_{r=r_0} = \frac{k \cos(kr_0 + \delta_k)}{\sin(kr_0 + \delta_k)} \simeq \frac{k}{\tan \delta_k} = \left. \frac{\chi'(r < r_0)}{\chi(r < r_0)} \right|_{r=r_0}. \tag{2.5}$$

Here, for the second approximate equality, we have used the low-energy property to approximate $kr_0 \approx 0$. Now, the question is, to determine the phase shift, do we need to know the full information of the wave function $\chi(r < r_0)$ inside $r < r_0$? Let us consider the situation that when $r < r_0$, the interaction potential changes very rapidly, and in this regime, the incoming energy E can be ignored compared with the strength of $V(\mathbf{r})$. Thus, it is reasonable to assume that the energy E dependence of $\chi(r < r_0)$ is insignificant for the low-energy states. Therefore, to the leading order, we can take $\chi'(r < r_0)/\chi(r < r_0)|_{r=r_0}$ simply as a constant denoted by $-1/a_s$, where a_s is called the *s-wave scattering length*. Thus we have

$$\frac{k}{\tan \delta_k} = -\frac{1}{a_s}. \tag{2.6}$$

With Eq. 2.6 the relation between δ_k and k is shown in Figure 2.1. It shows that for small k, δ_k linearly depends on momentum as $-ka_s$, and for large k (but still much smaller compared with $1/r_0$), δ_k saturates to $\pm\pi/2$. We should also note that for deriving Eq. 2.6, we only use the three conditions discussed at the beginning, and Eq. 2.6 is therefore valid for any value of a_s, including $a_s = \pm\infty$. As shown in Figure 2.1, when a_s becomes larger, the linear regime of δ_k becomes smaller. When $a_s = \pm\infty$, δ_k becomes $\mp\pi/2$ for any nonzero k.

We can further treat E as a perturbation for the Schrödinger equation in the regime $r < r_0$ and improve this expansion systematically. By expanding $k/\tan \delta_k$ to the next order in k^2, we obtain

$$\frac{k}{\tan \delta_k} = -\frac{1}{a_s} + \frac{1}{2}r_{\text{eff}}k^2 + \cdots, \tag{2.7}$$

where the coefficient defines an effective range r_{eff}. In most cases, the contribution from the r_{eff} term is negligible at low energy, and a_s is the most important parameter for describing low-energy two-body interactions. However, there are exceptions. For example, when a_s

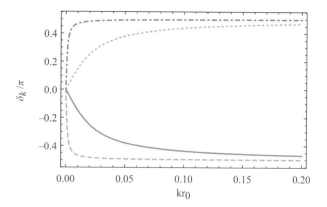

The s-wave phase shift. The phase shift δ_k/π as a function of kr_0 for different a_s/r_0; $a_s/r_0 = 50$ for the solid line, $a_s/r_0 = 10^3$ for the dashed line, $a_s/r_0 = -50$ for the dotted line, and $a_s/r_0 = -10^3$ for the dash-dotted line. A color version of this figure can be found in the resources tab for this book at cambridge.org/zhai.

approaches zero, the expansion equation 2.7 is not very appropriate because δ_k cannot be always zero for all k. In the limit $a_s \to 0$, it is better to expand [19]

$$-\frac{\tan \delta_k}{k} = \frac{1}{\frac{1}{a_s} - \frac{1}{2}r_{\text{eff}}k^2} \approx a_s + \frac{1}{2}r_{\text{eff}}a_s^2 k^2 + \ldots. \tag{2.8}$$

In fact, in the limit $a_s \to 0$, r_{eff} diverges such that $r_{\text{eff}}a_s^2$ remains finite, and $v = -r_{\text{eff}}a_s^2$ is called the *scattering volume*. That is to say, when a_s is finite, $\tan \delta_k$ linearly depends on k, and when a_s vanishes, the linear term vanishes and $\tan \delta_k$ depends on $\sim k^3$.

Considering two different short-range potentials $V_1(\mathbf{r})$ and $V_2(\mathbf{r})$, say, for two interaction potentials of two different atoms, the short-range wave functions $\chi_1(r)$ and $\chi_2(r)$ are also very different in the regime $r < r_0$. But as long as they give the same value of χ'/χ at $r = r_0$ and therefore the same phase shift, the low-energy physics of the two systems are identical, despite the very different behaviors of the short-range potentials. If we further focus on the situation that the effective range effect is negligible, then these two potentials share the same a_s, and this a_s is the only parameter that is needed for describing the low energy of two different microscopic potentials. This is so-called *universality*, which states that different systems with quite different microscopic details can be described universally by a few parameters.

The s-wave scattering length also possesses a clear geometric meaning. In the zero-energy limit, the s-wave wave function at $r > r_0$ can be expanded as

$$\chi(r) \propto \sin(kr + \delta_k) \approx \sin \delta_k + \cos \delta_k (kr)$$

$$\propto 1 + \frac{k}{\tan \delta_k}r = 1 - \frac{r}{a_s}. \tag{2.9}$$

It is clear that $\chi(r = a_s) = 0$; that is to say, a_s is the location of the node of the zero-energy radial wave function.

Let us consider a toy model with a finite range attractive square well potential $V(r) = -V_0$ ($V_0 > 0$) for $0 < r < r_0$, and $V(r) = 0$ for $r > r_0$, and we also consider a hard core boundary condition at $r = 0$. This simple toy model mimics the real interatomic potential. In this model, the zero-energy wave function for $0 < r < r_0$ is given by

$$\chi(r) = \sin\left(\sqrt{\frac{2V_0\bar{m}}{\hbar^2}}r\right). \tag{2.10}$$

It satisfies the hard core boundary condition at $r = 0$, and its slope at $r = r_0$ determines the wave function at the outside, whose node determines a_s. With this picture, it is easy to show how a_s changes as the depth V_0 of the attractive well increases. When the attractive well is shallow and $\sqrt{2V_0\bar{m}/\hbar^2}r_0 < \pi/2$, the situation is shown in the left of Figure 2.2(a), where the node of the wave function appears at a negative value, giving rise to a negative a_s. As V_0 increases, when $\sqrt{2V_0\bar{m}/\hbar^2}r_0$ approaches $\pi/2$, the slope for the zero-energy wave function approaches zero. As a result, a_s first approaches $-\infty$, and then jumps from $-\infty$ to $+\infty$, as shown in the middle of Figure 2.2(a). At this jump, the phase shift also jumps from $\pi/2$ to $-\pi/2$. Then, when V_0 further increases, the slope becomes negative and the node comes to a positive value, as shown in the right of Figure 2.2(a). As V_0 further increases, a_s decreases from $+\infty$ to finite positive value. This simple example shows that a_s can take any value from $-\infty$ to $+\infty$, as we shown in Figure 2.2(b).

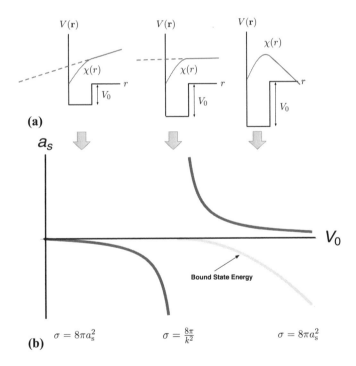

Figure 2.2 Geometric meaning of the scattering length. (a) The geometric meaning of the s-wave scattering length illustrated by a square well model. (b) The s-wave scattering length as a function of the depth of the attractive well. The low-energy bound state energy is also plotted. Different behaviors of the scattering amplitudes are marked in different regimes. A color version of this figure can be found in the resources tab for this book at cambridge.org/zhai.

In this simple toy model, it can also be shown that

$$-\frac{1}{a_s} = \frac{\chi'(r < r_0)}{\chi(r < r_0)}\bigg|_{r=r_0} = \frac{\sqrt{\frac{2V_0\bar{m}}{\hbar^2}}}{\tan\left(\sqrt{\frac{2V_0\bar{m}}{\hbar^2}}r_0\right)}. \tag{2.11}$$

It is clear that a_s will repeatedly change from ∞ to $+\infty$ as V_0 increases. That is to say, there will be a series of different values of V_0 that give the same value of a_s. With our discussion of universality above, this means that the low-energy physics is the same for these different values of V_0.

Finally, it is worth mentioning that the discussion of the phase shift can be generalized to other higher partial waves. One can show that for the lth partial wave, the corresponding phase shift $\delta_k \propto k^{2l+1}$. Therefore, for low-energy collision, the phase shifts for the higher partial waves are suppressed compared with the s-wave case. That means that at the lowest energy, the interaction effect is dominated by the s-wave channel, as long as the s-wave channel is not forbidden.

The Shallow Bound State. Above we have considered low-energy scattering states with $E > 0$, and now we turn to consider a bound state with negative energy $E < 0$. The difference between a scattering state and a bound state lies on the asymptotic behavior at large r. For scattering states, their wave functions keep oscillating with a fixed momentum at large r. The energy spectrum of scattering states is a continuum, and the short-range boundary condition determines the phase shift. But for bound states, their wave functions decay exponentially at large r, and the energy spectrum of bound states is discrete, which is determined by the short-range boundary condition. Explicitly, in the regime $r > r_0$, the radial wave function is given by

$$\chi = Ae^{-r\sqrt{2\bar{m}|E|/\hbar^2}}. \tag{2.12}$$

Similarly to the discussion of the low-energy scattering state, here we are concerned with the absolute value of the binding energy being much weaker than the strength of the potential, such that the wave function at the short distance $r < r_0$ is also insensitive to the bound state energy. Therefore, we can match the same boundary condition at $r = r_0$ for this bound state wave function and reach

$$\frac{\chi'}{\chi}\bigg|_{r=r_0} = -\sqrt{\frac{2\bar{m}|E|}{\hbar^2}} = -\frac{1}{a_s}. \tag{2.13}$$

Obviously, if $a_s < 0$, there is no solution for Eq. 2.13, which means that there is no shallow bound state for negative a_s. But for $a_s > 0$, we have a bound state solution with

$$E_b = -\frac{\hbar^2}{2\bar{m}a_s^2}, \tag{2.14}$$

which is shown in Figure 2.2(b).

Nevertheless, we should be very careful about the statement of no bound state with negative a_s. In fact, as shown in Figure 2.2, the first bound state appears at $\sqrt{2V_0\bar{m}/\hbar^2}r_0 = \pi/2$, and it will not disappear when V_0 further increases. However, there are regimes with larger V_0 where a_s is negative. That is to say, for these negative a_s, there actually exist bound states. To reconcile this fact with Eq. 2.13 above, we should note that Eq. 2.13 only applies to the low-energy bound state because of the assumption that boundary conditions are insensitive to energy. The absence of a solution for Eq. 2.13 only rules out the existence of the low-energy bound state but does not exclude the existence of deep bound states. In fact, it is easy to see that the bound state that emerged at $\sqrt{2V_0\bar{m}/\hbar^2}r_0 = \pi/2$ becomes a sufficient deep bound state when a_s becomes negative again. Moreover, even for positive a_s, Eq. 2.14 works only when $a_s/r_0 \gg 1$. The binding energy will deviate from this universal expression, and the short-range details will matter when the bound state is sufficiently deep.

Features of a Scattering Resonance. From above discussions, we can also find that the following three properties occur simultaneously, which we call an *s-wave scattering resonance*:

1. The *s*-wave scattering length jumps from $-\infty$ to $+\infty$.
2. The phase shift jumps by π.
3. A bound state appears at the threshold.

This connection between the jump of the phase shift and the existence of a zero-energy bound state is also related to Levinson's theorem in quantum mechanics.

Now we have introduced the scattering length a_s as a central concept for the *s*-wave scattering. Below we will address two important questions regarding how to interpret the physical meaning of a_s.

In What Sense Does a Larger $|a_s|$ Mean a Stronger Interaction? We have considered the two-body problem from the perspective of the eigenstate of the Schrödinger equation. Here we introduce another viewpoint. Considering an incoming wave e^{ikz} along the $\hat{z}$ direction that is scattered to an outgoing wave along the radial direction, the total wave function at large distance can be written as

$$\Psi = e^{ikz} + f(\theta)\frac{e^{ikr}}{r},\tag{2.15}$$

where $f(\theta)$ is called the *scattering amplitude*. To determine $f(\theta)$, we need to first rewrite Eq. 2.15 as

$$\Psi = \frac{1}{2ikr}\left[\sum_{l=0}^{+\infty}(2l+1)\mathcal{P}_l(\cos\theta)(e^{ikr-l\pi/2} - e^{-i(kr-\pi l/2)})\right] + f(\theta)\frac{e^{ikr}}{r}.\tag{2.16}$$

Because these two viewpoints should give the same results, by comparing this equation with Eq. 2.2 and Eq. 2.4, and focusing on the $l = 0$ channel, we can obtain the *s*-wave scattering amplitude as

$$f_s(\theta) = \frac{e^{2i\delta}-1}{2ik} = -\frac{1}{ik - k/\tan\delta} = -\frac{1}{1/a_s + ik},\tag{2.17}$$

which is independent of θ. For identical bosons, the scattering cross section σ is given by $8\pi|f_s|^2$. If $|ka_s| \ll 1$, we have $f(\theta) \sim -a_s$, and the scattering cross section is $\sigma = 8\pi a_s^2$. Thus, the larger $|a_s|$ is, the larger is the scattering cross section. In this sense, one can say that the absolute value of a_s represents the strength of the interaction. But this argument cannot be generalized to very large or even infinite a_s, because if $|ka_s| \gg 1$, $f(\theta)$ should be approximated as $-1/(ik)$, and then the scattering cross section becomes $8\pi/k^2$. It is interesting to note that in this regime, the scattering cross section strongly depends on momentum of particles under collision and does not depend on any other parameters. This is the so-called unitary regime. As already indicated in this formula of the scattering cross section, the interaction energy of a many-body system at the unitary regime only depends on density and temperature. These behaviors will be discussed in Chapters 5 and 6 in detail.

In What Sense Does $a_s > 0$ Mean Repulsive Interaction? When we talk about a pure short-range repulsive interaction, usually what we naturally have in mind is a hard core potential with size R_0. That is to say, $V(r) = +\infty$ for $r \leqslant R_0$ and $V(r) = 0$ for $r > R_0$, which forces the wave function to vanish at $r = R_0$. This leads to $\delta_k = -kR_0$. The interatomic interaction we considered here has an attractive well, and the microscopic potential form is very different from the hard core potential. However, the low-energy expansion of the phase shift given by Eq. 2.6 can agree with $\delta_k = -kR_0$, at least for a large range of small k, if a_s is positive and $a_s = R_0$. Therefore, as far as the low-energy phase shift of the scattering states is concerned, a positive a_s is equivalent to a hard core repulsive interaction. In other words, as we discussed at the beginning of this section, atoms in a dilute gas can only experience the phase shift δ_k, so they cannot distinguish the actual interatomic potential from a hard core potential for sufficiently low-energy atoms.

However, we shall emphasize that this equivalence is only valid for low-energy scattering states and small a_s such that $ka_s \ll 1$. There are several reasons. First, at large momentum, when ka_s is large, the phase shift given by Eq. 2.6 always saturates to $-\pi/2$, but the phase shift of a hard core potential keeps increasing linearly. Second, in order for the hard core model to be valid in a gas system, the hard core radius R_0 should be taken to be much smaller than the interparticle spacing, typically $1/k$. That also requires $a_s \sim R_0 \ll 1/k$. Third, as discussed above, for positive a_s, there is a low-energy bound state, and such a bound state is also absent in the hard core potential. Nevertheless, only when $ka_s \ll 1$, the absolute value of the binding energy $\hbar^2/(2\bar{m}a_s^2)$ is much larger than typical kinetic energy $\hbar^2 k^2/(2\bar{m})$, and this bound state is well beyond the low-energy regime. In summary, a positive a_s can be regarded as representing a repulsive interaction only when

1. $ka_s \ll 1$ and only the low-energy scattering states are considered.
2. the bound state is sufficiently deep that can be safely ignored for low-energy scattering.

How Short Range Is Short Ranged? So far, we have considered a finite range potential where the interaction is taken strictly as zero above a range r_0. But the actual Van der Waals potential is algebraically decaying one at large distance. Now we shall come back to briefly revisit how good this approximation is to replace an algebraically decaying potential as a strictly finite-range one.

Let us again first recall how we solve the finite-range potential. In the $r > r_0$ regime, the potential term vanishes and the Hamiltonian only contains the kinetic energy term. For a three-dimensional kinetic operator, there are two independent solutions for the s-wave

Box 2.1   **Meaning of Positive Scattering Lengths in This Book**

Since the main part of this book discusses interaction effects in ultracold atomic gases, here we shall make an important statement on modeling the interaction potential in different parts of this book. In Parts II and IV of this book, we always consider the situation that the scattering length is small compared with the interatomic distance. Hence, when we talk about positive a_s, by default, we consider that these two conditions are satisfied, and the interaction is taken as repulsive. In Part III of this book, we will consider the situation that a_s can go across infinity. In this case, these two conditions are not satisfied, and in particular, the role of the shallow bound state is very crucial. In this part, a positive a_s does not mean the interaction is repulsive.

channel, which can be taken as $\sin(kr)/(kr)$ and $\cos(kr)/(kr)$. The general wave function at large distance is a superposition of these two solutions, and the mixing angle of the superposition gives the phase shift, which should be determined by the short-range physics.

This strategy can be generalized straightforwardly to an algebraically decaying potential. The only difference is that one needs to find out the corresponding solutions for a $1/r^\alpha$ potential. It turns out that, for the s-wave case, one can also write down two independent solutions whose asymptotic behavior also approaches $\sin(kr)/(kr)$ and $\cos(kr)/(kr)$, respectively. Hence, the general wave function is a superposition of these two solutions, and the mixing angle of the superposition determines the phase shift in this case. In this case, the phase shift is also determined by the short-range physics. The similar treatment can be generalized to higher partial wave cases. In this way, one can show that for the lth partial wave, $\tan \delta_k \propto k^{2l+1}$ if $2l + 1 \leq \alpha - 2$ and $\tan \delta_k \propto k^{\alpha-2}$ if $2l + 1 \geq \alpha - 2$ [34]. Therefore, as far as the low-energy physics is concerned, if an algebraically decaying potential is considered to be equivalent to a finite-range one, at least the leading-order contribution to the low-energy phase shift has to be the same for these two potentials, which means $\alpha - 2$ has to be larger than $2l + 1$. Thus, for the s-wave channel, α should be greater than 3. For the realistic Van der Waals potential, $\alpha = 6$, and this means that for $l = 0, 1$ channels, it can be treated by the strictly finite-range approximation, and for $l \geq 2$, the algebraically decaying tail needs to be considered more seriously.

2.2 Zero-Range Models

We have discussed the low-energy physics for a two-body problem using a finite-range potential and shown that for most circumstances, the s-wave scattering length a_s is the only parameter that is required for describing the low-energy interaction in a dilute quantum gas. Here we would like to develop an effective model to describe the interaction effects in many-body systems, and we would like to require the following two features in our effective model:

- Inspired by the two-body discussion, we would like to use the s-wave scattering length a_s as the only parameter in our effective model and disregard the microscopic details of the interatomic potential.
- For the convenience of later studies of many-body theories, it will be useful that this effective model is a zero-range one, that is to say, two atoms interact only when they are exactly in the same spatial location. In Section 2.1, we always keep the interaction range r_0 finite, and in this section, we should be taking r_0 to zero.

Here, by *effective*, we mean that the low-energy scattering properties, including phase shift for the low-energy scattering state and the shallow bound state energy, can be well reproduced by this effective model. Below we present two different effective models, and both can achieve this goal.

Pseudopotential. The simplest form of a zero-range model is a delta-function potential $V(\mathbf{r}) \propto \delta(\mathbf{r})$. Obviously, for $r \neq 0$, $V(\mathbf{r}) = 0$ and $\chi(r) = \sin(kr + \delta_k)$ always satisfy the Schrödinger equation. The question is whether the Schrödinger equation can still be satisfied as $r \to 0$. However, as shown above, for zero energy, $\chi(r)$ behaves as $1 - r/a_s$ and therefore $\Psi(r)$ behaves as $1/r - 1/a_s$, which diverges as $1/r$ at the short distance. Therefore, a simple delta-function potential gives a divergent energy. We note that this $1/r$ divergence is not physical, because in the finite-range model discussed above, the free wave function terminates at r_0 and the even short-range wave function is determined by the microscopic potential. In other words, this $1/r$ divergence is an artifact arising from taking r_0 to zero. So a properly defined interaction potential should be able to eliminate this $1/r$ divergency at $r \to 0$ before taking the δ-function interaction.

Let us denote such a potential as $V(\mathbf{r}) = \delta(\mathbf{r})\hat{O}(r)$, and $V(\mathbf{r})$ should satisfy the Schrödinger equation as

$$\left[-\frac{\hbar^2}{2\bar{m}}\nabla^2 + V(\mathbf{r}) \right] \Psi = E\Psi, \qquad (2.18)$$

where $\Psi(r) = \sin(kr + \delta_k)/(kr)$. To focus on the $r \to 0$ limit, let us again consider the expansion of the wave function $\Psi(r)$ around $r = 0$ as

$$\Psi(r) = \frac{1}{r} - \frac{1}{a_s} + o(kr). \qquad (2.19)$$

It is straightforward to show that

$$[\partial_r \cdot r]\Psi(r) = -\frac{1}{a_s} + o(kr), \qquad (2.20)$$

which eliminates the short-range $1/r$ divergence. Hence, when $\hat{O}(r)$ takes the form [77]

$$\hat{O}(r) = \frac{2\pi\hbar^2 a_s}{\bar{m}}\partial_r \cdot r, \qquad (2.21)$$

we have

$$V(\mathbf{r})\Psi(r) = -\frac{2\pi\hbar^2}{\bar{m}}\delta(\mathbf{r}). \qquad (2.22)$$

And because in three dimensions,

$$-\frac{\hbar^2}{2\bar{m}}\nabla^2\Psi(r) = \frac{2\pi\hbar^2}{\bar{m}}\delta(\mathbf{r}), \qquad (2.23)$$

the Schrödinger equation is satisfied. This interaction potential is known as Fermi's pseudo-potential [77]. Furthermore, one can show that the energy of the low-energy bound state can also be reproduced.

Renomalizable Contact Potential. A pseudo-potential model can nicely reproduce the low-energy physics. However, it has a shortcoming that the operator is not Hermitian. Thus, it is not very convenient to use the pseudo-potential in many circumstances, in particular, when a second-quantized form of a many-body Hamiltonian is needed. For studying many-body physics, it is still convenient to use a delta-function contact potential as $V(\mathbf{r}) = g\delta(\mathbf{r})$. Though we have already known that it will cause a divergent problem at short distance, nevertheless, let us proceed further and see how serious the problem is and whether there are ways to fix the problem.

Here we consider spin-1/2 fermions with this delta-function interaction potential as an example. With a delta-function interaction potential, the second-quantized Hamiltonian for spin-1/2 fermions can be written as

$$\hat{\mathcal{H}} = \int d^3\mathbf{r} \left(\sum_\sigma \hat{\Psi}_\sigma^\dagger(\mathbf{r}) \left(-\frac{\hbar^2}{2m}\nabla^2 \right) \hat{\Psi}_\sigma(\mathbf{r}) + g\hat{\Psi}_\uparrow^\dagger(\mathbf{r})\hat{\Psi}_\downarrow^\dagger(\mathbf{r})\hat{\Psi}_\downarrow(\mathbf{r})\hat{\Psi}_\uparrow(\mathbf{r}) \right), \qquad (2.24)$$

where $\hat{\Psi}_\sigma^\dagger(\mathbf{r})$ and $\hat{\Psi}_\sigma(\mathbf{r})$ ($\sigma = \uparrow, \downarrow$) are creation and annihilation operators for fermions at position $\mathbf{r}$. In the momentum space, this Hamiltonian is given by

$$\hat{\mathcal{H}} = \sum_{\mathbf{k}\sigma} \frac{\hbar^2 k^2}{2m} \Psi_{\mathbf{k}\sigma}^\dagger \Psi_{\mathbf{k}\sigma} + \frac{g}{V} \sum_{\mathbf{k},\mathbf{k}_1,\mathbf{k}_2} \hat{\Psi}_{\frac{\mathbf{k}}{2}+\mathbf{k}_1,\uparrow}^\dagger \hat{\Psi}_{\frac{\mathbf{k}}{2}-\mathbf{k}_1,\downarrow}^\dagger \hat{\Psi}_{\frac{\mathbf{k}}{2}-\mathbf{k}_2,\downarrow} \hat{\Psi}_{\frac{\mathbf{k}}{2}+\mathbf{k}_2,\uparrow}, \qquad (2.25)$$

where V is the volume of the system. Here the second term represents scattering between atoms, with the center-of-mass momentum $\mathbf{k}$ conserved and the relative momenta changing from $\mathbf{k}_2$ to $\mathbf{k}_1$.

We first compute a two-body scattering T-matrix with Hamiltonian equation 2.25. We consider an on-shell scattering process with both incoming and outgoing states having the same energy E and the center-of-mass momentum equaling zero. Since the interaction vertex g is now a constant independent of momentum, the leading order diagram is a direct scattering from the incoming state to the outgoing state, whose contribution is g, as shown in Figure 2.3(a). The next-order diagram involves intermediate states, and the relative momentum $\mathbf{p}$ of the intermediate state can be taken at any momentum. Its contribution can be computed by the second-order processes as

$$\frac{1}{V} \sum_{\mathbf{p}} g \frac{1}{E - \frac{\hbar^2 p^2}{m} + i0^+} g, \qquad (2.26)$$

where $i0^+$ is a mathematical technicality necessary for the calculation of the integrals and is also a consequence of causality. Furthermore, one can systematically consider all the

T-matrix for two-body scattering. Ladder diagrams for two-body T-matrix (a) of the renormalizable contact potential model Eq. 2.24 and (b) for the two-channel model Eq. 2.66. The two-channel model will be discussed in Section 2.4.

higher order contributions by including more intermediate states, as illustrated by the so-called ladder diagram shown in Figure 2.3(a). It turns out that for two-body problems, unlike the many-body situation to be discussed in later chapters, the summation of the ladder diagram is an exact solution. The summation of the ladder diagram leads to the so-called *Schwinger–Dyson equation* given by

$$T_2(E) = g + \frac{1}{V} \sum_{\mathbf{p}} g \frac{1}{E - \frac{\hbar^2 \mathbf{p}^2}{m} + i0^+} g + \dots$$

$$= g + \frac{g}{V} \sum_{\mathbf{p}} \frac{1}{E - \frac{\hbar^2 \mathbf{p}^2}{m} + i0^+} T_2(E), \tag{2.27}$$

and thus

$$T_2(E) = \frac{g}{1 - \frac{g}{V} \sum_{\mathbf{p}} \frac{1}{E - \frac{\hbar^2 \mathbf{p}^2}{m} + i0^+}}. \tag{2.28}$$

Here it is important to notice that the summation over momentum in Eq. 2.28 behaves as $\int d^3\mathbf{p}(1/p^2)$ at large momentum and diverges at large momentum in three dimensions. This divergence comes from the upper limit of the energy integration and is called the *ultraviolet divergence*. As we discussed in Box 2.2, such an ultraviolet divergence means the short-range physics is not treated properly. Here, it means nothing but that the short-range $1/r$ behavior of the free wave function should not be taken to the $r \to 0$ limit, and the δ-function contact potential is not appropriate.

This divergence can also be viewed from the Hamiltonian in momentum space equation 2.25, where the scattering vertex is taken as independent of the momentum transfer, because the Fourier transformation of a δ-function potential is a constant. However, this is unphysical because in any physical model with finite range r_0, this scattering vertex always decays toward zero when the transferred momentum is much larger than $\hbar/r_0$. By taking this momentum dependence of the scattering vertex into account, the large momentum divergence in the summation of Eq. 2.28 can be avoided. Nevertheless, the momentum dependence of the scattering vertex at large momentum comes from the short-range structure of the microscopic potential, which is the nonuniversal physics that we do not want to include.

Hence, we encounter a dilemma. On one hand, we understand that the zero-range δ-function potential, or equivalently saying, a momentum-independent scattering vertex at large momentum, is unphysical, which causes ultraviolet divergence. On the other hand,

Quite often, we will encounter the situation that the integration over energy, or equivalently, the integration over momentum space, diverges. There are two different kinds of divergence. One is called the ultraviolet divergence, and the other is called the infrared divergence. The ultraviolet divergence is due to the upper limit of the energy integration taking to infinity, and the infrared divergence is due to the lower limit of the energy integration taking to zero. However, the physical quantity should always be finite. Thus, both divergences mean that something unphysical is mistaken. The ultraviolet divergence usually means that the high-energy physics, or equivalently, the short-range physics, is not treated properly. Here the delta-function potential is such an example. The infrared divergence usually means the low-energy, or equivalently, the large-scale structure, is mistaken. We will discuss an example of the infrared divergence in Section 3.4.

the details of the short-range potential, or the momentum dependence of the scattering vertex at large momentum, is nonuniversal, which we do not want to explicitly include. To overcome this problem, we will implement the idea of *renormalization*. We will still use the delta-function potential, but we will not treat interaction parameter g as a physical parameter. And we should find a way to properly renormalize the interaction parameter g and to relate it to the physical parameter a_s. Hence, let us rewrite

$$
\begin{aligned}
T_2(E) &= \frac{g}{1 - \frac{g}{V} \sum_{\mathbf{p}} \frac{1}{E - \hbar^2 \mathbf{p}^2/(m) + i0^+}} \\
&= \frac{1}{\frac{1}{g} + \frac{1}{V} \sum_{\mathbf{p}} \frac{1}{\hbar^2 \mathbf{p}^2/m} - \frac{1}{V} \sum_{\mathbf{p}} \left(\frac{1}{E - \frac{\hbar^2 \mathbf{p}^2}{m} + i0^+} + \frac{1}{\hbar^2 \mathbf{p}^2/m} \right)} \\
&= \frac{1}{\frac{1}{g} + \frac{1}{V} \sum_{\mathbf{p}} \frac{1}{\hbar^2 \mathbf{p}^2/m} + \frac{ikm}{4\pi \hbar^2}},
\end{aligned}
\tag{2.29}
$$

where $k = \sqrt{mE/\hbar^2}$. This two-body T-matrix should be related to the s-wave scattering amplitude of Eq. 2.17 determined by the two-body calculation above; therefore, we have

$$
T_2(E) = \frac{4\pi \hbar^2}{m} \frac{1}{\frac{1}{a_s} + ik} = \frac{1}{\frac{1}{g} + \frac{1}{V} \sum_{\mathbf{p}} \frac{1}{\hbar^2 \mathbf{p}^2/m} + \frac{ikm}{4\pi \hbar^2}}.
\tag{2.30}
$$

Hence, we reach the important renormalization identity that relates g to physical quantity a_s, that is,

$$
\frac{m}{4\pi \hbar^2 a_s} = \frac{1}{g} + \frac{1}{V} \sum_{\mathbf{p}} \frac{1}{\hbar^2 \mathbf{p}^2/m}.
\tag{2.31}
$$

To conclude, we will use Eq. 2.24 or Eq. 2.25 as our model for a many-body system. But one will often encounter an ultraviolet divergence problem when using this model. When the ultraviolet divergence is encountered, we should use Eq. 2.31 to replace g by the physical parameter a_s, and at the same time this replacement eliminates the divergency. This is an important result that will be repeatedly used in later chapters.

However, there is one important question. Here we obtain the renormalization condition by matching the two-body scattering amplitude. How can we be sure that this renormalization condition can work for a system with more than two particles? In general, the answer is that it may or may not work. If this works, the theory is called *renormalizable*. If this does not work, it means some extra high-energy scales emerge in few- or many-body systems, and these energy scales matter. In fact, as we will see in Chapter 5 and Chapter 6, theory for spin-1/2 fermion is renormalizable. But for spinless bosons, the renormalization condition actually does not work. This can be seen from Section 2.6, where we will discuss the three-body problem for bosons. We will see that an extra high-energy cutoff scale is required for the energy spectrum being bounded from below.

2.3 Spin-Dependent Interaction

In the discussion above, we do not explicitly include the role of the spin degree of freedom of the atoms under collision. From Section 1.1 we already know that atoms can have quite rich spin structures, and we have also discussed in Section 1.3 that in an optical trap, all spin components can be trapped. In fact, spins of atoms can play very important roles in two-body collisions, and their roles are different between the zero magnetic field limit and the finite magnetic field regime. In the zero magnetic field limit, the spin rotational symmetry is preserved, and the spin rotational symmetry imposes constraints on the form of two-body interactions, which will be discussed in this section. In the finite magnetic field regime, the spin rotational symmetry is broken by the Zeeman energy, but the Zeeman energy of spins can be used as a tool to tune the two-body interactions, which will be discussed in Section 2.4.

Alkali-Metal Atoms. Let us first consider the collision between two alkali-metal atoms with spin-f [2]. Here the spin refers to the total hyperfine spin. For simplicity, we take bosons with $f = 1$ as an example, which includes examples like the ground state of ^{87}Rb and ^{23}Na atoms. Due to the spin rotational symmetry, the total spin F of two atoms under collision should be conserved, and for the $f = 1$ case, the total spin of two atoms can therefore be either 0, 1, or 2. Thus, the interaction potential can be written in a diagonal form in the total spin bases as

$$\hat{V}(\mathbf{r}) = \frac{2\pi\hbar^2}{\bar{m}}(a_0\mathcal{P}_0 + a_2\mathcal{P}_2)\delta(\mathbf{r})\partial_r \cdot r, \tag{2.32}$$

where a_0 and a_2 denote the scattering length in the $F = 0$ and $F = 2$ channels, respectively. Here the $F = 1$ channel does not enter the s-wave scattering because the spin wave function is antisymmetric for total spin $F = 1$, and therefore the spatial wave function also has to be antisymmetric in order for the total wave function to be symmetric. Thus, the s-wave scattering is forbidden in this channel.

---

[2] Here we use little f to denote the spin of a single atom and capital F to denote the total spin of two atoms.

The projection operator $\mathcal{P}_F$ is to project the spin wave function of two atoms into subspace with total spin being F. To write $\mathcal{P}_F$ more explicitly, say, in terms of physical observables, we can make use of the following two identities. First, by definition, the identity operator can be written as

$$\sum_F \mathcal{P}_F = 1. \tag{2.33}$$

Second, we consider $\mathbf{f}_1 \cdot \mathbf{f}_2$, and because $\mathbf{f}_1 \cdot \mathbf{f}_2 = (\mathbf{F}^2 - \mathbf{f}_1^2 - \mathbf{f}_2^2)/2$, $\mathbf{f}_1 \cdot \mathbf{f}_2$ only depends on F as

$$\mathbf{f}_1 \cdot \mathbf{f}_2 = \sum_F \left(\frac{F(F+1)}{2} - f(f+1) \right) \mathcal{P}_F. \tag{2.34}$$

By further projecting both sides of Eq. 2.33 and Eq. 2.34 into the Hilbert space of symmetric total spin wave function, the term $\mathcal{P}_1$ can be dropped out. Therefore we have

$$\mathcal{P}_0 + \mathcal{P}_2 = 1 \tag{2.35}$$

$$-2\mathcal{P}_0 + \mathcal{P}_2 = \mathbf{f}_1 \cdot \mathbf{f}_2. \tag{2.36}$$

By solving these two equations, one can then express $\mathcal{P}_0$ and $\mathcal{P}_2$ in terms of identity operator and $\mathbf{f}_1 \cdot \mathbf{f}_2$, and one obtains [70, 130]

$$\hat{V}(\mathbf{r}) = \frac{2\pi \hbar^2}{\bar{m}} \left(a^{(n)} + a^{(s)} \mathbf{f}_1 \cdot \mathbf{f}_2 \right) \delta(\mathbf{r}) \partial_r \cdot r, \tag{2.37}$$

where

$$a^{(n)} = \frac{a_0 + 2a_2}{3} \tag{2.38}$$

$$a^{(s)} = \frac{a_2 - a_0}{3}. \tag{2.39}$$

Here $a^{(n)}$ and $a^{(s)}$ represent the density–density interaction and the spin-dependent interaction, respectively, and the latter is proportional to the difference in the scattering lengths between the $F = 0$ and $F = 2$ channels. When $a_0 = a_2$, the interactions are identical for different spin channels, and therefore $a^{(s)}$ vanishes. In this case, the interaction only depends on the total density, which is invariant under an arbitrary $SU(3)$ rotation of all three spin components. Therefore, the Hamiltonian is $SU(3)$ invariant instead of $SU(2)$ invariant.

Here we should also emphasize that one needs to carefully distinguish the high-spin representation of $SU(2)$ symmetry and the basic representation of $SU(N)$ symmetry. For the $SU(2)$ symmetry, there are only three generators, no matter how large S is, and the interaction is invariant under the rotation generated by these three generators. In the spin-S representation, these three generators are represented by $(2S+1) \times (2S+1)$ Pauli matrices. But for $SU(N)$ symmetry, there are in total $N^2 - 1$ generators, and the interaction is invariant under the rotation generated by all these $N^2 - 1$ generators.

In reality, for atoms like ^{87}Rb and ^{23}Na, the differences between a_0 and a_2 are actually quite small, and consequently, $a^{(s)}$ is only a few percent of $a^{(n)}$. Nevertheless, $a^{(s)}$ plays an important role for spin-1 alkali-metal atoms. Dynamically, this spin-dependent interaction

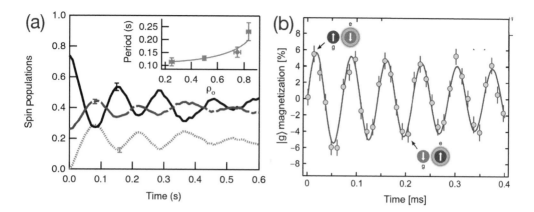

Figure 2.4 Interaction-induced spin exchanging dynamics. (a) The spin exchanging dynamics for spin-1 ^{87}Rb atom. The solid, dotted, and dashed lines are populations on $m_f = 0$, 1, and -1, respectively. Reprinted from Ref. [30]. (b) The nuclear spin exchanging between $|g\rangle$ and $|e\rangle$ states for ^{173}Yb atom. This line is a time-dependent nuclear spin polarization for atoms in $|g\rangle$ state, that is, the population difference between $|g\rangle|\uparrow\rangle$ and $|g\rangle|\downarrow\rangle$. Reprinted from Ref. [27]. A color version of this figure can be found in the resources tab for this book at cambridge.org/zhai.

can lead to a spin exchanging process. Because there are terms $f_1^+ f_2^- + f_1^- f_2^+$ in $\mathbf{f}_1 \cdot \mathbf{f}_2$, two incoming atoms with $f_z = 0$ can be scattered into one in $f_z = 1$ and the other in the $f_z = -1$ state. This process has been observed experimentally, and one of the examples is shown in Figure 2.4(a). Moreover, $a^{(s)}$ determines the spin structure of the Bose–Einstein condensate of spin-1 atoms, as we will see in Section 4.3.

Alkaline-Earth-Metal Atoms. As we have discussed in Section 1.1, for the ground state, the spin of alkaline-earth-metal atoms is purely nuclear spin I. Because the electron spin is zero, there is no coupling between the nuclear spin and the electronic degree of freedom. The nuclear spin is nonzero only for the fermionic alkaline-earth-metal atom because all the bosonic isotopes of alkaline-earth-metal atoms have zero nuclear spin. On the other hand, for fermionic isotopes, the nuclear spin usually can be quite large, and because of the decoupling between the nuclear spin and the electronic degree of freedom, the interaction between two ground-state alkaline-earth-metal atoms is nearly independent of nuclear spin I [193]. Therefore, all the scattering lengths between any two components are all identical, and the interaction only depends on the total density. Such an interaction term possesses $SU(2I + 1)$ symmetry.

Another interesting aspect of alkaline-earth-metal atoms is the interaction between the ground state $^1S_0$ (usually denoted by $|g\rangle$) and the clock state $^3P_0$ (usually denoted by $|e\rangle$). As discussed in Section 1.1, the clock state has long enough lifetime whose single-particle decay can be safely ignored in practice. In literature, these two states are also referred to as a doublet of the "orbital" degree of freedom.[3] Here we consider the interaction between two fermionic alkaline-earth-metal atoms, one in $|g\rangle$ state and the other in $|e\rangle$ state, and

---

[3] Note that here "orbital" labels an internal degree of freedom for atoms, that is, one of the valance electrons is excited to the excited p-orbit.

they can be in two (among $2I + 1$) different nuclear spin states, denoted by $| \uparrow \rangle$ and $| \downarrow \rangle$, respectively. Because the total wave function has to be antisymmetric, and because we consider the s-wave interaction that requires spatial wave function to be symmetric, the internal wave function has to be antisymmetric, which limits the internal Hilbert space to either the orbital triplet and the nuclear spin singlet or the orbital singlet and the nuclear spin triplet, which are

$$|+\rangle = \frac{1}{2} (|g\rangle|e\rangle + |e\rangle|g\rangle) \otimes (| \uparrow \rangle | \downarrow \rangle - | \downarrow \rangle | \uparrow \rangle), \tag{2.40}$$

$$|-,0\rangle = \frac{1}{2} (|g\rangle|e\rangle - |e\rangle|g\rangle) \otimes (| \uparrow \rangle | \downarrow \rangle + | \downarrow \rangle | \uparrow \rangle), \tag{2.41}$$

$$|-,1\rangle = \frac{1}{\sqrt{2}} (|g\rangle|e\rangle - |e\rangle|g\rangle) \otimes | \uparrow \rangle | \uparrow \rangle, \tag{2.42}$$

$$|-,-1\rangle = \frac{1}{\sqrt{2}} (|g\rangle|e\rangle - |e\rangle|g\rangle) \otimes | \downarrow \rangle | \downarrow \rangle, \tag{2.43}$$

where $\pm$ refers to orbital triplet and singlet, respectively, and $0, \pm 1$ in Eq. 2.41–2.43 refers to the z-component of the total nuclear spin. Here we have ignored two orbital triplet states where both atoms are in ground states or both atoms are in the clock state, because here we are interested in the interorbital interaction.

Above we have discussed that the interactions between atoms in the $^1S_0$ state have $SU(2I+1)$ symmetry. For atoms in the clock state $^3P_0$, because the total electronic angular momentum is also zero, the nuclear spin is still decoupled from the electronic degree of freedom, and therefore the interorbital interactions between the $^1S_0$ state and $^3P_0$ state also possess the $SU(2I+1)$ symmetry. Here, if we only consider two out of $2I+1$ nuclear spin components, the interactions possess an $SU(2)$ nuclear spin rotational symmetry. On the other hand, because the "orbital" degree of freedom is just a label of two different states, which is similar to the pseudo-spin-1/2 discussed in Box 2.3, there is no rotational symmetry requirement in the orbital space. This symmetry requirement leads to the following: (1) the interaction is diagonal in the bases of Eq. 2.40–2.43 listed above and (2) the $|+\rangle$ channel has one scattering length, and all three $|-\rangle$ channels share another different scattering length.

Denoting $\mathcal{P}_+$ as projection operator to $|+\rangle$ state, and $\mathcal{P}_-$ as projection operator to the Hilbert space spanned by three $|-\rangle$ ($m_n = 0, \pm 1$) states, we have

$$\mathcal{P}_+ = |+\rangle\langle+| \tag{2.44}$$

$$\mathcal{P}_- = |-,0\rangle\langle-,0| + |-,-1\rangle\langle-,-1| + |-,+1\rangle\langle-,+1|. \tag{2.45}$$

The interaction can be written as

$$\hat{V}(\mathbf{r}) = \frac{2\pi \hbar^2}{\bar{m}} \left(\sum_{\pm} a_{\pm} \mathcal{P}_{\pm} \right) \delta(\mathbf{r}) \partial_r \cdot r, \tag{2.46}$$

where $a_{\pm}$ are two different scattering lengths. For this interaction form, when one atom in $|g\rangle | \uparrow \rangle$ state collides with another atom in $|e\rangle | \downarrow \rangle$ state, there is a channel in which the outcoming atoms are one in $|g\rangle | \downarrow \rangle$ state and the other in $|e\rangle | \uparrow \rangle$ state. In other words,

Spin and Spin Rotational Symmetry

In ultracold atom literatures, "spin" can have different meanings. In some cases, spin means the total hyperfine spin of an atom, as discussed in this section and in spinor condensate discussed in Section 4.3. In such cases, interaction between different spin components should obey the spin rotational symmetry at zero field. In this case, the accuracy of the spin rotational symmetry is guaranted by the fact that the collision energy between two atoms is much weaker compared with the hyperfine coupling of a single atom.

In some other cases, spin actually means the pseudo-spin, which are essentially two or more eigenstates of the total spin Hamiltonian, including both hyperfine coupling and the Zeeman field, as discussed in Section 1.2. In such cases, it is not necessary for interactions between different pseudo-spin components to obey the spin rotational symmetry. In this context, we can have pseudo-spin-$1/2$ Bose gas, which is not possible with real spins. For pseduo-spin-$1/2$ bosons, the two intracomponent interaction parameters and the intercomponent interaction parameter can in principle take arbitrary values. Therefore, there is no spin $SU(2)$ symmetry. The spin-orbit coupled Bose condensate discussed in Section 4.5 is such an example.

For pseudo-spin-$1/2$ Fermi gas, at the lowest order, there are intercomponent s-wave interactions and two intracomponent p-wave interactions, because the intracomponent s-wave interactions vanish due to the Fermi statistics. In general, the two intracomponent p-wave interactions are different, especially when one of them possesses a p-wave Feshbach resonance. However, away from the high-partial wave Feshbach resonances, the high-partial wave interaction can be safely ignored at ultra low temperature compared with the s-wave interaction, as discussed in Section 2.1, and we only need to retain the intercomponent interaction. Under this situation, the interaction again possesses an emergent $SU(2)$ symmetry. In this case, the accuracy of this emergent $SU(2)$ symmetry is guaranteed by the fact that the high-partial wave interaction energy is much weaker compared with the intercomponent s-wave interaction. The spin-$1/2$ Fermi gas discussed in Part III of this book, as well as the Fermi–Hubbard model discussed in Section 8.2, belongs to this case. In particular, we will emphasize the role of the $SU(2)$ spin rotational symmetry in the discussion of the Fermi–Hubbard model.

the nuclear spin between two different orbital states can be exchanged during the collision. This spin exchanging interaction strength is proportional to the difference between a_+ and a_-. This spin exchanging processes have also been observed in experiments [152, 27], as shown in Figure 2.4(b).

Such a spin-exchanging process can find broad applications in quantum simulation of many-body physics, for instance, in simulating the famous Kondo physics with ultracold atoms. The Kondo physics in condensed matter system arises from a localized magnetic impurity embedded in metal, and this magnetic impurity can exchange spin with itinerant electrons. Here, because the scalar polarizabilities are different between atoms in the $|g\rangle$ state and atoms in the $|e\rangle$ state, with the optical lattice scheme discussion in Section 1.3, one can create a situation that atoms in $|e\rangle$ state are localized by a deep potential, and atoms in $|g\rangle$ state experience a shallow potential and remain itinerant [62, 190, 149]. Thus, the atoms in $|e\rangle$ state act as localized impurities embedded in a Fermi sea of the itinerant atoms in $|g\rangle$ state, and the spin exchanging interaction between them can realize the Kondo effect [62, 190].

2.4 Feshbach Resonance

The discussion in Section 2.1 has established that the scattering length is an important quantity for describing the interatomic interaction. Can we tune the scattering length experimentally? In Section 2.1, using the square well potential as an example, we have also shown that the scattering length can be changed by changing the depth of a square well potential. However, in practice, it is hard to vary the strength of the Van der Waals potential over a large energy range. Nevertheless, the discussion in Section 2.1 gives an important hint, that is, if one can tune the energy of a bound state to be close to the scattering threshold, it can strongly affect the scattering length. This is essentially the key idea behind all tunable scattering resonances. Here we should first discuss a magnetic field–tunable *Feshbach resonance*.

The discussion of magnetic Feshbach resonance involves the internal spin structure of atoms in a magnetic field. We should recall that in Section 2.3, we have discussed the role of internal spin structure for two-body collision. The difference between the discussion here and that in Section 2.3 is that here we consider the regime where the effect of an external Zeeman field is strong enough. In Section 2.3 we focus on the zero-field regime where the spin rotational symmetry plays an important role, and we have discussed how the spin rotational symmetry imposes constraints on the form of interaction. But here the presence of a finite Zeeman field breaks the spin rotational symmetry, and therefore such a constraint no longer exists.

Now let us be more specific. We consider interaction between two alkali-metal atoms whose internal spin structure in a Zeeman field has been discussed in Sec 1.2. Let us label each internal spin eigenstate of a single atom in a Zeeman field by $|q\rangle$, which has a well-defined quantum number F_z. For instance, for ^{6}Li, the internal spin eigenstates are shown in Figure 2.5(b). Though there is no $SU(2)$ spin rotational symmetry because of the Zeeman field, there is still a spin rotational symmetry along the field direction, and thus F_z is still a good quantum number. When two atoms are far from each other, they are in the eigenstate of $|q_1\rangle \otimes |q_2\rangle$. Now we can introduce the concepts of two scattering channels. One is the called the *open channel* and the other is called the *closed channel*. These channels are defined as eigenstates when two atoms are far separated. Here are a few remarks about these two channels:

- Quantum Number: Due to the rotational symmetry along $\hat{z}$, the total $F_z^1 + F_z^2 + L_z$ is conserved. Here $F_z^{i=1,2}$ is the z-component of the hyperfine spin of these two atoms, respectively, and L_z is the $\hat{z}$-component of their relative angular momentum. Here, for simplicity, we only consider the s-wave states in both the open and the closed channels, and $L_z = 0$. With this simplification, the total $F_z^1 + F_z^2$ of the open channel should equal to that of the closed channel. For instance, for ^{6}Li, if the open channel is taken as $|a\rangle \otimes |b\rangle$

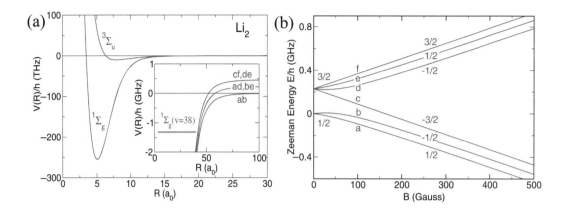

Figure 2.5 Interaction channels between two alkali-metal atoms. (a) The electronic spin singlet and triplet interaction potential of Li_2 at short interatomic separation. The inset shows a zoom-in plot of the interaction potential of two ^6Li atoms at large interatomic separation. The five pairs of states all have total $F_z = 0$. The horizontal line shows a bound state in the closed channel. (b) The internal eigenstate labeled from a to f of ^6Li. The number on each curve is the value of F_z for each state. Reprinted from Ref. [34]. A color version of this figure can be found in the resources tab for this book at cambridge.org/zhai.

with total $F_z = 0$, there are in total five combinations that have total $F_z = 0$, and the other four are $|a\rangle \otimes |d\rangle$, $|b\rangle \otimes |e\rangle$, $|c\rangle \otimes |f\rangle$, and $|d\rangle \otimes |e\rangle$.

- Closed versus Open: When two atoms are far separated, the energy difference between the open and the closed channel is set by the Zeeman energy, which is normally much higher than the kinetic energy. Therefore, when two atoms collide from the low-energy scattering state of the open channel, they cannot be scattered into scattering states of the closed channel. That is why these channels are called "closed channels." As shown in the inset of Figure 2.5(a), all the other four combinations can be taken as closed channels when $|a\rangle \otimes |b\rangle$ is chosen as the open channel.

- Energy Tunability: Usually when the open channel is chosen as the low-lying hyperfine spin state, as the magnetic field increases, the energy of the open channel decreases with respect to the closed channel. Thus, it is conceivable that, as magnetic field increases, the scattering threshold can approach a bound state in the closed channel from above.

- Coupling between Channels: When two atoms are close to each other, the inter-atomic potential between two atoms mostly depends on the electronic degree of freedom of two atoms. Here, since each alkali-metal atom has one electron, the interatomic potential depends on whether their total electronic spin is singlet or triplet,[4] an example of which is shown in Figure 2.5(a). For instance, considering the open channel $|a\rangle \otimes |b\rangle$, their electron spins are polarized by the magnetic field, and their total electronic spin is more close to a triplet. However, the hyperfine coupling mixes in electron spin singlet

---

[4] In practice, the Van der Waals part is the same for electron spin singlets and triplets, but the short-range repulsive part depends on electron spin.

component. Hence, the short-range potential couples different channels, though the coupling is usually weak.

Coupled-Channel Model. With these features of the two channels discussed above, we can consider a simplified coupled-channel model to demonstrate how the scattering length can be changed by the magnetic field [34]. The model is schematically illustrated in Figure 2.6(a). The major considerations are as follows:

- At the distance $r > r_0$, two channels are decoupled, and they are respectively denoted by the open channel $|o\rangle$ and the closed channel $|c\rangle$. Because the energy of the closed

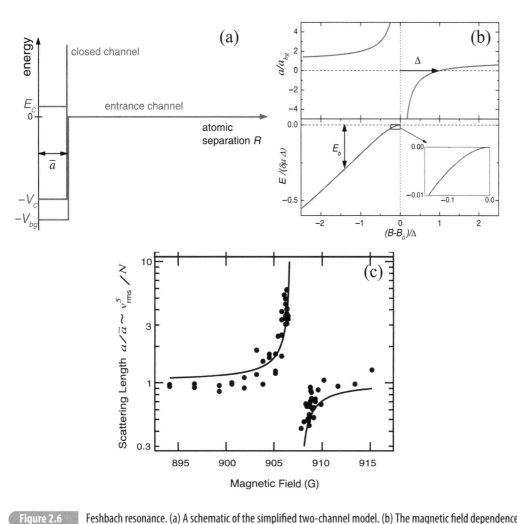

Figure 2.6 Feshbach resonance. (a) A schematic of the simplified two-channel model. (b) The magnetic field dependence of the scattering length and the bound state energy, where B_0 is B_{res} in the text. (c) The first experimental observation of a Feshbach resonance in ^{23}Na. (a) and (b) are reprinted from Ref. [34], and (c) is reprinted from Ref. [79]. A color version of this figure can be found in the resources tab for this book at cambridge.org/zhai.

channel is much higher than the typical kinetic energy of incoming scattering states of the open channel, the wave function of low-lying scattering states only exists in the open channel $|o\rangle$, and the low-energy s-wave wave function is given by $\Psi = \chi/r$ and

$$\chi = \sin(kr + \delta_k)|o\rangle. \tag{2.47}$$

- At the distance $r < r_0$, the wave function is diagonalized in the $|+\rangle$ and $|-\rangle$ bases, written as

$$\chi = \chi_+|+\rangle + \chi_-|-\rangle. \tag{2.48}$$

Here $|\pm\rangle$ are superposition of $|o\rangle$ and $|c\rangle$ as

$$|+\rangle = \cos\theta|o\rangle + \sin\theta|c\rangle \tag{2.49}$$
$$|-\rangle = -\sin\theta|o\rangle + \cos\theta|c\rangle. \tag{2.50}$$

Without loss of generality, we consider θ as spatially independent and quite small.

The wave function in the $r < r_0$ regime can now be written as

$$\chi = (\chi_+\sin\theta + \chi_-\cos\theta)|c\rangle + (\chi_+\cos\theta - \chi_-\sin\theta)|o\rangle. \tag{2.51}$$

To match the boundary conditions in $r = r_0$, we obtain

$$\chi_+\sin\theta + \chi_-\cos\theta\bigg|_{r=r_0} = 0 \tag{2.52}$$

$$\frac{\chi_+'\cos\theta - \chi_-'\sin\theta}{\chi_+\cos\theta - \chi_-\sin\theta}\bigg|_{r=r_0} = \frac{k}{\tan\delta_k} \equiv -\frac{1}{a_s}. \tag{2.53}$$

Eq. 2.52 comes from that the closed channel wave function vanishes at $r = r_0$, and Eq. 2.53 determines the phase shift in the open channel scattering wave function. As discussed in Section 2.1, here we have assumed that both $\chi_\pm$ is independent of energy and r_0 is a small value. These two equations give

$$-\frac{1}{a_s} = \frac{\chi_+'}{\chi_+}\bigg|_{r=r_0}\cos^2\theta + \frac{\chi_-'}{\chi_-}\bigg|_{r=r_0}\sin^2\theta. \tag{2.54}$$

Since θ is usually quite small, the second term on the r.h.s. of Eq. 2.54 is usually insignificant. In that case, the scattering length is provided by the $|+\rangle$ channel only, and we denote

$$\frac{\chi_+'}{\chi_+}\bigg|_{r=r_0} = -\frac{1}{a_{bg}}, \tag{2.55}$$

where a_{bg} is called the background scattering length. Now we have

$$-\frac{1}{a_s} = -\frac{1}{a_{bg}}\cos^2\theta + \frac{\chi_-'}{\chi_-}\bigg|_{r=r_0}\sin^2\theta. \tag{2.56}$$

Again because θ is small, we can approximate $\cos^2\theta \approx 1$ and $\sin^2\theta \approx \theta^2$, and the second term can give rise to a significant contribution only when $\chi_-'/\chi_-|_{r=r_0}$ is very large. As we will see, this means nothing but a bound state appearing nearby the threshold in the $|-\rangle$ channel.

For instance, let us consider the $|-\rangle$ channel as a square well with $V(r) = -V_0$ ($V_0 > 0$) for $r < r_0$, as shown in Figure 2.6(a). In this case, when the energy of the scattering state can be ignored compared with V_0, we have $\chi_- = \sin(q_1 r)$ and $q_1 = \sqrt{mV_0/\hbar^2}$, therefore,

$$\left.\frac{\chi_-'}{\chi_-}\right|_{r=r_0} = \frac{q_1 \cos(q_1 r_0)}{\sin(q_1 r_0)}. \tag{2.57}$$

If there is a bound state with energy E_c, then the bound state wave function is $\chi_-(r) = \sin(q_2 r)$ and $q_2 = \sqrt{m(V_0 + E_c)/\hbar^2}$, and to zeroth order of θ, $|-\rangle$ channel connects to $|c\rangle$ channel at $r = r_0$ and $\sin(q_2 r_0) = 0$. When E_c is small, we can expand q_1 around q_2 and obtain

$$\left.\frac{\chi_-'}{\chi_-}\right|_{r=r_0} \approx \frac{q_1}{(q_1 - q_2)r_0} \approx \frac{2q_1^2}{(q_1^2 - q_2^2)r_0} = -\frac{2\hbar^2 q_1^2}{mr_0 E_c}. \tag{2.58}$$

In fact, although we derive Eq. 2.58 using a square well potential, it holds for a general potential that $\chi_-'/\chi_-|_{r=r_0}$ is inversely proportional to E_c. Denoting $\gamma = 2\hbar^2 q_1^2 \theta^2/(mr_0)$, Eq. 2.56 can be rewritten as

$$\frac{1}{a_s} = \frac{1}{a_{bg}} + \frac{\gamma}{E_c}. \tag{2.59}$$

Here it is important to note that γ depends on θ, which is the coupling between two channels. Eq. 2.59 gives

$$a_s = a_{bg}\left(1 - \frac{\gamma a_{bg}}{E_c + \gamma a_{bg}}\right). \tag{2.60}$$

In the presence of a magnetic field, the threshold energies of the open and the closed channel change as $-\mu_o B$ and $-\mu_c B$, respectively. In most cases, $\mu = \mu_o - \mu_c > 0$. E_c is replaced by $E_c - \mu_c B + \mu_o B = E_c + \mu B$. Defining $\Delta = \gamma a_{bg}\mu^{-1}$ and $B_{res} = -\mu^{-1}E_c - \Delta$, Eq. 2.60 can be rewritten as

$$a_s = a_{bg}\left(1 - \frac{\Delta}{B - B_{res}}\right). \tag{2.61}$$

This result shows that, usually for $\mu > 0$, a_s diverges to $+\infty$ when $B \to B_{res}$ from below and diverges to $-\infty$ when $B \to B_{res}$ from above, as shown in Figure 2.6(b). B_{res} denotes the magnetic field for a scattering resonance, which is close to the position with $E_c = 0$ but is shifted away by Δ. Δ defines the width of a resonance. From Eq. 2.61, one can see that $a_s = \infty$ when $B = B_{res}$ and $a_s = 0$ when $B = B_{res} + \Delta$, and the latter is known as the zero crossing. Thus, Δ measures the distance between the magnetic field for resonant scattering and the magnetic field for the zero crossing. Figure 2.6(c) shows the first experimental observation of a Feshbach resonance in ^{23}Na [79]. Later Feshbach resonances are found in almost all alkali-metal and magnetic atoms, which have become the most important tools for controlling interaction in ultracold atomic physics.

Wide versus Narrow Resonance. One can further show that at finite energy,

$$a_s(E) = a_{bg} \left(1 - \frac{\mu\Delta}{\mu(B - B_{res}) - E} \right). \tag{2.62}$$

Expanding $-1/a_s(E) = -1/a_s + r_{eff}k^2/2$, with $E = \hbar^2 k^2/m$, one obtains the effective range as

$$r_{eff} = -\frac{2\hbar^2\Delta}{\mu m a_{bg}(B - B_{res} - \Delta)^2} \approx -\frac{2\hbar^2}{\mu\Delta m a_{bg}}, \tag{2.63}$$

where the second approximate equality is valid nearby the resonance. This equation shows that the effective range depends on Δ; that is, it depends on γ or θ. This is a major difference between the single-scattering channel model discussed in Section 2.1 and the two-channel model discussed here. In the single-channel model, as discussed in Section 2.1, one can also fine-tune the potential such that there is a bound state at the threshold, and such a resonance is also called a *shape resonance*. Usually for an *s*-wave shape resonance, r_{eff} is usually negligible. But for the two-channel model, depending on how strong the mixing between the open and the closed channel is, the effective range can be tuned over a wide range from very small to quite large, and the sign of r_{eff} depends on the sign of a_{bg}. That is to say, only when Δ in the Feshbach resonance is large enough that r_{eff} is sufficiently small, a Feshbach resonance in the two-channel model is equivalent to a shape resonance in a single-channel model. To characterize the role of the effective range in a many-body system of degenerate Fermi gas, a dimensionless quantity $k_F r_{eff}$ is introduced as

$$k_F r_{eff} = \frac{4E_F}{\mu\Delta(k_F a_{bg})}. \tag{2.64}$$

If $k_F r_{eff} \ll 1$, we call it a *wide resonance*, and if $k_F r_{eff} \gg 1$, we call it a *narrow resonance*. For a narrow resonance, effectively, the scattering length varies a lot over the energy range of E_F; thus the many-body system cannot be described by a single energy-independent parameter of the scattering length a_s. In Chapters 5 and 6, when we discuss the many-body physics of ultracold Fermi gases across a Feshbach resonance, we focus on the wide resonances.

Zero-Range Two-Channel Model. In Section 2.2, we have introduced a zero-range model to describe a single-channel scattering problem. We emphasize that a renormalization condition has to be introduced in order to remove the artificial short-range divergency when taking the range of potential to zero. Above we have introduced a coupled two-channel scattering problem, and we have also noted that the two-channel model is not always equivalent to the single-channel model when the energy dependence of the scattering length has to be taken into account for narrow resonances. Hence, it is desirable to introduce a zero-range version of the two-channel model, which can describe both the wide and the narrow Feshbach resonances. As we will see, here we also need to be careful about the renormalization of the model parameters.

Here, similarly to in Section 2.2, we consider two-component fermions as an example. To capture the two-channel nature of the problem, we explicitly introduce a bosonic $\hat{b}$

field to describe the two-body bound state in the closed channel, which is also called the molecular state. Now the Hamiltonian is written as

$$\hat{\mathcal{H}} = \sum_{\mathbf{k}\sigma} \frac{\hbar^2 \mathbf{k}^2}{2m} \hat{\Psi}_{\mathbf{k}\sigma}^\dagger \hat{\Psi}_{\mathbf{k}\sigma} + \sum_{\mathbf{k}} \left(\frac{\hbar^2 \mathbf{k}^2}{4m} + \nu \right) \hat{b}_{\mathbf{k}}^\dagger \hat{b}_{\mathbf{k}}$$
$$+ \frac{g}{V} \sum_{\mathbf{k},\mathbf{k}_1,\mathbf{k}_2} \hat{\Psi}_{\frac{\mathbf{k}}{2}+\mathbf{k}_1,\uparrow}^\dagger \hat{\Psi}_{\frac{\mathbf{k}}{2}-\mathbf{k}_1,\downarrow}^\dagger \hat{\Psi}_{\frac{\mathbf{k}}{2}-\mathbf{k}_2,\downarrow} \hat{\Psi}_{\frac{\mathbf{k}}{2}+\mathbf{k}_2,\uparrow}$$
$$+ \frac{\alpha}{\sqrt{V}} \sum_{\mathbf{k},\mathbf{k}_1} \hat{\Psi}_{\frac{\mathbf{k}}{2}+\mathbf{k}_1,\uparrow}^\dagger \hat{\Psi}_{\frac{\mathbf{k}}{2}-\mathbf{k}_1,\downarrow}^\dagger \hat{b}_{\mathbf{k}} + \hat{b}_{\mathbf{k}}^\dagger \hat{\Psi}_{\frac{\mathbf{k}}{2}-\mathbf{k}_1,\downarrow} \hat{\Psi}_{\frac{\mathbf{k}}{2}+\mathbf{k}_1,\uparrow}, \tag{2.65}$$

where $\hat{\Psi}_\sigma^\dagger$ and $\hat{\Psi}_\sigma$ are the creation and annihilation operators for scattering states in the open channels. The last term denotes the conversion between the open channel scattering states and the closed channel molecular state, with the strength given by α. Here ν is the detuning of the molecular state in the closed channel, and g is the bare interaction between open channel atoms themselves. This model is a zero-range model because both g and α are chosen as momentum independent. Here, for the reason discussed above, we do not include the scattering states in the closed channel. To find the renormalization relations for ν, α, and g, similarly to our calculation done in Section 2.2, we can sum over the ladder diagrams for the two-channel model to obtain the two-body scattering T-matrix. The ladder diagram for the two-channel model is shown in Figure 2.3(b), compared with the ladder diagrams in the single-channel model. Here we obtain the T_2 as

$$T_2(E) = \frac{g + \frac{|\alpha|^2}{E-\nu}}{1 - \left(g + \frac{|\alpha|^2}{E-\nu} \right) \frac{1}{V} \sum_{\mathbf{k}} \frac{1}{E - \hbar^2 \mathbf{k}^2/(m)}}. \tag{2.66}$$

By comparing $T_2(E = 0) = 4\pi \hbar^2 a_s/m$ with a_s given by Eq. 2.61, we can obtain the renormalization conditions that

$$\frac{1}{g} = \frac{m}{4\pi \hbar^2 a_{bg}} - \Lambda, \tag{2.67}$$

$$\frac{1}{\alpha} = \left(1 - \frac{4\pi \hbar^2 a_{bg}}{m} \Lambda \right) \sqrt{\frac{m}{4\pi \hbar^2 a_{bg} \mu \Delta}}, \tag{2.68}$$

$$\nu = \mu(B - B_{res}) + \frac{\Lambda}{1 - \frac{4\pi \hbar^2 a_{bg}}{m} \Lambda} \frac{4\pi \hbar^2 a_{bg} \mu \Delta}{m}, \tag{2.69}$$

where Λ denotes

$$\Lambda = \frac{1}{V} \sum_{\mathbf{k}} \frac{1}{\hbar^2 \mathbf{k}^2/m}. \tag{2.70}$$

General Schemes of the Feshbach Resonance. From the discussion above, we can summarize the following three key ingredients in order to support a Feshbach resonance:

- For $r > r_0$, atoms stay in the single-particle eigenstates, and the different quantum numbers of the single-particle eigenstates define "channels."

- The energy spacing between different channels can be tuned by an external parameter.
- The short-range potential at $r < r_0$ does not respect the good quantum number of the single-particle Hamiltonian and thus mixes different channels.

For the magnetic Feshbach resonance of alkali-metal atoms discussed above, these three conditions are satisfied as follows:

- The channel is defined in terms of the spin quantum number of a single atom in the presence of a magnetic field, that is, the eigenstate of both the hyperfine interaction and the Zeeman field.
- The energy splitting between two channels can therefore be tuned by the Zeeman energy.
- The short-range potential largely depends on the total electron spin of two atoms being singlet or triplet, which does not conserve the spin quantum number of the single atom.

In the same spirit, we can also have several different types of Feshbach resonance. One is the optical Feshbach resonance. Here we briefly introduce how the optical Feshbach resonance satisfies the three ingredients:

- For $r > r_0$, the atoms are labeled by the electronic quantum number of a single atom. Taking an alkali-metal atom as an example, for the open channel, two atoms are both in the $^2S_{1/2}$ ground state, and for the closed channel, one atom is still in the $^2S_{1/2}$ state and the other atom is in the excited $^2P_{1/2}$ state. Here we should note that, although there presents a laser field, the laser frequency is far detuned from the single-particle transition, and to very good approximation, the single-particle electronic states are not affected by the laser when two atoms are far separated.
- In the presence of light, and by rotating wave approximation as discussed in Section 1.3, the effective energy difference between two channels is the excitation energy subtracted by the single photon energy. Thus, the energy spacing between two channels can be tuned by the laser frequency. When the laser frequency is detuned to be resonant with a bound state energy in the closed channel, the bound state is effectively tuned to the threshold of the open channel, at which a scattering resonance occurs.
- Since the laser is tuned to be resonant with a bound state in the closed channel, the two channels are coupled by the laser at the short distance when the molecular wave function is concentrated.

The optical Feshbach resonances have great advantages that they can provide very fast temporal control and small spatial resolution control of interactions, because the laser can be turned on and off much more rapidly than the magnetic field, and the laser intensity can be varied on the spatial scale of less than 1 μm. However, the disadvantage is that the excited state (such as $^2P_{1/2}$ of alkali-metal atoms) usually has finite lifetime due to the spontaneous emission. The loss, as well as the heating due to the loss, can be quite significant, preventing the system from reaching equilibrium in the regime nearby a resonance. In fact, a better stratagem is to combine the optical control with the magnetic Feshbach resonance, such that one can take the advantages of temporal and spatial control and can also avoid the heating problem.

Another example is the orbital Feshbach resonance in alkaline-earth-metal atom, which has been first theoretically predicted [189] and then experimentally observed in ^{173}Yb [133, 75] and in ^{171}Yb [18]. As we have seen, the electron spin plays an important role in the magnetic Feshbach resonance of alkali-metal atoms, because the short-range potentials are labeled by the total electronic spin singlet and triplet. Because the electron spin of the ground state ($^1S_0$) alkaline-earth-metal atom is zero, the short-range potentials do not have the choice of the total electron spin being singlet or triplet, and therefore, the mechanism for the magnetic Feshbach resonance in alkali-metal atoms does not hold for alkaline-earth-metal atoms. Nevertheless, let us recall that in Section 1.1, we have discussed that alkaline-earth-metal atoms have a long-lived clock state $^3P_0$, and in Section 2.3, we have discussed the collision between two different nuclear spin states ($|\uparrow\rangle$ and $|\downarrow\rangle$) of a fermionic alkaline-earth-metal atom, with one in the ground state ($^1S_0$ denoted by $|g\rangle$) and the other in the clock state ($^3P_0$ denoted by $|e\rangle$). In Section 2.3, we focus on the zero magnetic field limit, and here we consider the presence of finite magnetic field. One crucial fact is that the nuclear spin Landé g-factor for $|e\rangle$ state is slightly larger than that of the $|g\rangle$ state [20]. This is because, as we have discussed in Section 1.1, $^3P_0$ state possesses certain coupling to $^3P_1$ state through the hyperfine coupling, which can be further coupled to $^1P_1$ state. The small but finite coupling to the electronic spin gives rise to a slightly larger g-factor of $^3P_0$ compared with $^1S_0$ state. With this in mind, let us briefly introduce how these three conditions can be satisfied in alkaline-earth-metal atoms [191]:

- For $r > r_0$, atoms stay in the single-particle spin eigenstates in the presence of a magnetic field. Here, for the open channel, one atom stays in $|g\downarrow\rangle$ and the other atom stays in $|e\uparrow\rangle$, and the wave function under antisymmetrization reads

$$|o\rangle = \frac{1}{\sqrt{2}} \left(|g\downarrow\rangle|e\uparrow\rangle - |e\uparrow\rangle|g\downarrow\rangle \right). \tag{2.71}$$

For the closed channel, one atom stays in $|g\uparrow\rangle$ and the other atom stays in $|e\downarrow\rangle$, and the wave function under antisymmetrization reads

$$|c\rangle = \frac{1}{\sqrt{2}} \left(|g\uparrow\rangle|e\downarrow\rangle - |e\downarrow\rangle|g\uparrow\rangle \right). \tag{2.72}$$

- As mentioned above, because the $|g\rangle$ state and $|e\rangle$ state have slightly different g-factors, the energy difference between the open and the closed channels can in principle be tuned by the magnetic field. However, also because this g-factor difference is quite small, the range of tunability is also rather small. Typically, changing the magnetic field by 1 gauss, the Zeeman energy between two channels changes about $2\pi\hbar \times 100$ Hz. Note that for alkali-metal atoms, for the same amount of magnetic field, the change of Zeeman energy between channels is about five orders of magnitude larger. With such limited Zeeman energy tunability, it is hard to access a bound state with an accessible magnetic field range in the laboratory. But fortunately, nature is very kind. For both ^{173}Yb and ^{171}Yb atoms, there exists quite a shallow bound state in the interaction potential, which can be accessed even with this narrow tunable energy window.

- As we discussed in Section 2.3, for the four states mentioned above, the short-range potential is diagonal in the bases labeled by $|+\rangle$ and $|-, 0\rangle$, as shown in Eq. 2.40 and

Eq. 2.41. Because $|+\rangle$ and $|-,0\rangle$ can be written as $(|o\rangle \pm |c\rangle)/\sqrt{2}$, respectively, this short-range potential mixes the open and the closed channels.

With these three conditions satisfied, a magnetic field–tunable Feshbach resonance can also be reached in the alkaline-earth-metal atoms. However, the role of electronic spin in the alkali-metal case is now replaced by the so-called *orbital* degree of freedom that labels $^1S_0$ and $^3P_0$. To highlight this difference, the new Feshbach resonance is named as the *orbital Feshbach resonance*. There is a major physical difference between the magnetic Feshbach resonance and the orbital Feshbach resonance. In the former, as we repeatedly emphasized, the energy difference between two channels is much larger than the kinetic energy such that the closed channel cannot be populated by scattering states. But for the latter, this energy difference is reduced by five orders of magnitude, and therefore it is no longer much larger than the kinetic energy. Hence, the so-called closed channel can be populated by low-energy scattering states in a many-body system, and it is no longer closed [189]. This difference can manifest significantly in a strongly interacting Fermi gas nearby these resonances [189, 191].

2.5 Confinement-Induced Resonance

When a strong one- or two-dimensional confinement potential is applied, such a geometric confinement can reduce a three-dimensional system to a quasi-two- or quasi-one-dimensional one. In this section, we will discuss how to deduce the effective interaction strength for scattering in lower dimensions, starting from the original three-dimensional scattering problem with confinement potentials. We will show that the effective interaction strength in lower dimensions can diverge even when the original s-wave scattering length in three dimensions is finite. This is known as the *confinement-induced resonance* [131].

Here, as an example, we consider the quasi-one-dimensional situation; that is, a strong harmonic trap in the transverse xy plane is applied to a three-dimensional system, and the system remains uniform along the $\hat{z}$ direction. Note that the center of mass and relative motions are still separable with the presence of a harmonic trap, and the Schrödinger equation for the relative motion between two atoms is written as

$$\left[\frac{\hat{p}_z^2}{2\bar{m}} + \frac{\hat{p}_x^2 + \hat{p}_y^2}{2\bar{m}} + \frac{\bar{m}\omega_\perp^2(x^2+y^2)}{2} + V(\mathbf{r})\right]\Psi(\mathbf{r}) = E\Psi(\mathbf{r}) \tag{2.73}$$

where $V(\mathbf{r})$ is the interatomic potential. Similarly to discussion in Section 2.1, when $r > r_0$, we can ignore the interaction potential, and the wave function is determined by the free Hamiltonian. Note that the transverse mode has energy $(n_x + n_y + 1)\hbar\omega_\perp$ $(n_x, n_y \geq 0)$. Here we focus on the energy range $\hbar\omega_\perp < E < 2\hbar\omega_\perp$. In this energy range, if the atoms are in the lowest transverse mode, they can be in scattering state along the longitudinal direction. And if atoms are in the transverse excited states, they can only be in the bound state along the longitudinal direction. Hence, the general form of the wave function can be written as

$$\Psi = (e^{ik_z z} + f_{\text{even}} e^{ik_z |z|}) \varphi_0(x) \varphi_0(y) + \sum_{n_x + n_y \neq 0} \alpha_{n_x, n_y} \varphi_{n_x}(x) \varphi_{n_y}(y) e^{-\kappa_{n_x n_y} |z|}, \quad (2.74)$$

where $E = \hbar \omega_\perp + \hbar^2 k_z^2 / (2\bar{m}) = \hbar \omega_\perp (n_x + n_y + 1) - \hbar^2 \kappa_{n_x n_y}^2 / (2\bar{m})$. Here f_{even} is the even parity scattering amplitude, and odd parity scattering amplitude vanishes because of the requirement of wave function continuity at $z = 0$. Here φ_n is the eigen-mode of a one-dimensional harmonic oscillator. Because the second term in the wave function Eq. 2.74 vanishes at large z, the asymptotic form of this scattering wave function is given by the first term as

$$\Psi(z, \rho) \to (e^{ik_z z} + f_{\text{even}} e^{ik_z |z|}) \varphi_0(x) \varphi_0(y). \quad (2.75)$$

For $r < r_0$, the single-particle energy can be ignored, and the wave function is determined by the short-range interaction potential. Similarly, for the s-wave channel, we can match the boundary condition by requiring $(r\Psi)'/(r\Psi)\big|_{r=r_0} = -1/a_s$, and for a higher partial wave channel, we assume the interaction effects are negligible. Nevertheless, the difficulty here is that the short-range boundary condition is spherical symmetrical but the wave function Eq. 2.74 is cylindrical symmetrical. After some quite involved calculation using the frame transformation [187], one finally reaches [131]

$$f_{\text{even}}(k_z \to 0) = -\frac{1}{1 - \frac{ik_z a_\perp}{2} \left(\frac{a_\perp}{a_s} + \mathcal{C} \right)}, \quad (2.76)$$

where $a_\perp = \sqrt{\hbar / \bar{m} \omega}$ and $\mathcal{C} \approx -1.46$.

Next we consider a real one-dimensional case. We shall also model the one-dimensional scattering process in terms of a zero-range potential. Unlike the three-dimensional case, the one-dimensional wave function does not display any singularity when $z \to 0$, and therefore, a δ-function potential is regular in one dimension. Hence, we write down the Hamiltonian with δ-function interaction as

$$\left[-\frac{\hbar^2}{2\bar{m}} \frac{\partial^2}{\partial z^2} + g_{1d} \delta(z) \right] \Psi(z) = E\Psi(z). \quad (2.77)$$

When $z \neq 0$, the wave function of the kinetic energy eigenstate is generally written as

$$\Psi = e^{ik_z z} + f_{\text{even}} e^{ik_z |z|}, \quad (2.78)$$

where $E = \hbar^2 k_z^2 / (2\bar{m})$. For a δ-function potential, we can use the continuity condition that $\Psi'(0^+) - \Psi'(0^-) = 2\bar{m} g_{1d} \Psi(0)/\hbar^2$ to determine f_{even}, which gives rise to

$$ik_z f_{\text{even}} = \frac{\bar{m} g_{1d}}{\hbar^2} (1 + f_{\text{even}}); \quad (2.79)$$

that is

$$f_{\text{even}} = -\frac{1}{1 - i\frac{\hbar^2}{\bar{m} g_{1d}} k_z}. \quad (2.80)$$

Introducing "one-dimensional scattering length" a_{1d} as

$$g_{1d} = -\frac{\hbar^2}{\bar{m} a_{1d}}, \quad (2.81)$$

we can write

$$f_{\text{even}} = -\frac{1}{1 + ika_{1\text{d}}}. \tag{2.82}$$

To determine the effective one-dimensional interaction scattering length $a_{1\text{d}}$, one requires that the scattering amplitude f_{even} obtained from the one-dimensional model Eq. 2.82 reproduce f_{even} of Eq. 2.76 obtained from the full three-dimensional calculation with confinement potential. As we emphasized at the beginning of this chapter, since the ultra-cold atomic systems are dilute and the typical interatomic separation is much larger than the range of potential, and the collision energy is also very small compared with the interaction potential, the interaction mostly manifests in the asymptotic wave function. Therefore, if these two situations give the same asymptotic wave functions, we consider this $g_{1\text{d}}$ as a faithful representation of the interaction in the reduced dimension. Hence, by matching Eq. 2.82 with Eq. 2.76, we obtain

$$\frac{a_{1\text{d}}}{a_\perp} = -\frac{1}{2}\left(\frac{a_\perp}{a_s} + \mathcal{C}\right). \tag{2.83}$$

This shows that when $a_\perp/a_s = -\mathcal{C}$, $a_{1\text{d}} = 0$ and $g_{1\text{d}}$ diverges, which is known as the confinement-induced resonance [131].

Although the discussion of the confinement-induced resonance appears quite different from the discussion of Feshbach resonance in Section 2.4, it can be essentially understood in the same way as a Feshbach resonance [17]. In Section 2.4, we established three points as the key ingredients for a Feshbach resonance, and here we can show that the confinement-induced resonance can also be understood in terms of these three points.

- When two atoms are separated, atoms stay in the single-particle eigenstates. Here we use different eigenstates in the transverse direction to label "channels." For the open channel, both atoms are in the transverse ground state. For the closed channel, atoms are in the transverse excited states
- The energy difference between the open and the closed channels is given by the transverse confinement energy and can be tuned by the external confinement potential.
- The single-particle eigenstate has cylindrical symmetry, but the short-range potential has spherical symmetry. The incompatibility of two symmetries naturally leads to coupling between channels.

With this understanding, resonance occurs when a bound state in the closed channel matches the scattering threshold of an open channel. Here, the energy offset between the closed channel and the open channel is typically $\hbar\omega = 2\hbar^2/(ma_\perp^2)$, and the bound state energy in three dimensions is estimated by $-\hbar^2/(ma_s^2)$. Thus, the resonance condition can be roughly estimated as

$$\hbar\omega - \frac{\hbar^2}{ma_s^2} = 0, \tag{2.84}$$

which leads to

$$\frac{a_\perp}{a_s} = \sqrt{2}. \tag{2.85}$$

This is not that different from the exact results in Eq. 2.83 where $\sqrt{2}$ is replaced by $1.46\dots$. A similar argument can be applied to confinement into quasi two dimensions, or mixed dimensions. Here mixed dimensions means that one atom is confined to the d_1 dimension and the other atom is confined to the d_2 dimension, where both d_1 and d_2 can take a value between 0 and 3.[5]

2.6 Efimov Effect

In the above sections, we discussed different methods, as well as a general framework, to tune the two-body interaction potential to a scattering resonance. A quantum many-body system with such a resonant interaction potential has many intriguing properties, as we will discuss in Chapter 5 and Chapter 6. Here, before studying many-body physics, we first study a manifestation of resonant interaction in a three-body system. This problem can be generally solved by a so-called hyper-spherical coordinate approach [21, 22], but the calculation is quite involved. Here, to illustrate the essential physics, we take a simpler case of one light atom interacting with two heavy atoms, and we can utilize the Born–Oppenheimer approximation to simplify the calculation [141].

Born–Oppenheimer Approximation. First of all, we fix the positions of two heavy atoms with mass M at $\mathbf{R}/2$ and $-\mathbf{R}/2$, respectively, and study the motion of the light atom with mass m in the presence of these two heavy atoms. The Hamiltonian for the light atom therefore reads

$$\hat{H} = -\frac{\hbar^2 \nabla^2}{2m} + \frac{2\pi \hbar^2 a_s}{m} \delta(\mathbf{R}_+) \frac{\partial}{\partial |\mathbf{R}_+|} |\mathbf{R}_+| + \frac{2\pi \hbar^2 a_s}{m} \delta(\mathbf{R}_-) \frac{\partial}{\partial |\mathbf{R}_-|} |\mathbf{R}_-|, \qquad (2.86)$$

where $\mathbf{R}_\pm = \mathbf{r} \pm \mathbf{R}/2$, and we take $m \ll M$ such that the reduced mass is simplified as m. In the regime $\mathbf{r} \neq \pm\mathbf{R}/2$, let us consider the following three requirements: (i) the wave function should be an eigenstate of the kinetic operator; (ii) we consider that the light atom forms a bound state around both the two heavy atoms; and (iii) the wave function is symmetric or antisymmetric with respect to exchanging $\mathbf{R}_+$ and $\mathbf{R}_-$. Thus, we can write down the wave function as

$$\Psi_\pm(\mathbf{r}) \propto \frac{\exp\{-\kappa|\mathbf{R}_+|\}}{|\mathbf{R}_+|} \pm \frac{\exp\{-\kappa|\mathbf{R}_-|\}}{|\mathbf{R}_-|}, \qquad (2.87)$$

where κ is real and positive. The energy of this wave function is $-\hbar^2\kappa^2/(2m)$.

Expanding the wave function around either $|\mathbf{R}_+|$ or $|\mathbf{R}_-|$ yields

$$\Psi_\pm \propto \frac{1}{|\mathbf{R}_\pm|} - \kappa \pm \frac{e^{-\kappa R}}{R} + \dots, \qquad (2.88)$$

---

[5] The exceptions are that they cannot both be 0 where no scattering state can be defined, and they cannot both be 3 when no confinement is applied at all.

where $R = |\mathbf{R}|$, and the pseudo-potential requires the short-range behavior of the wave function to be

$$\Psi_{\pm} \propto \frac{1}{|\mathbf{R}_{\pm}|} - \frac{1}{a_s}. \tag{2.89}$$

Thus, it leads to

$$\kappa \mp \frac{e^{-\kappa R}}{R} = \frac{1}{a_s}. \tag{2.90}$$

Clearly, the term with a plus sign in the l.h.s. of Eq. 2.90 has no solution for negative and infinite a_s. So we consider the equation with a minus sign, resulting from the symmetric wave function in Eq. 2.87. The solution in general has the form

$$\kappa = \frac{1}{R} f\left(\frac{R}{a_s}\right), \tag{2.91}$$

where $f(y)$ is the solution to the equation $x - e^{-x} = y$. One can see that at unitarity with $a_s = \infty$ and $y = 0$, $f(0)$ is a constant. Therefore, $\kappa \sim 1/R$, and the energy is proportional to $-\hbar^2/(mR^2)$.

Continuous and Discrete Scaling Symmetry. With the help of the Born–Oppenheimer approximation, we have found that, at two-body resonance, the light atom induces an effective potential $\sim -\hbar^2/(mR^2)$ between two heavy atoms. Then, the Schrödinger equation for two heavy atoms is given by

$$\left(-\frac{\hbar^2 \nabla_{\mathbf{R}}^2}{M} - \frac{\hbar^2 c_0^2}{mR^2}\right) \Psi = E\Psi, \tag{2.92}$$

where c_0^2 is a constant. The most important feature of this equation is that the interaction energy scales the same way as the kinetic energy under a scaling transformation $\mathbf{R} \to \lambda \mathbf{R}$. Therefore, it looks as though, by applying this scale transformation, if E is an eigenenergy, E/λ^2 is also an eigenenergy. This works for any λ, which is known as the *continuous scaling symmetry*. However, if this is true, that also implies that the energy spectrum of this Hamiltonian is not bound from below. Hence, we need to apply an extra short-range cutoff to bound the spectrum from below. This short-range boundary condition can be a nonuniversal one depending on short-range details. And the fact that an extra nonuniversal high-energy cutoff is required also means that a theory with zero-range interaction potential is not renormalizable in this case.

Here we explicitly show how the extra short-range boundary condition affects the scaling symmetry. Since here we are interested in a three-body bound state, and since above we have considered that the light atom already forms a bound state with both heavy atoms, we now need only consider the bound state solution between these two heavy atoms. In the spherical coordinate of $\mathbf{R}$, we write $\Psi(\mathbf{R}) = \chi(R)/R$, as we now only consider the s-wave solution when two heavy atoms are bosons or distinguish particles. The Schrödinger equation for $\chi(R)$ is written as

$$\left[-\frac{\hbar^2}{M}\frac{d^2}{dR^2} - \frac{\hbar^2 c_0^2}{mR^2}\right] \chi = E\chi. \tag{2.93}$$

Now we consider a zero-energy solution, or alternatively speaking, we consider the wave function at a distance when $E \ll 1/R^2$. Because of the scaling symmetry, we can assume $\chi = R^s$, and by setting $E = 0$, Eq. 2.93 gives

$$s(s-1) + \frac{c_0^2 M}{m} = 0. \tag{2.94}$$

This leads to $s = 1/2 \pm i s_0$, where $s_0 = \sqrt{\frac{c_0^2 M}{m} - \frac{1}{4}}$, and we consider $M \gg m$ such that s_0 is always real. Thus, the two independent solutions can be written as

$$\chi_{\pm} = \sqrt{R} R^{\pm i s_0} = \sqrt{R} e^{\pm i s_0 \ln R}. \tag{2.95}$$

Each of $\chi_{\pm}$ is still invariant under a continuous scaling transformation, but because of the short-range boundary condition, the general wave function should be a superposition of both χ_+ and χ_- to satisfy the boundary condition. Note that the two solutions can also be written as $\sqrt{R} \cos(s_0 \ln R)$ and $\sqrt{R} \sin(s_0 \ln R)$, and so a general solution can be constructed as

$$\chi(R) = \sqrt{R} \cos(s_0 \ln R + \theta), \tag{2.96}$$

where θ should be determined by the short-range boundary condition. Clearly the wave function Eq. 2.96 is no longer invariant under a continuous scaling transformation, but if the scaling factor $\lambda = e^{\pi n/s_0}$, where n is an integer, the wave function is still invariant. This is known as the *discrete scaling symmetry*, because the scaling factor can only take values in a set of discrete numbers. Under the discrete scaling transformation, the energy becomes $E \to E e^{-2\pi/s_0}$. That is to say, if E_0 denotes the lowest-energy bound state, and E_n denotes the nth bound state counting from below, then there is an infinite number of bound states, and their binding energies satisfy

$$E_n = E_{n-1} e^{-2\pi/s_0}. \tag{2.97}$$

Note that the solutions are actually the binding energies of the three-atom bound state, which means that the three-body bound state energies obey a geometric sequence. This result was first obtained by Efimov from solving the problem of three identical bosons nearby a two-body resonance and thus is named the *Efimov effect*. The Efimov effect in a three-body system was first experimentally found in cold ^{133}Cs gas of identical bosons [94], and later was also found between two bosons and a third distinguishable atom, or three distinguishable atoms [21, 22]. The discrete scaling symmetry has also been confirmed experimentally [176, 143].

Here we highlight that, from the symmetry perspective, the defining property of the Efimov effect is the discrete scaling symmetry with a universal scaling factor, which resulted from a Hamiltonian with continuous scaling symmetry plus a nonuniversal short-range boundary condition. We emphasize that this defining property has at least two nontrivial points:

- In many cases, a short-range boundary condition completely breaks the continuous scaling symmetry, but in this case, it still leaves a discrete scaling symmetry. Mathematically, it happens when Eq. 2.94 for s has a pair of conjugate solutions.

• Although the short-range boundary condition is nonuniversal, the scaling factor is actually universal. In this case, one can see that although θ is a nonuniversal value depending on the details of the short-range boundary condition, and the exact value of the lowest binding energy E_0 is also nonuniversal, the scaling factor e^{π/s_0} is a constant only depending on the mass ratio and does not depend on the short-range details.

This definition of the Efimov effect from the symmetry perspective allows one to generalize this effect beyond few-body physics and to find more intriguing manifestations of this effect in many-body systems. One such example is the quantum many-body expansion dynamics of a scaling-invariant quantum gas in a specially designed expanding harmonic trap, which follows the same symmetry definition and is named the *Efimovian expansion* [48].

Finally, let us briefly discuss how these three-body bound states behave when the interaction is tuned away from the resonance. It turns out that when a_s is negative, the effective attraction is weaker than $\sim -1/R^2$ at large distance, which first affects these shallow bound states whose wave functions are more extended and have more weight on the long-range part. The energies of these bound states will increase as the interaction is tuned away from the resonance to the negative side, and they will in turn merge into the three-body continuum, as shown in Figure 2.7. When one of the three-body bound states meets the three-body threshold, it yields a three-body scattering resonance. When a_s is positive, the effective attraction is deeper than $\sim -1/R^2$. However, on this side, there also exists a two-body bound state, and as we have discussed in Section 2.1, the dimer energy is $-\hbar^2/(ma_s^2)$. Hence, the atom-dimer threshold energy is $-\hbar^2/(ma_s^2)$. It turns out that the increasing of the three-body binding energy is slower than the increasing of the two-body binding energy

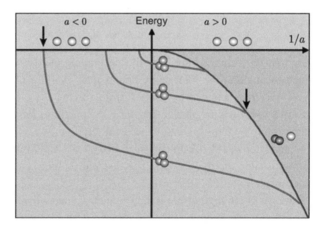

Figure 2.7   Three-body Efimov effect. The energy of three-body bound state of spinless bosons as a function of two-body scattering length a_s (a in the figure). The three-atom threshold is always $E = 0$. The atom-dimer threshold behaves as $-\hbar^2/(ma_s^2)$ in the positive a_s side. Three typical spectrum lines for the three-body bound state energy are shown. At resonance when $1/a_s = 0$, the binding energies form a geometric sequence. Two arrows label examples of three-atom scattering resonance and atom-dimer scattering resonance, respectively. Reprinted from Ref. [55]. A color version of this figure can be found in the resources tab for this book at cambridge.org/zhai.

as $1/a_s$ increases; therefore, the three-body bound state will in turn merge into the atom-dimer continuum when the interaction is tuned away from the resonance to the positive side, as also shown in Figure 2.7. When one of the three-body bound states meets the atom-dimer threshold, it yields an atom-dimer scattering resonance. Both the three-atom resonance and the atom-dimer resonance will manifest in the loss rate of atoms, which can be experimentally measured as evidence of the Efimov effect [94, 176, 143].

2.7 From Few to Many

In this chapter, we have discussed two- and three-body problems. In the next chapter, we will start to discuss many-body physics in ultracold atomic gases. Here we would like to point out that there are many connections between few- and many-body problems in ultracold atomic physics. First of all, few-body problems help us to build up the right model for many-body physics, as we have discussed in Section 2.2 and Section 2.4. Second, few-body problems help us to locate the parameter regimes where the many-body physics can be interesting. We discussed in Section 2.4 and Section 2.5 how to tune the elastic scattering to be very strong. In addition, there is another important aspect that we do not discuss in this book, which is about the inelastic part of the scattering process. The inelastic part of the scattering process leads to atom loss. A strong inelastic scattering can lead to strong loss and, therefore, a short lifetime of the many-body system. Hence, in order that intriguing many-body physics takes place, we not only require the elastic scattering to be strong enough but also require the inelastic scattering not to be too strong. We need the solutions of the few-body problem to help us locate such regimes.

Third, few-body problems provide an alternative way to analyze correlations in a quantum many-body system. Generally speaking, there are two different approaches to studying many-body corrections, which are known as the top-down approach and the bottom-up approach. Here the top-down means starting from large-scale, long-wave length, or low-energy structures. Various kinds of mean-field theories that we will discuss in the next chapter belong to the top-down approach. In contrast, the bottom-up approach means understanding correlations in a many-body system from its microscopic building blocks, that is to say, from two-body, three-body, and then gradually adding more particles. The advances in ultracold atomic experiments allow us to control atom number very precisely, which makes this approach even experimentally possible. In experiments, one can observe how the many-body correlation gradually builds up by adding to the atom number one by one [197]. Theoretically, one systematic method to carry out this bottom-up approach is in fact the high-temperature expansion. This expansion uses the fugacity as a small parameter, and therefore it also works in the resonance when the interaction is very strong. Below we will briefly discuss this approach.

We consider the partition function $\mathcal{Z}$ at high temperature. At high temperature, μ is very negative and the fugacity $z = e^{\mu/(k_B T)}$ is very small. Hence we can use z as a small parameter to expand $\mathcal{Z}$ as

$$\mathcal{Z} = \mathrm{Tr} e^{-(H-\mu N)/(k_B T)} = 1 + z \sum_{n_1} e^{-E_{n_1}/(k_B T)} + z^2 \sum_{n_2} e^{-E_{n_2}/(k_B T)} + \ldots, \qquad (2.98)$$

where we have taken the $N = 1$ in the second term of Eq. 2.98 and n_1 denotes quantum numbers of all single-particle eigenstates, and $N = 2$ for the third term in Eq. 2.98 and n_2 denotes the quantum numbers of all two-particle eigenstates. For a uniform system,

$$\sum_{n_1} e^{-E_{n_1}/(k_B T)} = \sum_{\mathbf{k}} e^{-\hbar^2 \mathbf{k}^2/(2mk_B T)} = V \left(\frac{mk_B T}{2\pi \hbar^2} \right)^{3/2} = \frac{V}{\lambda^3}, \qquad (2.99)$$

where $\lambda = \sqrt{2\pi \hbar^2/(mk_B T)}$ is the thermal de Broglie wavelength. E_{n_2} contains the center-of-mass motion $\mathbf{K}^2/(4m)$ and the relative motion with eigenenergies denoted by ϵ_{rel}. Since the center-of-mass and relative coordinates are separable, we have

$$\sum_{n_2} e^{-E_{n_2}/(k_B T)} = V \left(\frac{\sqrt{2}}{\lambda} \right)^3 \sum_{\epsilon_{rel}} e^{-\epsilon_{rel}/(k_B T)}. \qquad (2.100)$$

Thus, the solution of the two-body problem allows us to obtain the partition function to the order of z^2. Furthermore, with the solutions of the three-body problem, we can obtain information on the partition function up to z^3, and this expansion can be systematically carried on. Here, for simplicity, we only consider the z^2 order.

Up to the z^2 order, we can therefore rewrite the partition function as

$$\mathcal{Z} = \mathcal{Z}^0 + V z^2 \left(\frac{\sqrt{2}}{\lambda} \right)^3 b_2, \qquad (2.101)$$

where $\mathcal{Z}^0$ is the partition function in the absence of interactions, and

$$b_2 = \sum_{\epsilon_{rel}} \left(e^{-\epsilon_{rel}/(k_B T)} - e^{-\epsilon_{rel}^0/(k_B T)} \right). \qquad (2.102)$$

Here b_2 is called the second *virial coefficient*, and ϵ_{rel}^0 is the eigenstate for relative motion in the absence of interactions. Below we shall discuss how to compute b_2 with the knowledge of two-body problem discussed in Section 2.1 [86].

For the reason we discussed in Section 2.1, we ignore the interaction effect in all high partial wave channels and only consider the interaction effect in the s-wave channel. Note that b_2 can be rewritten as

$$b_2 = \sum_{n_b} e^{-E_{n_b}/(k_B T)} + \int_0^{+\infty} dk (g(k) - g_0(k)) e^{-\hbar^2 k^2/(mk_B T)}, \qquad (2.103)$$

where the first contribution comes from bound states due to interactions and E_{n_b} denotes binding energies, and the second contribution comes from all scattering states in the s-wave channel; $g(k)dk$ and $g_0(k)dk$ denote the number of eigenstates with wave vector between k and $k + dk$ for interacting systems and noninteracting systems, respectively. As we have shown in Section 2.1, the wave function for the relative motion between two particles in the s-wave channel can be written as

$$\Psi = \frac{\sin(kr + \delta_k)}{r}. \qquad (2.104)$$

Considering a spherical box with radius size R, the wave function has to satisfy the boundary condition at $r = R$, which yields

$$kR + \delta_k = s\pi, \tag{2.105}$$

where s is an integer. Eq. 2.105 gives

$$\left(R + \frac{d\delta_k}{dk}\right)\Delta k = \pi \Delta s. \tag{2.106}$$

The number of eigenstates increases by 1 when Δs increases by 1, which requires Δk increasing by

$$\Delta k = \frac{\pi}{R + \frac{d\delta_k}{dk}}. \tag{2.107}$$

Thus, we have

$$g(k)dk = \frac{1}{\pi}\left(R + \frac{d\delta_k}{dk}\right)dk, \tag{2.108}$$

and for the noninteracting case, $g_0(k)dk = Rdk/\pi$. Therefore, b_2 can be written as

$$b_2 = \sum_{n_b} e^{-E_{n_b}/(k_BT)} + \frac{1}{\pi}\int_0^\infty \frac{d\delta_k}{dk}e^{-\hbar^2k^2/(mk_BT)}dk. \tag{2.109}$$

Using $\tan\delta_k = -ka_s$, one can obtain

$$\int_0^\infty \frac{d\delta_k}{dk}e^{-\hbar^2k^2/(mk_BT)}dk = -\text{sgn}(a_s)\int_0^\infty \frac{|a_s|}{(k|a_s|)^2 + 1}e^{-\hbar^2k^2/(mk_BT)}dk$$
$$= -\text{sgn}(a_s)\frac{\pi}{2}\text{Erfc}[\alpha]e^{\alpha^2}, \tag{2.110}$$

where sgn is the sign function, Erfc is the complementary error function, and $\alpha = \lambda/(\sqrt{2\pi}|a_s|)$. Hence, if one excludes the contribution from the bound state,

$$b_2 = -\text{sgn}(a_s)\frac{1}{2}\text{Erfc}[\alpha]e^{\alpha^2}. \tag{2.111}$$

As shown in Figure 2.8(a), b_2 decreases from zero to $-1/2$ if a_s increases from zero to positive infinite and increases from zero to $1/2$ if a_s decreases from zero to negative infinite. This jump of unity at resonance can be exactly compensated by a zero-energy bound state contribution. Including the contribution from the bound state, b_2 becomes a smooth function when a_s changes from negative infinite to positive infinite, and b_2 monotonically increases as $-\lambda/a_s$ decreases.

With the help of the partition function, one can show that the total energy can be deduced as [72]

$$\mathcal{E} = \frac{3nk_BT}{2}\left(1 + \frac{n\lambda^3}{2^{7/2}}\right) + \mathcal{E}_{\text{int}} = \mathcal{E}_{\text{kin}} + \mathcal{E}_{\text{int}} \tag{2.112}$$

and

$$\mathcal{E}_{\text{int}} = \frac{3nk_BT}{2}(n\lambda^3)\left[-\frac{b_2}{\sqrt{2}} + \frac{\sqrt{2}}{3}T\frac{\partial b_2}{\partial T}\right], \tag{2.113}$$

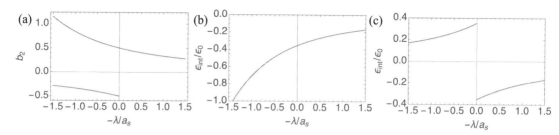

Figure 2.8 High-temperature expansion. (a) The second virial coefficient b_2 as a function of $-\lambda/a_s$. For positive a_s, the positive b_2 branch includes the contribution from the shallow bound state and smoothly connects to the negative a_s side, and the negative b_2 branch excludes the contribution from the shallow bound state. (b–c) The interaction energy $\mathcal{E}_{\text{int}}$ in units of $\mathcal{E}_0 = 3nk_BT(n\lambda^3)/2$ as a function of $-\lambda/a_s$ (b) with and (c) without the bound state contribution.

where n is the density of the system. With b_2, one can straightforwardly obtain the interaction energy with or without the contribution from the bound state, as shown in Figures 2.8(b) and (c), respectively. One can see that, including the bound state contribution, the interaction energy is always negative, consistent with the fact that the underlying potential is attractive. For positive scattering length, when the bound state contribution is excluded, the interaction energy is positive, which is called the *upper branch*. When the bound state contribution is included, the interaction energy is negative, which is called the *lower branch*. We will come back to revisit this physics in the discussion of polarons in Section 5.2. When the bound state contribution is excluded, one can see that the interaction energy becomes small when a_s is small, consistent with our discussion in Section 2.1 that the amplitude of a_s characterizes interaction strength when $|a_s|$ is small. One can also see that the interaction energy remains finite even when a_s is infinite at resonance, and the interaction energy becomes proportional to the thermal kinetic energy at resonance. That the interaction energy scales the kinetic energy characterizes strong interaction effects, as we will discuss again in Chapter 6.

Exercises

2.1 Calculate the scattering length for a three-dimensional square well interaction potential $V(r) = 0$ for $r > r_0$, $V(r) = -V_0$ for $r_0 > r > 0$ with $V_0 > 0$, and $V(r) = \infty$ for $r = 0$. Discuss how the scattering length changes as a function of V_0, and discuss when the binding energy satisfies the relation $E = -\hbar^2/(2\bar{m}a_s^2)$.

2.2 Calculate the scattering length for a three-dimensional hard core potential $V(r) = 0$ for $r > r_0$, $V(r) = V_0$ for $r_0 > r > 0$ with $V_0 > 0$, and $V(r) = \infty$ for $r = 0$. Discuss how the scattering length changes as a function of V_0 and the difference from the square well potential above.

2.3 Show that for a finite range interaction $V(r) \simeq 0$ for $r > r_0$, the phase shift for the lth partial wave $\delta_l \propto k^{2l+1}$.

2.4 Show that the bound state wave function $\Psi = \chi/r$ with χ given by Eq. 2.12 also satisfies the Schrödinger equation 2.18 with $V(r)$ given by $\delta(r)\hat{O}(r)$ and $\hat{O}(r)$ given by Eq. 2.21.

2.5 Analytically show that

$$\frac{1}{V}\sum_{\mathbf{p}}\left(\frac{1}{E - \hbar^2\mathbf{p}^2/m + i0^+} + \frac{1}{\hbar^2\mathbf{p}^2/m}\right) = -\frac{ikm}{4\pi\hbar^2}, \qquad (2.114)$$

where $k = \sqrt{mE/\hbar^2}$.

2.6 Derive the general interaction form between two spin-2 atoms.

2.7 Show Eq. 2.62 for a finite energy scattering state using the simplified two-channel model discussed in this chapter.

2.8 (1) Show that the two-body T-matrix for the two-channel model is given by Eq. 2.66, following the same method of summing up the ladder diagram shown in Figure 2.3(b). (2) Verify the normalization conditions of Eq. 2.67–2.69 by comparing the two-body T-matrix (Eq. 2.66) with $T_2(E = 0) = 4\pi\hbar^2 a_s/m$ and a_s given by Eq. 2.61.

2.9 Use a variational wave function to show that the lowest eigenenergy of the Hamiltonian Eq. 2.92 is not bound from below if no short-range cutoff is imposed.

2.10 With the help of Eq. 2.90, discuss the effective three-body interaction potential when a_s is away from infinite.

2.11 Compute the chemical potential and the pressure up to z^2 order by using the high-temperature expansion.

Part II

Interacting Bose Gas

3 Interaction Effects

Learning Objectives

- Define Bose–Einstein condensation in an interacting boson system in terms of macroscopic occupation.
- Introduce the Gross–Pitaevskii equation and its applicable conditions.
- Derive the hydrodynamic equations and various physical consequences of the hydrodynamics with or without harmonic trap.
- Introduce sound waves and emphasize the difference between wave behavior and diffusive behavior.
- Introduce the criterion for superfluidity and determine the superfluid critical velocity.
- Introduce two-fluid hydrodynamics.
- Introduce the Bogoliubov theory.
- Introduce various concepts related to quasi-particles, including ground state as a vacuum of quasi-particles, the zero-point energy of quasi-particles, quasi-particle lifetime, and vacuum fluctuation.
- Introduce healing length as the characteristic length scale for low-energy physics of an interacting Bose condensate.
- Summarize the difference between an interacting and noninteracting Bose condensate.
- Introduce Bragg spectroscopy.
- Show that the Bogoliubov theory fails in one dimension due to the infrared divergence.
- Discuss the basic idea of the Bethe-ansatz solution and the Tonks–Girardeau limit.
- Introduce scale-invariant quantum gases.
- Introduce the Josephson effect and self-trapping as two different transport phenomena driven by phase coherence.
- Discuss the conjugate relation between the relative phase fluctuation and the relative particle number fluctuation and show that strong enough phase fluctuation renders the system into a Fock state.
- Discuss an example of quantum measurement that can project two independent condensates to a coherent superposition state.
- Discuss interference effects in density and density–density correlations.
- Discuss the relation between the instability of the Schödinger Cat state and the stability of symmetry breaking.

3.1 Bose–Einstein Condensation

Before we discuss an interacting system, let us first review the concept of Bose–Einstein condensation (BEC) in a free boson system. Considering free bosons with dispersion $\epsilon_{\mathbf{k}} = \hbar^2 k^2 / (2m)$ in three dimensions, at high temperatures, each mode with a given wave vector $\mathbf{k}$ is populated by

$$n_{\mathbf{k}} = \frac{1}{e^{(\epsilon_{\mathbf{k}} - \mu)/(k_B T)} - 1}, \tag{3.1}$$

where μ is the chemical potential and is negative at high temperature. In statistical mechanics, we always consider the thermodynamic limit that $N \to \infty$ and $V \to \infty$ with density $n = N/V$ fixed. It is important to note that the population fraction $n_{\mathbf{k}}/V$ at each mode vanishes in the thermodynamic limit. This is always true as long as the chemical potential μ is negative. As temperature decreases, μ increases. Then the question is whether there exists a critical temperature, denoted by T_c, at which the chemical potential μ will increase to zero. If such a T_c exists, by setting $\mu = 0$, T_c is given by

$$N = \frac{V}{(2\pi)^3} \int d^3\mathbf{k} \frac{1}{e^{\epsilon_{\mathbf{k}}/(k_B T_c)} - 1} = \frac{V(mk_B T_c)^{3/2}}{\sqrt{2}\pi^2\hbar^3} \int_0^\infty \frac{\sqrt{z}dz}{e^z - 1}. \tag{3.2}$$

Because the integration in the r.h.s. of Eq. 3.2 is finite, Eq. 3.2 is equivalent to $1/\lambda^3 \sim n \sim 1/d^3$, where d is the mean interparticle spacing and $\hbar^2/(m\lambda^2) = k_B T/(2\pi)$ defines the thermal de Broglie wave length. The physical meaning of Eq. 3.2 is that the thermal de Broglie wave length is comparable to the mean inter-particle distance d, or equivalently speaking, $k_B T_c$ is comparable to the degenerate energy $\hbar^2/(md^2)$. It is the same condition as fermions entering quantum degeneracy.

Below T_c, μ cannot further increase to be positive for free bosons, and μ should retain zero. In this case, the occupation at $\mathbf{k} = 0$ mode N_0 should be considered separately, that is,

$$N = N_0 + \frac{V}{(2\pi)^3} \int d^3\mathbf{k} \frac{1}{e^{\epsilon_{\mathbf{k}}/(k_B T)} - 1}. \tag{3.3}$$

It is easy to show that $N_0/N \neq 0$ in the thermodynamic limit, and this is taken as the Bose–Einstein condensate (BEC). That a population fraction at a certain mode is nonzero in the thermodynamic limit is called the *macroscopic occupation*. From the discussion of BEC in a free bosons system, we learn that the Bose condensation is a transition from one situation that population fractions at all modes vanish in the thermodynamic limit to another situation that at least one mode is macroscopically occupied. In fact, as we will summarize at the end of Section 3.3, many properties are different between an interacting BEC and a noninteracting BEC. In order to define BEC in an interacting system, it is important to first identify which property of a noninteracting BEC should be regarded as the essential defining property of a BEC, and we should capture and generalize this defining property to an interacting system. Here we argue that one should take this macroscopic occupation as the defining property of a BEC. Nevertheless, in a free system, it is straightforward to define the occupation in single-particle eigenmodes. Hence, the question is how to properly define occupation in an interacting system, and the answer to this question leads to the concept of the off-diagonal long-range order (ODLRO) [183].

General Definition of BEC. Now we present a general definition of BEC in an interacting system. First of all, let us introduce the concept of the density matrix. When the many-body system is in a pure state with the many-body wave function $\Psi(\mathbf{r}_1, \mathbf{r}_2, \ldots, \mathbf{r}_N)$, the one-body density matrix is defined as

$$\rho(\mathbf{r}, \mathbf{r}') = N \int \Psi^*(\mathbf{r}, \mathbf{r}_2, \ldots, \mathbf{r}_N)\Psi(\mathbf{r}', \mathbf{r}_2, \ldots, \mathbf{r}_N)d^3\mathbf{r}_2 \ldots d^3\mathbf{r}_N. \tag{3.4}$$

If the system is in a mixed state with probability p_s in a many-body wave function $\Psi_s(\mathbf{r}_1, \mathbf{r}_2, \ldots, \mathbf{r}_N)$, the one-body density matrix is defined as

$$\rho(\mathbf{r}, \mathbf{r}') = N \int \sum_s p_s \Psi_s^*(\mathbf{r}, \mathbf{r}_2, \ldots, \mathbf{r}_N)\Psi_s(\mathbf{r}', \mathbf{r}_2, \ldots, \mathbf{r}_N)d^3\mathbf{r}_2 \ldots d^3\mathbf{r}_N. \tag{3.5}$$

Alternatively, in a second quantized form, it can be equivalently defined as

$$\rho(\mathbf{r}, \mathbf{r}') = \langle \hat{\Psi}^\dagger(\mathbf{r})\hat{\Psi}(\mathbf{r}') \rangle, \tag{3.6}$$

where $\hat{\Psi}^\dagger(\mathbf{r})$ and $\hat{\Psi}(\mathbf{r})$ are boson creation and annihilation operators at position $\mathbf{r}$, respectively. This one-body density matrix $\rho(\mathbf{r}, \mathbf{r}')$ can be diagonalized and decomposed as

$$\rho(\mathbf{r}, \mathbf{r}') = \sum_i N_i \psi_i^*(\mathbf{r})\psi_i(\mathbf{r}'). \tag{3.7}$$

Therefore, the eigenvector ψ_i defines the wave function of each mode, and N_i defines the single-particle occupation of each mode. With this definition of occupation of single-particle modes, we can now introduce the definitions of BEC in an interacting system as follows:

- If, for all N_i, $\lim_{N \to \infty} N_i/N = 0$, we call it a *normal phase*.
- If there is one and only one N_i, $\lim_{N \to \infty} N_i/N \neq 0$, we call it a simple *BEC*.
- If there are more than one N_i, $\lim_{N \to \infty} N_i/N \neq 0$, we call it a *fragmented BEC*.

For the latter two cases, we say the system has ODLRO [183].

Similarly, we can introduce a higher-order density matrix. For instance, a two-body density matrix can be defined as

$$\rho(\mathbf{r}_1, \mathbf{r}_2, \mathbf{r}_1', \mathbf{r}_2') = \langle \hat{\Psi}^\dagger(\mathbf{r}_1)\hat{\Psi}^\dagger(\mathbf{r}_2)\hat{\Psi}(\mathbf{r}_1')\hat{\Psi}(\mathbf{r}_2') \rangle. \tag{3.8}$$

Usually we consider the situation that $\mathbf{r}_1 \approx \mathbf{r}_2 \approx \mathbf{r}$ and $\mathbf{r}_1' \approx \mathbf{r}_2' \approx \mathbf{r}'$; we can similarly decompose the two-body density matrix as

$$\rho(\mathbf{r}_1, \mathbf{r}_2, \mathbf{r}_1', \mathbf{r}_2') = \sum_i N_i \psi_i^*(\mathbf{r}_1 \approx \mathbf{r}_2 \approx \mathbf{r})\psi_i(\mathbf{r}_1' \approx \mathbf{r}_2' \approx \mathbf{r}'). \tag{3.9}$$

In this way, we can define a *boson pair condensate* when there exists one or more N_i that satisfy $\lim_{N \to \infty} N_i/N \neq 0$. In many cases, a fragmented BEC defined by the one-body density matrix can be a simple BEC defined by the higher order density matrix [123]. We will come back to revisit this in Section 3.5.

In ultracold atom experiments, a BEC is achieved after the laser cooling and the evaporative cooling. The first measurement proving Bose condensation is the time-of-flight measurement of the momentum distribution [5, 44], which experimentally determines the onset of BEC by this property of macroscopic occupation. When the harmonic trap is suddenly released, atoms acquire a velocity $\mathbf{v} = \mathbf{k}/m$ where $\mathbf{k}$ is the momentum of atoms before turning off the trap. Afterward, these atoms flight in free space with this velocity.[1] After long enough time, when the size of the initial cloud can be ignored compared with the distance that atoms have traveled, the initial momentum distribution can be revealed by measuring the distance that atoms have flighted and dividing the distance by the time that they have flighted. In other words, the spatial distribution after the time of flight, as shown in Figure 3.1 as an example, reveals the momentum distribution before turning off the trap. Two typical sets of measurements from the first BEC experiments are shown in Figure 3.1. At higher temperature, this reveals the momentum distribution of thermal bosons. And below certain temperature, there is a sudden onset of peak which expands much slower

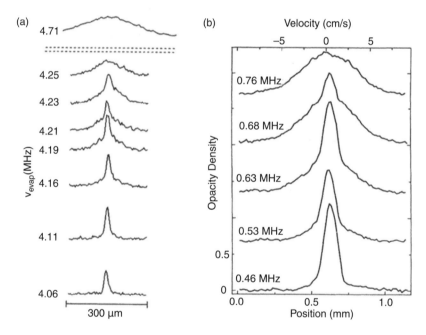

Figure 3.1   Time-of-flight measurements of momentum distribution of a Bose gas. The onset of a sharp peak in the momentum distribution marks the Bose–Einstein condensation transition. The number at each curve is called ν_{evap}, which labels the radio-frequency used in the evaporative cooling. The smaller this number is, the cooler is the gas. (a) is reprinted from Ref. [5], and (b) is reprinted from Ref. [44].

---

[1]   If there are other lasers to create optical lattices, or spin-orbit coupling, or other effects, as we will discuss in later chapters, they will also be turned off at the same time when the trapping laser is turned off. This flighting velocity $\mathbf{k}/m$ is not necessarily the same as the velocity of atoms before turning off the trap, and the latter is defined as $\mathbf{v} = \partial \mathcal{E}_{\mathbf{k}}/\partial \mathbf{k}$, where $\mathcal{E}_{\mathbf{k}}$ is the single-particle dispersion in the presence of optical lattices, or spin-orbit coupling or other effects. That is to say, the time-of-flight measures initial momentum distribution not the initial velocity distribution.

compared with the thermal components, because these atoms are the condensed part and their expansion dynamics are no longer driven by the kinetic energy. The expansion dynamics of the condensed part will be discussed in Section 3.2. As shown in Figure 3.1, below T_c, this component increases, and the atom number inside this peak becomes a fraction of the total number of atoms. This macroscopically occupied component is considered as the Bose condensate, and the ratio between atoms in this component and the total atom number is called the *condensate fraction*.

Below we will first focus on a simple BEC. For a simple BEC, since one eigenvalue of the density matrix (say, N_0) is much larger than any others, we can make a bold approximation for the density matrix as

$$\rho(\mathbf{r}, \mathbf{r}') \simeq N_0 \psi^*(\mathbf{r}) \psi(\mathbf{r}'). \tag{3.10}$$

Here N_0 is of the order of N, and $\psi(\mathbf{r})$ is the corresponding eigenvector. Eq. 3.10 means a long-range correlation because $\rho(\mathbf{r}, \mathbf{r}')$, or equivalently speaking, the correlation function $\langle \hat{\Psi}^\dagger(\mathbf{r}) \hat{\Psi}(\mathbf{r}') \rangle$, does not vanish even when $|\mathbf{r} - \mathbf{r}'|$ is taken to infinity. Here $\sqrt{N_0} \psi(\mathbf{r})$ is called the *condensate wave function*. The focus below is to discuss the equations that govern the ground state and the dynamical behaviors of this condensate wave function. To this end, we need to work out a microscopic theory, and we should make a proper approximation for the underlying many-body wave function. Since here we consider our approximation equation 3.10 of $\rho(\mathbf{r}, \mathbf{r}')$ is the key for BEC, let us ask an inverse question of what kind of many-body microscopic state can give rise to such a one-body density matrix. In fact, at least we can come up with the following two ways to satisfy Eq. 3.10.

- We can assume the system is a pure state and the many-body wave function is a product state as

$$\Psi(\mathbf{r}_1, \ldots, \mathbf{r}_N) = \prod_{i=1}^{N} \psi(\mathbf{r}_i). \tag{3.11}$$

It is easy to verify that the wave function of Eq. 3.11 reproduces the density matrix given by Eq. 3.10.

- We can assume that the wave function is a coherent state given by

$$|\Psi\rangle = e^{\int d^3\mathbf{r} \sqrt{N_0} \psi(\mathbf{r}) \hat{\Psi}^\dagger(\mathbf{r})} |0\rangle, \tag{3.12}$$

and using the property of the coherent state, we have $\hat{\Psi}(\mathbf{r})|\Psi\rangle = \sqrt{N_0} \psi(\mathbf{r})|\Psi\rangle$. Then it is straightforward to show that

$$\langle \hat{\Psi}^\dagger(\mathbf{r}) \hat{\Psi}(\mathbf{r}') \rangle = N_0 \psi^*(\mathbf{r}) \psi(\mathbf{r}'). \tag{3.13}$$

Before ending this part, we shall make two remarks:

- Though these two microscopic descriptions look quite different, they give the same one-body density matrix with ODLRO. As we will show in the following two sections, our derivation of the hydrodynamic theory ultilizes the first description, and the Bogoliubov theory is based on the second description. We will see that these two different theories give the same low-energy excitation spectrum.

- The density matrix itself does not infer the global phase of the condensate wave function. However, once a condensate wave function is chosen, we have to choose a fixed global phase. Nevertheless, the energies of wave functions with different global phases are degenerate, and we will discuss the connection to the gapless phonon mode in the next section.

3.2 Hydrodynamic Theory

Now we consider an interacting many-boson system, and we start from the pseudo-potential model described in Section 2.2:

$$\hat{\mathcal{H}} = \sum_{i=1}^{N} \left(\frac{\hat{\mathbf{p}}_i^2}{2m} + V(\mathbf{r}_i) \right) + \sum_{i<j} \frac{4\pi \hbar^2 a_s}{m} \delta(\mathbf{r}_{ij}) \frac{\partial}{\partial r_{ij}} r_{ij}. \tag{3.14}$$

We evaluate the energy expectation value of this Hamiltonian under the wave function Eq. 3.11. Here we should note that the wave function Eq. 3.11 remains regular at short-distance between two atoms, and it does not obey the short-range behavior of two-body wave function as we discussed in Sec 2.2, which requires that the wave function diverges as $1/r_{ij}$ when $\mathbf{r}_i$ approaches $\mathbf{r}_j$. The question is that how we reconcile the inconsistency between this many-body wave function of Bose condensation and the requirement for two-body wave function at short distance. To this end, we should emphasize that the assumption of Bose condensation wave function is only able to reproduce the approximate form of the density matrix Eq. 3.10, which ignores all the modes that are not macroscopically occupied. In other words, this microscopic wave function only captures the mode that is macroscopically occupied. Since this macroscopically occupied mode must be a low-lying mode, this is equivalent to say that this trial wave function only captures the low-energy and the long wavelength physics, and the wavelength should be much larger than the interparticle spacing. Therefore, it is natural that this wave function fails to capture short-range and high energy physics at the scale comparable or shorter than the interparticle spacing. We shall always keep this in mind that the following theory based on this assumption can only be applied to the length scale larger than interparticle spacing.

Since the wave function Eq. 3.11 is regular at short distance between any two particles, $\frac{\partial}{\partial r_{ij}} r_{ij}$ will not play any role and it is straightforward to calculate the energy as

$$\frac{\mathcal{E}}{N} = \int d^3\mathbf{r} \left[\psi^*(\mathbf{r}) \left(-\frac{\hbar^2 \nabla^2}{2m} + V(\mathbf{r}) \right) \psi(\mathbf{r}) + \frac{(N-1)}{2} \frac{4\pi \hbar^2 a_s}{m} |\psi(\mathbf{r})|^4 \right]. \tag{3.15}$$

Minimizing the energy with respect to ψ^*, and redefining $\sqrt{N}\psi$ as ψ, one obtains

$$\left(-\frac{\hbar^2 \nabla^2}{2m} + V(\mathbf{r}) \right) \psi + U|\psi|^2 \psi = \mu \psi, \tag{3.16}$$

where μ is the chemical potential. The dynamical version of this equation is given by

$$\left(-\frac{\hbar^2 \nabla^2}{2m} + V(\mathbf{r})\right)\psi + U|\psi|^2\psi = i\hbar\frac{\partial \psi}{\partial t}. \tag{3.17}$$

Here U denotes $4\pi\hbar^2 a_s/m$. Throughout this and next chapter, we always consider positive a_s and $U > 0$, unless specifically stated. Here $U > 0$ means repulsive interaction between atoms, as we have discussed in Box 2.1. Eq. 3.16 and Eq. 3.17 are known as the *Gross–Pitaevskii equation*, often short-noted as the GP equation. These equations are also called nonlinear Schrödinger equations. Mathematically, they differ from the single-particle Schrödinger equation because of the presence of the nonlinear term $U|\psi|^2\psi$. Here we should also remark the physical difference between this equation and the single-particle Schrödinger equation, although both of them are equations for single-particle wave function.[2] The single-particle Schrödinger equation describes a system with only one particle alone, and here the nonlinear Schrödinger equation describes a system with macroscopic number of particles occupying a single-particle mode, where the nonlinear term presents the interaction effects between these atoms. Below, we will mostly focus on the interaction effects. First of all, for a uniform system with $V(\mathbf{r}) = 0$, the density of the ground state should be uniform, therefore we have the chemical potential $\mu = Un_0$, where $n_0 = |\psi|^2$ is the mean condensate density. This already differs from the non-interacting case where μ is always zero for a BEC. To further discuss the dynamical behaviors and excitations, we proceed to introduce the hydrodynamic equations.

Hydrodynamic Equation. We decompose ψ as $\sqrt{n}e^{i\theta}$, where n is the density and θ is the phase. Both n and θ are functions of space and time. Substituting $\psi = \sqrt{n}e^{i\theta}$ into the time-dependent GP equation Eq. 3.17, and after eliminating $e^{i\theta}$ from both sides, the real part of the equation gives

$$\hbar\frac{\partial \theta}{\partial t} = -\left[-\frac{\hbar^2}{2m}\frac{1}{\sqrt{n}}\nabla^2\sqrt{n} + \frac{1}{2}m v_s^2 + V(\mathbf{r}) + Un\right], \tag{3.18}$$

where $v_s = \hbar\nabla\theta/m$ is the superfluid velocity. Taking the derivative ∇ at both sides, we obtain

$$m\frac{\partial v_s}{\partial t} = -\nabla\left[-\frac{\hbar^2}{2m}\frac{1}{\sqrt{n}}\nabla^2\sqrt{n} + \frac{1}{2}m v_s^2 + V(\mathbf{r}) + Un\right]. \tag{3.19}$$

The first term in the bracket of the r.h.s. of Eq. 3.19 is called the quantum pressure, and other terms are in turn the kinetic energy, the trapping energy and the interaction energy. Since the gradient of energy is force, the physical meaning of Eq. 3.19 is nothing but $\mathbf{F} = m\mathbf{a}$ in the classical mechanics. Thus Eq. 3.19 is also called the *Newton equation*. The imaginary part gives

$$\frac{\partial \sqrt{n}}{\partial t} = -\frac{\hbar}{2m}\left(2(\nabla\sqrt{n})\cdot(\nabla\theta) + \sqrt{n}\nabla^2\theta\right), \tag{3.20}$$

and by using the definition of v_s, it leads to

$$\frac{\partial n}{\partial t} + \nabla\cdot(nv_s) = 0. \tag{3.21}$$

---

[2] Here single-particle wave function means the wave function has a single spatial coordinate $\mathbf{r}$ as its variable.

This equation means that the change in the local density should be equal to the net flux of current. This equation is also called the *continuity equation*. Eq. 3.19 and Eq. 3.21 form the zero-temperature superfluid hydrodynamic equations.

Hydrodynamic equations refer to equations that govern dynamics of a fluid. Here we should make an important remark between the hydrodynamics and the BEC. We have shown that a BEC with interaction naturally leads to the hydrodynamic behavior, but on the other hand, hydrodynamics does not always require Bose condensation. Strong interactions can also lead to the hydrodynamic behavior. In normal fluids the strong interaction due to high density leads to hydrodynamic behavior. In dilute gas strong interaction can arise from Feshbach resonances. Experiments have also observed the hydrodynamic behavior in strongly interacting ultracold atomic gases nearby a Feshbach resonance, even when the system is not cold enough to become a superfluid [129].

Sound Velocity. Let us first consider a uniform system with $V(\mathbf{r}) = 0$, the ground state has a uniform density n_0 and $\mathbf{v}_s = 0$. We can expand $n = n_0 + \delta n$, and simplify the hydrodynamic equations by focusing on near-equilibrium and low-energy dynamics. Because of being near equilibrium, we can only keep the leading order of δn and $\mathbf{v}_s$, and because of being low-energy dynamics, we only need to keep the leading order of k and ignore the higher-order derivative terms. Thus, the hydrodynamic equations can be greatly simplified as

$$m\frac{\partial \mathbf{v}_s}{\partial t} = -U\boldsymbol{\nabla}\delta n \qquad (3.22)$$

$$\frac{\partial \delta n}{\partial t} = -\boldsymbol{\nabla}(n_0\mathbf{v}_s). \qquad (3.23)$$

Here the density fluctuation and the phase fluctuation are locked together,[3] which gives rise to a single low-energy mode described by

$$\frac{\partial^2 \delta n}{\partial t^2} = \frac{Un_0}{m}\nabla^2\delta n. \qquad (3.24)$$

This equation contains both second-order time and spatial derivatives, which is a *wave equation*.

Here we should note the general difference between a wave equation and a *diffusion equation*. Generally, considering an observable $\mathcal{W}$,

$$\frac{\partial^2 \mathcal{W}}{\partial t^2} = c^2\nabla^2\mathcal{W}, \qquad (3.25)$$

is called a wave equation, such as Eq. 3.24, where c is a constant velocity. It is easy to show that if $\mathcal{W}$ obeys such a wave equation, when a profile is created in $\mathcal{W}(\mathbf{r})$, the shape of the profile will keep unchanged as time evolves, and its center propagates in space with the velocity c. That is also the reason why waves can carry information. In nature, sound wave, electromagnetic wave and gravitational wave all obey wave equations, and they all

---

[3] It will be different in the relativistic case where density and phase modes are decoupled at the lowest order, as we will discuss in Section 8.1.

can carry information. In contrast, a diffusive equation contains a first-order time derivative and a second-order spatial derivative, which can be generally written as $\mathcal{W}$,

$$\frac{\partial \mathcal{W}}{\partial t} = \mathcal{D}\nabla^2 \mathcal{W}, \tag{3.26}$$

where $\mathcal{D}$ is usually called diffusion constant. If $\mathcal{W}$ obeys such a diffusion equation, when a profile is created in $\mathcal{W}(\mathbf{r})$, the shape of the profile will gradually smear out as time evolves. In a normal state, heat and Brownian motion are examples of diffusive motions. Below, when we discuss the two-fluid hydrodynamics, we will discuss that the propagation of the heat changes from the diffusive behavior to the wave behavior across the superfluid transition.

The wave equation Eq. 3.24 gives rise to sound wave, or phonon mode, as the low-energy excitations, and the phonon mode has a linear dispersion as

$$\omega = \sqrt{Un_0/m}|\mathbf{k}|, \tag{3.27}$$

where the phonon velocity $c = \sqrt{Un_0/m}$. The sound wave describes the propagation of a density deviation from the equilibrium density. Experimentally, one can use a focused laser beam to create either a local density dip or a density hump at the center of a BEC, and then watch the motion of this density dip or hump [8]. The results are shown in Figure 3.2. From this measurement one can see that the shape of the density dip or hump does not change and its location moves from the center toward the edge of the cloud, from which one can deduce the sound velocity. By repeating such measurements with different densities, one can obtain the relation between the sound velocity and the density, as shown in Figure 3.2(d), which verifies Eq. 3.27.

This low energy gapless excitation with linear dispersion is also called *the Goldstone mode*. Note that the density and the phase are coupled through the continuity equation, the phonon mode can be also be viewed as an excitation of spatially twisting phase of the condensate wave function. In the long wave length limit by taking $k \to 0$, the excitation turns to a uniform rotation of the phase of the condensate wave function. As discussed in Section 3.1, since two condensate wave functions with different global phases are degenerate, the excitation energy vanishes in the long wave length limit.

Superfluidity and Critical Velocity. The phonon mode with linear dispersion is the only mode at the lowest energy. As we will discuss here, this has very dramatic consequence. Considering an impurity with mass m_0 moving inside a condensate with velocity $\mathbf{v_i}$, friction occurs when this impurity can be scattered to another velocity $\mathbf{v}_f$ and the momentum is transferred into an excitation of condensate with momentum $\mathbf{q}$. Here we remark that, for a system with Galilean invariance, this is equivalent to a system moving with velocity $\mathbf{v}_i$ in the presence of a static impurity. However, for a systems without Galilean invariance, the two cases are not equivalent, which leads to two distinct critical velocities. We will discuss such an example in the spin-orbit coupled BEC in Section 4.5.

Let us now focus on the situation with a moving impurity in a static BEC. When this linearly dispersive mode is the only low-lying excitation, the momentum conservation and the energy conservation together give

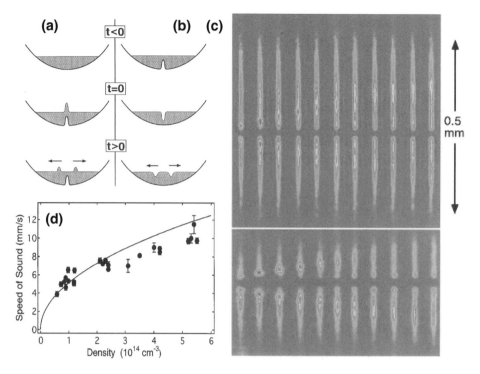

Figure 3.2 Measurement of the phonon velocity in a BEC. (a–b) A local density dip or a hump created by a focused laser beam in the center of a BEC, which propagates toward the edge of a BEC. (c) Propagating of the local density modulation at different times, from which one can deduce the sound velocity. (d) Measured sound velocity as a function of atom density. Reprinted from Ref. [8]. A color version of this figure can be found in the resources tab for this book at cambridge.org/zhai.

$$m_0\mathbf{v}_i = m_0\mathbf{v}_f + \mathbf{q} \tag{3.28}$$

$$\frac{m_0\mathbf{v}_i^2}{2} = \frac{m_0\mathbf{v}_f^2}{2} + c|\mathbf{q}|. \tag{3.29}$$

Replacing $\mathbf{v}_f = \mathbf{v}_i - \mathbf{q}/m_0$ in Eq. 3.29, we obtain

$$\mathbf{v}_i \cdot \mathbf{q} - c|\mathbf{q}| = \frac{\mathbf{q}^2}{2m_0}. \tag{3.30}$$

Therefore, if $|\mathbf{v}_i| < c$, Eq. 3.30 cannot be satisfied. That means if the velocity of a moving impurity is smaller than the phonon velocity, it cannot be scattered and there is no friction for its motion. With the Galilean invariance, it is equivalent to say, when the fluid moves with a velocity smaller than c, there is no friction. This phenomenon is known as *superfluidity*. The upper bound v_c for the velocity of a moving fluid without friction is called the *superfluid critical velocity*. Thus, the sound velocity here equals to the superfluid critical velocity. For a more general isotropic quasi-particle dispersion $\mathcal{E}(|\mathbf{q}|)$, the similar argument leads to a general condition for the critical velocity given by

$$v_c = \min\left(\frac{\mathcal{E}(|\mathbf{q}|)}{|\mathbf{q}|}\right). \tag{3.31}$$

This is known as the Landau criterion for the superfluid critical velocity. From the Landau criterion it is also very clear that for a noninteracting BEC, the dispersion remains quadratic and the critical velocity is zero. In other words, a noninteracting BEC is *not* a superfluid. Bose condensation and interaction together lead to superfluidity. We will apply this criterion again when we discuss the Fermi superfluid in Section 6.2.

Soon after the BEC is achieved, the existence of a critical velocity is also experimentally confirmed [146]. Experimentally, they scan a focused beam back and forth in a BEC, as shown in Figure 3.3(a). Then, after a certain scanning time, they turn off the scan and let the condensate thermalize again. After that they measure the increasing of the thermal fraction in the BEC. If there exists a critical velocity, because there is no friction and no excitation when the scanning velocity is below the critical velocity, the thermal fraction will not change. But when the scanning velocity is above the critical velocity, thermal fraction increases as the velocity increases. In this experiment [146], they have tried different scanning frequencies and amplitudes. When they plot the thermal fraction as a function of the scanning velocity, as shown in Figure 3.3(b), the measurements with different frequencies collapse nearly into the same curve, which clearly displays the behavior of a critical velocity.

Thomas–Fermi Distribution. Now we consider the physical consequences of the hydrodynamic equation in the presence of a trapping potential $V(\mathbf{r})$. By ignoring the quantum

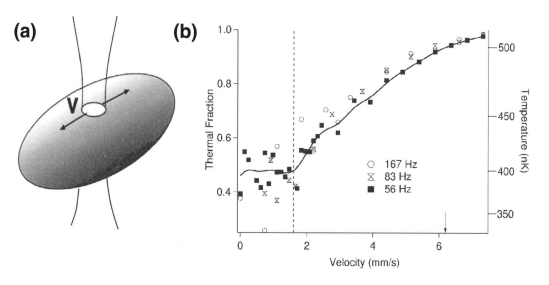

Figure 3.3   Experimental measurement of the critical velocity of a BEC. (a) Schematic of a local stirring of the condensate with a fixed velocity v. (b) Thermal fraction after certain duration of stirring as a function of the velocity, for different stirring amplitudes and frequencies. The dashed line indicates the critical velocity. Reprinted from Ref. [146].

pressure term in the Newton equation, the equilibrium solution gives $\mathbf{v}_s = 0$ and the equilibrium density distribution

$$n_0(\mathbf{r}) = \frac{\mu - V(\mathbf{r})}{U} \tag{3.32}$$

for the regime with $\mu > V(\mathbf{r})$, and $n_0(\mathbf{r}) = 0$ for $\mu < V(\mathbf{r})$. The equation $\mu = V(\mathbf{R})$ determines the boundary of a BEC. For a harmonic trap, the density profile is an inverted parabolic function, which is also called the *Thomas–Fermi distribution*. Note that $\mu = U n_0$ is the equation-of-state for a uniform system, the result of Eq. 3.32 can be interpreted as the local density approximation, that is, we simply replace the chemical potential μ in the equation-of-state of a uniform system by a "local chemical potential" $\mu(\mathbf{r}) = \mu - V(\mathbf{r})$. Furthermore, the total number conservation equation is

$$\int_{\mathbf{r} \subset \mathbf{R}} d^3\mathbf{r}\, n_0(\mathbf{r}) = N, \tag{3.33}$$

which determines μ for each given N, and fixes the entire density profile. For instance, considering an isotropic harmonic trap $m\omega_0^2 r^2/2$, one can find $4\pi m\omega_0^2 R^5 = 15UN$, and thus $R \propto N^{1/5}$.

Ignoring the quantum pressure term is crucial for obtaining the Thomas-Fermi distribution. Is this approximation justified, or consistent with the resulting density profile? Here, with Eq. 3.32, let us do a "back-of-envelope" estimation. Using the result $R \propto N^{1/5}$, we can estimate that the total interaction energy is proportional to $UnN \propto UN^2/R^3 \propto UN^{7/5}$, and the total harmonic trapping energy is $\sim R^2 N \propto N^{7/5}$. For large number of atoms, since these two terms have the same dependence on N, they can balance each other. However, the quantum pressure term is originated from the density inhomogeneity, and is given by $N/R^2 \propto N^{3/5}$. Thus, when N is sufficiently large, the quantum pressure energy is always much smaller compared with both the interaction energy and the harmonic trapping energy, and therefore, the quantum pressure term can be safely ignored in the hydrodynamic equation. This is a significant difference between the interacting and the noninteracting cases. In the noninteracting case, the ground state density profile is determined by the balance between the harmonic trap potential and the kinetic energy, which gives a Gaussian distribution. However, in the interacting case with large number of bosons, the ground state density profile is determined by the balance between the interaction energy and the harmonic trapping energy, which results in an inverted parabola.

Low-Energy Modes in a Harmonic Trap. Now we consider the low-energy excitations describing the density fluctuation on top of the equilibrium Thomas-Fermi distribution. Let us write

$$n(\mathbf{r}, t) = n_0(\mathbf{r}) + \delta n(\mathbf{r}, t) = \frac{\mu - V(\mathbf{r})}{U} + \delta n(\mathbf{r}, t), \tag{3.34}$$

and to the linear order of δn, one obtains

$$\frac{\partial^2 \delta n}{\partial t^2} = \nabla \left[\left(\frac{\mu - V(\mathbf{r})}{m} \right) \nabla \delta n \right] \tag{3.35}$$

$$= \nabla \left(\frac{\mu - V(\mathbf{r})}{m} \right) \nabla \delta n + \frac{U n_0(\mathbf{r})}{m} \nabla^2 \delta n, \tag{3.36}$$

where the first term in Eq. 3.36 varies in space of the order of $1/R$. Considering δn varying in space with a typical wave vector k, and if $k \gg 1/R$, the first term in Eq. 3.36 can be ignored compared with the second term in Eq. 3.36, then, it yields a sound mode with $\omega = c(\mathbf{r})k$, where $c(\mathbf{r}) = \sqrt{Un_0(\mathbf{r})/m}$. This is consistent with the local density approximation where sound propagates with a local sound velocity.

When $k \sim 1/R$, the first term becomes important. In this regime, the collective mode becomes a global motion of entire condensate. For simplicity, let us consider a three-dimensional isotropic trap as $V(\mathbf{r}) = m\omega_0^2 \mathbf{r}^2/2$, we can expand δn as $\delta n = \mathcal{P}_l^{2n_r}(r/R)r^l Y_{lm}(\theta, \phi)$. Here $\mathcal{P}_l^{2n_r}(r/R)$ are polynomials of degree $2n_r$, which only contain even-power terms and satisfy the orthogonality condition. This polynomial contains n nodes in the radial direction. The solution of frequencies can be obtained as [167, 137]

$$\omega(n_r, l) = \omega_0 \sqrt{2n_r^2 + 2n_r l + 3n_r + l}. \tag{3.37}$$

This should be compared with the spectrum of noninteracting harmonic oscillator where the excitation spectrum is given by $\omega(n_r, l) = \omega_0(2n_r + l)$. The difference is another manifestation of the interaction effects.

Surface modes. Modes with $n_r = 0$ are called the surface modes because the density change δn mostly manifests itself around the boundary, as shown in Figure 3.4 (a) and (b). Their frequencies are given by $\omega = \omega_0\sqrt{l}$. For $l = 1$, the density changes as $\delta n = rY_{1m} = x \pm y, z$. Adding such δn into the Thomas-Fermi distribution in a harmonic trap, it is easy to see that this mode corresponds to a global shift of the density distribution, as shown in Figure 3.4(a). It describes the center-of-mass oscillation of the entire condensate in the harmonic trap. This is called the *dipole mode* and its frequency is the same as trap frequency independent of the interactions. Figure 3.4(b) shows the case with $l = 2$ called the *quadrupole mode*, which causes a quadrupole deformation of the surface. In Figure 3.5 we show experimental observation of the quadrupole mode, together with the hexadecapole mode with $l = 4$. Other than $l = 1$, the frequencies of all surface modes are smaller than that of their noninteracting counterpart.

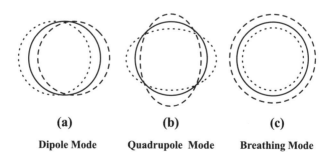

       (a)   **(b)**   **(c)**

   Dipole Mode   **Quadrupole Mode**   **Breathing Mode**

Figure 3.4   Schematic of the collective modes of a BEC. (a) The dipole model with $n_r = 0, l = 1$. (b) The quadrupole mode with $n_r = 0, l = 2$. (c) The breathing mode with $n_r = 1, l = 0$. The solid line denotes the equilibrium boundary, and the dashed and dotted lines denote the boundary at $T/4$ and $3T/4$, respectively. T is a period of the collective mode oscillation.

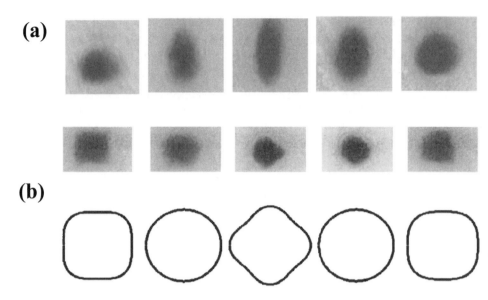

The observation of the surface modes. (a) Density profile at different times of the oscillation. The upper panel is the quadrupole mode with $l = 2$, and the lower panel is the hexadecapole mode with $l = 4$. (b) Schematic of the $l = 4$ mode. Reprinted from Ref. [132].

Compressional modes. Modes with $n_r \neq 0$ are called the *compressional mode*. In particular, δn for the mode with $n_r = 1$ and $l = 0$ has one node in the radial direction, which corresponds to either increasing density at the center and decreasing the density at the edge, or vice versa. This periodic change in the condensate size is called the *breathing mode*, as shown in Figure 3.4(c). The frequency of this breathing mode is $\sqrt{5}\omega_0$.

Anisotropic Expansion. Now let us consider the time-of-flight expansion of a Bose condensate. As we have discussed above, the time-of-flight expansion of the normal component is a ballistically expansion with a velocity $\hbar\mathbf{k}/m$. However, the expansion of the Bose condensed component follows the hydrodynamic equations. From Eq. 3.19, one can see that, when the trap is turned off, the acceleration of the velocity in the initial stage of this expansion is mostly determined by the gradient of the local internal energy $\nabla(Un_0)$. Considering an anisotropic trap

$$V(\mathbf{r}) = \frac{1}{2}m\omega_z^2 z^2 + \frac{1}{2}m\omega_\perp^2 (x^2 + y^2), \tag{3.38}$$

and if $\omega_z \ll \omega_\perp$, then the size R_z along $\hat{z}$ is much longer than the size $R_\perp$ in the transverse plane, that is,

$$R_z = \sqrt{\frac{2\mu}{m\omega_z^2}} \gg R_\perp = \sqrt{\frac{2\mu}{m\omega_\perp^2}}. \tag{3.39}$$

Therefore, it is also easy to see that the gradient term $\partial_z n(\mathbf{r})$ along $\hat{z}$ is smaller than the gradient terms $\partial_x n(\mathbf{r})$ or $\partial_y n(\mathbf{r})$ in the transverse plane. According to Eq. 3.19, a larger gradient term will give rise to a faster acceleration, and therefore, a larger cloud size after expansion. Thus, the aspect ratio of the cloud size is inverted before and after the time-of-flight expansion. This is sharply in contrast to the thermal cloud where the cloud size will finally become isotropic after long time time-of-flight, because the momentum distribution of the thermal atoms are isotropic. This inverted aspect ratio is a hallmark of the hydrodynamic behavior, which has also been observed in the experiments of BEC [84, 163], and later also in strongly interacting Fermi gases nearby a Feshbach resonance [129].

Two-Fluid Hydrodynamics. Now we return to the uniform situation but we concern about the finite temperature effect. Let us now ask ourselves following question. From the previous discussion of the hydrodynamic equations, we have obtained a sound velocity $c = \sqrt{Un_0/m}$, where n_0 is the condensate density. If we simply extend this formula to finite temperature, such a velocity will vanish at the transition temperature T_c. On the other hand, we know from our daily physical intuition that there is always a sound velocity even in normal gas above T_c, and such velocity should always be finite at T_c. How shall we reconcile this contradiction ? This contradiction seems to indicate that we have missed a sound mode below T_c. To solve this problem, we need to introduce a hydrodynamic theory for the superfluid at finite temperature, which is known as the *two-fluid hydrodynamics*.

Here two fluids means one superfluid component describing the macroscopically occupied mode and another normal component describing particles in all the other excited modes. Landau formulated and solved the two-fluid equations in the non-dissipative limit. The two-fluid model is described by several variables including the superfluid density n_s, the superfluid velocity $\mathbf{v}_s$, the normal density n_n, the normal velocity $\mathbf{v}_n$ and entropy S. These quantities satisfy following four two-fluid hydrodynamic equations [100]

$$\frac{\partial n}{\partial t} + \nabla\mathbf{j} = 0, \tag{3.40}$$

$$\frac{\partial S}{\partial t} + \nabla(S\mathbf{v}_n) = 0, \tag{3.41}$$

$$\frac{\partial \mathbf{v}_s}{\partial t} = -\nabla\left(\mu + \frac{1}{2}\mathbf{v}_s^2\right), \tag{3.42}$$

$$\frac{\partial j_i}{\partial t} + \frac{\partial \Pi_{ij}}{\partial x_j} = 0. \tag{3.43}$$

Here the total density n is a sum of both the superfluid component and the normal component as $n = n_s + n_n$. n_s vanishes at the transition temperature, and n_n vanishes at zero temperature. $\mathbf{j}$ is the total current given by $\mathbf{j} = n_s \mathbf{v}_s + n_n \mathbf{v}_n$. Eq. 3.40 is an equation for total number conservation. Eq. 3.41 is the entropy conservation because the nondissipative limit is considered here. Here the entropy is only carried by the normal component, and therefore the entropy flows with the velocity $\mathbf{v}_n$. Eq. 3.42 governs the motion of the superfluid velocity $\mathbf{v}_s$, which in fact takes the same form as Eq. 3.19. Eq. 3.43 governs the motion of the total current, where Π_{ij} is momentum flux density tensor given by

$$\Pi_{ij} = P\delta_{ij} + n_s v_{si}v_{sj} + n_n v_{ni}v_{nj}, \tag{3.44}$$

where P is the pressure, i and j label spatial directions, and v_{si} and v_{ni} are the $i = x, y, z$ component of $\mathbf{v_s}$ and $\mathbf{v_n}$, respectively.

We will not further proceed the detailed derivation of solving the two-fluid hydrodynamics. Here we just briefly mention that the consequence of the two-fluid hydrodynamics is the existence of another sound mode. The physical meaning is that, the entropy, or the heating, also propagates as a wave in the superfluid phase. These two waves of the heat and the density in general are hybridized, which gives rise to the first and the second sounds in a superfluid. This is a highly unconventional phenomenon and is sharply in contrast to our daily intuition from a normal fluid, where the motion of heat is diffusive. Approaching T_c from below, one of the sound waves becomes density wave, which survives above T_c. The other sound wave becomes heat wave, which undergoes a transition from the wave behavior to the diffusive behavior across T_c.

3.3 Bogoliubov Theory

Now we take an alternative approach to treat the excitations of a BEC. In the Bogoliubov theory, we start with a second quantized Hamiltonian with renormalizable contact potential introduced in Section 2.2:

$$\hat{H} = \sum_{\mathbf{k}} (\epsilon_{\mathbf{k}} - \mu)\hat{a}^\dagger_{\mathbf{k}}\hat{a}_{\mathbf{k}} + \frac{g}{2V} \sum_{\mathbf{k_1 k_2 k_3 k_4}} \hat{a}^\dagger_{\mathbf{k_1}} \hat{a}^\dagger_{\mathbf{k_2}} \hat{a}_{\mathbf{k_3}} \hat{a}_{\mathbf{k_4}}, \tag{3.45}$$

with $\mathbf{k_1} + \mathbf{k_2} = \mathbf{k_3} + \mathbf{k_4}$, where $\epsilon_{\mathbf{k}} = \hbar^2\mathbf{k}^2/(2m)$. The Bogoliubov theory is a perturbative approach. Let us first consider the noninteracting limit that bosons are all condensed in the zero-momentum state. Following our discussion in Section 3.1, we take the unperturbed wave function as the coherent state given by

$$|G_0\rangle \propto e^{\sqrt{N_0}\hat{a}^\dagger_{\mathbf{k}=0}}|0\rangle, \tag{3.46}$$

and under this wave function, we have

$$\langle G_0|\hat{a}_{\mathbf{k}=0}|G_0\rangle = \langle G_0|\hat{a}^\dagger_{\mathbf{k}=0}|G_0\rangle = \sqrt{N_0}. \tag{3.47}$$

Hence, under this wave function, we can replace both the operator $\hat{a}^\dagger_{\mathbf{k}=0}$ and $\hat{a}_{\mathbf{k}=0}$ as $\sqrt{N_0}$. Since $\sqrt{N_0}$ is a large number, in the Bogoliubov theory, we perform a systematical expansion of the interaction term in terms of $\sqrt{N_0}$. The leading order is that all four $\mathbf{k_1}, \ldots, \mathbf{k_4}$ equal zero and all operators in the interaction term are replaced by $\sqrt{N_0}$, which gives rise to condensate mean-field energy

$$\frac{\mathcal{E}}{V} = -\mu n_0 + \frac{g}{2}n_0^2, \tag{3.48}$$

where $n_0 = N_0/V$. Thus, it gives $\mu = gn_0$. Next, if one sets three of four momenta $\mathbf{k_1}, \ldots, \mathbf{k_4}$ to zero, the last momentum also has to be zero. Therefore, it does not lead to new term.

In the next order, one can take two of four momenta $\mathbf{k}_1, \ldots, \mathbf{k}_4$ to be zero, and replace the operator $\hat{a}^\dagger_{\mathbf{k}=0}$ or $\hat{a}_{\mathbf{k}=0}$ as $\sqrt{N_0}$. We shall also use $\mu = gn_0$.[4] Thus, the Bogoliubov Hamiltonian for finite momentum modes becomes

$$\hat{H} = \sum_{\mathbf{k}\neq 0}(\epsilon_{\mathbf{k}} + gn_0)\hat{a}^\dagger_{\mathbf{k}}\hat{a}_{\mathbf{k}} + \frac{gn_0}{2}\sum_{\mathbf{k}\neq 0}\left(\hat{a}^\dagger_{\mathbf{k}}\hat{a}^\dagger_{-\mathbf{k}} + \hat{a}_{-\mathbf{k}}\hat{a}_{\mathbf{k}}\right). \tag{3.49}$$

Note that the zero-momentum state has been excluded from the momentum summation in the Bogoliubov Hamiltonian. The second term is the most important process. It appears that the total number of atoms are not conserved. In fact, it is the total number of atoms in the finite momentum states that is not conserved, because we have treated the atoms in the zero-momentum state separately as a condensate. The last term describes that a pair of bosons can be scattered into finite momentum states from the zero momentum condensate, or a pair of bosons with finite but opposite momenta can annihilate into the zero momentum condensate.

It is straightforward to show that by introducing

$$\hat{a}_{\mathbf{k}} = u_{\mathbf{k}}\hat{\alpha}_{\mathbf{k}} - v_{\mathbf{k}}\hat{\alpha}^\dagger_{-\mathbf{k}}, \tag{3.50}$$

$$\hat{a}^\dagger_{-\mathbf{k}} = -v_{\mathbf{k}}\hat{\alpha}_{\mathbf{k}} + u_{\mathbf{k}}\hat{\alpha}^\dagger_{-\mathbf{k}}, \tag{3.51}$$

and

$$u^2_{\mathbf{k}} = \frac{1}{2}\left(\frac{\epsilon_{\mathbf{k}} + gn_0}{\mathcal{E}_{\mathbf{k}}} + 1\right), \tag{3.52}$$

$$v^2_{\mathbf{k}} = \frac{1}{2}\left(\frac{\epsilon_{\mathbf{k}} + gn_0}{\mathcal{E}_{\mathbf{k}}} - 1\right), \tag{3.53}$$

the Bogoliubov Hamiltonian can be diganoalized as

$$\hat{H} = \sum_{\mathbf{k}\neq 0}\mathcal{E}_{\mathbf{k}}\hat{\alpha}^\dagger_{\mathbf{k}}\hat{\alpha}_{\mathbf{k}} + \frac{1}{2}\sum_{\mathbf{k}\neq 0}(\mathcal{E}_{\mathbf{k}} - (\epsilon_{\mathbf{k}} + gn_0)), \tag{3.54}$$

with

$$\mathcal{E}_{\mathbf{k}} = \sqrt{(\epsilon_{\mathbf{k}} + gn_0)^2 - (gn_0)^2} = \sqrt{\epsilon_{\mathbf{k}}(\epsilon_{\mathbf{k}} + 2gn_0)}. \tag{3.55}$$

Here we should note that Eq. 3.50 and Eq. 3.51 are not a unitary transformation, since a unitary transformation of $\hat{a}_{\mathbf{k}}$ and $\hat{a}^\dagger_{-\mathbf{k}}$ does not keep the boson commutative relation, but Eq. 3.50 and Eq. 3.51 can.

Here we should also remark that it is a subtle issue of how to relate g to a_s in such a perturbative calculation. The relation Eq. 2.31 between g and a_s derived from the two-body calculation in Section 2.2 is valid up to all orders. When this relation is applied to a many-body calculation, it should be used in a way that is consistent with the degree of approximation in the many-body calculation. Here at the leading order, the order of approximation in the many-body calculation corresponds to the first order in the ladder summation of the two-body calculation discussed in Section 2.2, and therefore we simply take $g = U = 4\pi\hbar^2 a_s/m$. Then, the spectrum becomes

$$\mathcal{E}_{\mathbf{k}} = \sqrt{\epsilon_{\mathbf{k}}(\epsilon_{\mathbf{k}} + 2Un_0)}. \tag{3.56}$$

[4] This treatment is to enforce the Hugenholtz–Pines relation such that the excitation remains gapless [78].

Here we should introduce the concept of *quasi-particle*, which is one of the most fundamental concepts in quantum many-body physics. In a uniform system, "quasi-particle" refers to eigenmodes with well-defined energy-momentum dispersion relation. Because of interactions, a quasi-particle is normally different from the constitution particle of the system. For example, in this case the operator for the constitution particle of this system is $\hat{a}_{\mathbf{k}}$ but the operator for quasi-particle is $\hat{\alpha}_{\mathbf{k}}$. Because $\mathcal{E}_{\mathbf{k}}$ is always positive for all $\mathbf{k}$, the ground state should be a vacuum of all quasi-particles, that is, $\hat{\alpha}_{\mathbf{k}}|G\rangle = 0$. This leads to the ground state $|G\rangle$ that is modified from $|G_0\rangle$ by interactions, and

$$|G\rangle = e^{\sqrt{N_0}\hat{a}_{\mathbf{k}=0}^{\dagger}} e^{-\sum_{\mathbf{k}\neq 0} \frac{v_{\mathbf{k}}}{u_{\mathbf{k}}} \hat{a}_{\mathbf{k}}^{\dagger}\hat{a}_{-\mathbf{k}}^{\dagger}}|0\rangle. \tag{3.57}$$

Up to this level of the Bogoliubov Hamiltonian, $\hat{\alpha}_{\mathbf{k}}^{\dagger}|G\rangle$ creates an eigenmode with well-defined dispersion relation $\mathcal{E}_{\mathbf{k}}$. In contrast, the original boson operator $\hat{a}_{\mathbf{k}}^{\dagger}$ does not generate an eigenmode, and $|G\rangle$ is also not the vacuum of $\hat{a}_{\mathbf{k}}$. That is to say, there will always be finite population on these finite momentum states, which is called the *quantum depletion* as we will discuss below. The last term in Eq. 3.54 represents a zero-point energy associated with the quasi-particle at each momentum, and this vacuum energy will lead to the *Lee–Huang–Yang correction* [105] discussed below.

We should also note that there are also residual interactions between quasi-particles, which will give rise to a finite lifetime $\tau_{\mathbf{k}}$ for each quasi-particle. Thus, in order for a quasi-particle to be well defined, we require $\hbar/\tau_{\mathbf{k}} \ll \mathcal{E}_{\mathbf{k}}$. The finite lifetime leads to a broadening of quasi-particle energy, which is given by $\hbar/\tau_{\mathbf{k}}$. If this broadening is comparable or even larger than the excitation energy itself, the excitation energy will be smeared out by the broadening. Another way to understand the well-defined quasi-particle is to consider a superposition between a vacuum state and a state with one quasi-particle, which will undergo Rabi oscillation with period $2\pi\hbar/\mathcal{E}_{\mathbf{k}}$. $\hbar/\tau_{\mathbf{k}} \ll \mathcal{E}_{\mathbf{k}}$ also means that $\tau_{\mathbf{k}} \gg 2\pi\hbar/\mathcal{E}_{\mathbf{k}}$. That is to say, for a well-defined quasi-particle, one can observe many periods of Rabi oscillations before it damps out. In the opposite limit when $\hbar/\tau_{\mathbf{k}} \gg \mathcal{E}_{\mathbf{k}}$, the oscillation decays even before finishing one period, and the quasi-particle is no longer well defined. Such a system without quasi-particle description remains as a theoretical challenge in modern quantum many-body theory.

The Healing Length. As shown in Figure 3.6, the quasi-particle dispersion Eq. 3.56 exhibits a linear dispersion $\hbar ck$ at small k, consistent with what we have derived from the hydrodynamic theory in Section 3.2. In this regime, $u_{\mathbf{k}}^2 \sim v_{\mathbf{k}}^2 \sim Un_0/(2\hbar ck)$, which means that each quasi-particle mode contains equal contribution of particle and hole. This regime is called the phonon regime where the excitation is collective. At large momentum the asymptotic behavior of $\mathcal{E}_{\mathbf{k}}$ is $\epsilon_{\mathbf{k}} + Un_0$, which is simply a Hatree–Fock energy shift to the free-boson dispersion. At the same time $u_{\mathbf{k}} \to 1$ and $v_{\mathbf{k}} \to 0$, and $\hat{\alpha}_{\mathbf{k}}$ approaches $\hat{a}_{\mathbf{k}}$. This regime is therefore called the free-particle regime where the excitation behaves more like a free-particle. There exists a characteristic momentum regime where the quasi-particle dispersion undergoes a crossover from the phonon regime to the free-particle regime. Roughly speaking, it is determined by $\hbar^2 k_0^2/(2m) = \hbar ck_0$ and $k_0 = 2mc/\hbar$. This sets a length scale $\xi = \sqrt{2}/k_0 = \hbar/(\sqrt{2}mc)$ and ξ is called *the healing length*. We have $\hbar^2/(2m\xi^2) = Un_0$. In other words, for $k \ll 1/\xi$, interaction energy dominates over the kinetic energy and that is why the excitation is collective. And for $k \gg 1/\xi$ the kinetic energy dominates over the

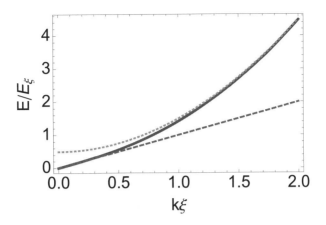

Figure 3.6 The Bogoliubov excitation spectrum. Eq. 3.56 is shown by the solid line and is compared with the linear dispersion $\hbar c k$ (dashed line) at small momentum and the free-particle dispersion with a constant offset $\epsilon_{\mathbf{k}} + U n_0$ (dotted line) at large momentum. Here ξ is the healing length, and $E_\xi = \hbar^2/(m\xi^2)$. A color version of this figure can be found in the resources tab for this book at cambridge.org/zhai.

interaction energy so that the excitation becomes free-particle like. The healing length ξ is an important quantity that has a few important physical meanings. We will encounter it again below.

The Lee–Huang–Yang Correction. The summation of the second term in Eq. 3.54 represents the zero-point energy of each quasi-particle mode. For a single harmonic oscillator, even without any excitation, there is a zero-point energy that is half of the harmonic frequency. Here it can be viewed as that there is a harmonic oscillator on each $\mathbf{k}$ mode, and the summation of all zero-point energies gives rise to the so-called *vacuum energy*. Here, this summation is given by

$$\frac{1}{2}\sum_{\mathbf{k}\neq 0}(\mathcal{E}_{\mathbf{k}} - \epsilon_{\mathbf{k}} - gn_0) = \frac{1}{2}\sum_{\mathbf{k}}\frac{-(gn_0)^2}{\mathcal{E}_{\mathbf{k}} + \epsilon_{\mathbf{k}} + gn_0}. \tag{3.58}$$

In the micro-canonical ensemble, the total energy density up to this order is given by

$$\frac{\mathcal{E}}{V} = \frac{gn^2}{2} + \frac{1}{2V}\sum_{\mathbf{k}}\frac{-(gn)^2}{\mathcal{E}_{\mathbf{k}} + \epsilon_{\mathbf{k}} + gn}, \tag{3.59}$$

where the first term comes from the mean-field interaction energy of the zero-momentum condensate. Here we have used the fact that, for weakly interacting regime, the quantum depletion is small as we will show below, and we take $n_0 \approx n$. We can see that the second term diverges as $\sum_{\mathbf{k}} m/(\hbar^2 \mathbf{k}^2)$. Therefore, we need to use the renormalization condition to eliminate this divergency. However, since this energy contribution is only a perturbative

correction to the next order of mean-field, we shall only use the renormalization condition
up to the second-order scattering processes. More explicitly, we should use

$$g = U + \frac{U^2}{V} \sum_{\mathbf{k}} \frac{1}{2\epsilon_{\mathbf{k}}} \tag{3.60}$$

in the first term of Eq. 3.59 and replace g with U in the second term of Eq. 3.59. This leads
to the total energy density given by

$$\frac{\mathcal{E}}{V} = \frac{Un^2}{2} + \frac{(Un)^2}{2V} \sum_{\mathbf{k}} \left(\frac{1}{2\epsilon_{\mathbf{k}}} - \frac{1}{\mathcal{E}_{\mathbf{k}} + \epsilon_{\mathbf{k}} + Un} \right). \tag{3.61}$$

The first term of Eq. 3.61 is the mean-field energy. The summation in the second term of
Eq. 3.61 now converges, which gives rise to the leading-order correction to the mean-field
energy. This result was first obtained by Lee, Huang and Yang using the pseudo-potential
model [105] and is now named as the Lee–Huang–Yang correction, and

$$\frac{\mathcal{E}_{\mathrm{LHY}}}{V} = \frac{Un^2}{2} \frac{128}{15\sqrt{\pi}} \sqrt{na_s^3}. \tag{3.62}$$

Hence, it is now clear that the LHY correction comes from the zero-point quantum fluctuation of all quasi-particles modes. To observe the LHY correction, the dimensionless
parameter na_s^3 cannot be too small. Using the tunability of the scattering length by Feshbach resonance, one can make the LHY correction visible. In the past years, the LHY
correction has been observed in several ultracold atom experiments [126].

Here let us make a short comment on more general role of such a correction. When
the ground state is unique at the mean-field level, the correction is quantitative. Nevertheless, there are also cases that the correction is qualitative rather than quantitative. It
happens when the mean-field ground states have degeneracy, but such degeneracy is not
protected by an exact symmetry of the full Hamiltonian. This usually occurs in many *frustrated models*. When this happens, the mean-field energy alone cannot determine a unique
ground state, and this fluctuation energy will play a crucial role in selecting out the actual
ground state, because the fluctuation energy usually can lift the degeneracy between different degenerate mean-field states. Usually, the intuition is that the fluctuation energy is
smaller if the sound velocity is smaller. This is because for smaller sound velocity, the excitation energy of the phonon mode increases slower, and a smaller excitation energy also
means a smaller contribution of the zero-point energy. This phenomenon of selecting out
the ground state by fluctuation energy is the *order by disorder* mechanism. Here the word
"order" refers to that a unique mean-field state is selected out, and the word "disorder"
actually means the fluctuation energy.

The Quantum Depletion. Below we will check two issues in order for the Bogoliubov
theory to be a self-consistent theory. First, we have mentioned that the Bogoliubov theory
is a perturbative approach, which treats the noninteracting Bose condensate as the unperturbed wave function. We should first check that the condensate remains as the dominate
component when interactions are turn on. In other words, when interactions are turned on,
there exists population on the finite momentum state. The total number of atoms populated

at finite momentum states is called the quantum depletion. In order for the Bogoliubov theory to be valid, the quantum depletion should remain small compared with the total number of atoms.

To compute the quantum depletion N_{dp}, we note that the ground state $|G\rangle$ is a vacuum of quasi-particle operator $\hat{\alpha}_{\mathbf{k}}$ but not for atom operator $\hat{a}_{\mathbf{k}}$. Using Eq. 3.50 and 3.51, it can be straightforwardly calculated that

$$N_{dp} = \langle G| \sum_{\mathbf{k}\neq 0} \hat{a}^{\dagger}_{\mathbf{k}} \hat{a}_{\mathbf{k}} |G\rangle = \sum_{\mathbf{k}} v^2_{\mathbf{k}}, \qquad (3.63)$$

and the density of quantum depletion $n_{dp} = N_{dp}/V$ is given by

$$n_{dp} = \frac{1}{V} \sum_{\mathbf{k}\neq 0} v^2_{\mathbf{k}} = \frac{1}{3\pi^2} \left(\frac{mc}{\hbar}\right)^3 \propto \frac{1}{\xi^3}. \qquad (3.64)$$

That is to say, the depleted bosons is of the order one in a volume of ξ^3, which gives another physical meaning for the healing length. Thus, the density of quantum depletion is much smaller than the total density when $\xi \gg d$, where d is the mean interparticle distance. In other words, the ratio between the density of quantum depletion to the total density is given by

$$\frac{n_{dp}}{n} = \frac{8}{3\sqrt{\pi}} (na^3_s)^{1/2}, \qquad (3.65)$$

and $n_{dp}/n \ll 1$ in the weakly interacting regime when $na^3_s \ll 1$. Hence, this assumption of the Bogouliubov theory is valid in the weakly interacting regime.

The Beliaev–Landau Damping. Another thing to check is whether the Bogoliubov quasi-particles are really well-defined quasi-particles. To this end, we note that the Bogoliubov Hamiltonian only retains the quadratic terms in terms of the creation and annihilation operators, and at this order the quasi-particles are well-defined eigenmodes. But all these quasi-particles acquire a finite lifetime when we go beyond this order. Here, we only replace one of momentum of $\mathbf{k}_1, \ldots, \mathbf{k}_4$ as zero momentum, which gives

$$V_{int} = \frac{g\sqrt{N_0}}{V} \sum_{\mathbf{k}_1\mathbf{k}_2\mathbf{k}_3} \hat{a}^{\dagger}_{\mathbf{k}_1} \hat{a}^{\dagger}_{\mathbf{k}_2} \hat{a}_{\mathbf{k}_3} + \hat{a}^{\dagger}_{\mathbf{k}_1} \hat{a}_{\mathbf{k}_2} \hat{a}_{\mathbf{k}_3}. \qquad (3.66)$$

At this order, the Bogoliubov quasi-particle interact with each other.

Rewritting the boson operator into quasi-particle operators, V_{int} contains four different types of terms like (i) $\hat{\alpha}^{\dagger}_{\mathbf{k}_1} \hat{\alpha}^{\dagger}_{\mathbf{k}_2} \hat{\alpha}_{\mathbf{k}_3}$, (ii) $\hat{\alpha}^{\dagger}_{\mathbf{k}_1} \hat{\alpha}_{\mathbf{k}_2} \hat{\alpha}_{\mathbf{k}_3}$, (iii) $\hat{\alpha}^{\dagger}_{\mathbf{k}_1} \hat{\alpha}^{\dagger}_{\mathbf{k}_2} \hat{\alpha}^{\dagger}_{\mathbf{k}_3}$ and (iv) $\hat{\alpha}_{\mathbf{k}_1} \hat{\alpha}_{\mathbf{k}_2} \hat{\alpha}_{\mathbf{k}_3}$. The first term (i) describes one quasi-particle decays into two, and the second term (ii) describes two quasi-particles scatter and merge into one quasi-particle. The terms (iii) and (iv) describe simultaneously creation or annihilation three quasi-particles. The energy conserved on-shell processes (i) and (ii) give rise to a damping rate and a finite lifetime of the quasi-particles. The off-shell processes (iii) and (iv) do not satisfy energy conservation, but they together can produce a second-order perturbation. For instance, three quasi-particles can first be simultaneously created from the vacuum and then all of them are annihilated together, which gives a correction to the quasi-particle energy. This effect is called the *vacuum polarization*.

Here, it is inspiring to compare the discussion here with the discussion of the hydrogen atom in atomic physics.

- When we only consider the Coulomb interaction, as we have discussed in Section 1.1, the energy levels are exact eigenstates. This is at the same level as the Bogoliubov Hamiltonian.
- When we consider the spontaneous emission, electron can decay from one excited state to a lower energy state by simultaneously emitting a photon, and such processes give rise to level broadening for all excited states. This spontaneous emission process is very much like processes (i) discussed here, which gives rise to a finite lifetime for all quasi-particles.
- When we further consider the quantum electrodynamics, the vacuum polarization means that an electron-positron pair can be created from the vacuum and then annihilate each other. This process can modify the Coulomb energy and gives the famous Lamb shift for the energy level of the hydrogen atom. The shift of quasi-particle energy due to the second-order process combining (iii) and (iv) is similar as the effect of the vacuum polarization.

Here let us focus on the quasi-particle lifetime due to processes (i) and (ii). First, let us consider the zero-temperature case. We consider that a quasi-particle with momentum $\mathbf{k}$ is excited on top of the vacuum, and then we monitor how long it will decay. In this case, since there is no other quasi-particle, only the first term contributes to the quasi-particle lifetime. This is known as the *Beliaev damping*. For instance, let us consider a term as

$$\mathcal{M}_{\mathbf{q},\mathbf{k}-\mathbf{q},\mathbf{k}}\hat{\alpha}^{\dagger}_{\mathbf{k}-\mathbf{q}}\hat{\alpha}^{\dagger}_{\mathbf{q}}\hat{\alpha}_{\mathbf{k}}, \tag{3.67}$$

where $\mathcal{M}$ is a matrix element made of u's and v's resulting from the expansion, and the damping rate for quasi-particle with momentum $\mathbf{k}$ is given by the Fermi-Gorden rule as

$$\frac{1}{\tau} = \frac{2\pi}{\hbar} \sum_{\mathbf{q},\mathbf{k}-\mathbf{q}} |\mathcal{M}_{\mathbf{q},\mathbf{k}-\mathbf{q},\mathbf{k}}|^2 \delta(\mathcal{E}_{\mathbf{k}} - \mathcal{E}_{\mathbf{q}} - \mathcal{E}_{\mathbf{k}-\mathbf{q}}). \tag{3.68}$$

Straightforward calculation of this integration will lead to $\hbar/\tau \propto k^5$ when the momentum k is in the phonon regime. Of course, the proportional constant depends on the interaction strength. This strong suppression at small momentum is partly because of the restriction of the phase space due to the energy conservation constraint, that is to say, due to the energy conservation, $|\mathbf{q}|$ must be smaller than $|\mathbf{k}|$. Here the most important point is that k^5 is much smaller than linear k for small k, which means that the energy level broadening due to quasi-particle lifetime is much smaller than the excitation energy itself, that is, $\hbar/\tau_{\mathbf{k}} \ll E_{\mathbf{k}}$. This satisfies the condition for a quasi-particle to be well defined as discussed above. Experimentally, the Bogoliubov quasi-particles with a fixed momentum can be first excited by the two-photon Bragg pulse, as we will discuss below. And then by monitoring the number of excited atoms, the collision section can be determined, and the experimental results are shown in Figure 3.7(a). Indeed, one can see a strong suppression at low momentum.

At finite temperature, the second term (ii) can also contribute to damping. This is known as the *Landau damping*. In fact, each term in (ii) can find a conjugate process in (i), thus,

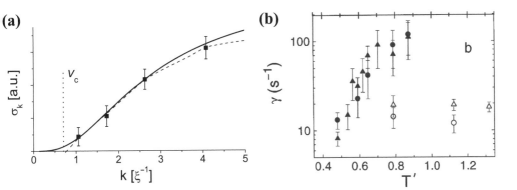

Figure 3.7   Experiments on the quasi-particle damping rate. (a) At low temperature, damping is dominated by the Beliaev damping. The suppression of quasi-particle collision at low momentum is measured. Here the quasi-particles are first excited by the Bragg pulse. Reprinted from Ref. [87]. (b) The damping rate of a low-energy collective mode as a function of temperature. Reprinted from Ref. [84].

we consider a term like $\mathcal{K}_{\mathbf{k+q,q,k}}(\hat{\alpha}^{\dagger}_{\mathbf{k+q}}\hat{\alpha}_{\mathbf{q}}\hat{\alpha}_{\mathbf{k}} + \text{h.c.})$. On one hand, a quasi-particle at momentum $\mathbf{k}$ can decay to another momentum $\mathbf{k+q}$ by merging another quasi-particle at momentum $\mathbf{q}$, and this rate is given by

$$\frac{df_{\mathbf{k}}}{dt} = -\frac{2\pi}{\hbar}\sum_{\mathbf{q}}|\mathcal{K}_{\mathbf{k+q,q,k}}|^2 f_{\mathbf{k}}f_{\mathbf{q}}(1+f_{\mathbf{k+q}})\delta(\mathcal{E}_{\mathbf{k+q}} - \mathcal{E}_{\mathbf{q}} - \mathcal{E}_{\mathbf{k}}), \qquad (3.69)$$

where $f_{\mathbf{k}}$ is the boson population at momentum $\mathbf{k}$ and $1+f_{\mathbf{k+q}}$ is due to the boson enhancement factor. On the other hand, another quasi-particle with momentum $\mathbf{k+q}$ can decay into two quasi-particles with momentum $\mathbf{k}$ and $\mathbf{q}$, and the rate for this inverse process is given by

$$\frac{df_{\mathbf{k}}}{dt} = \frac{2\pi}{\hbar}\sum_{\mathbf{q}}|\mathcal{K}_{\mathbf{k+q,q,k}}|^2(1+f_{\mathbf{k}})(1+f_{\mathbf{q}})f_{\mathbf{k+q}}\delta(\mathcal{E}_{\mathbf{k+q}} - \mathcal{E}_{\mathbf{q}} - \mathcal{E}_{\mathbf{k}}). \qquad (3.70)$$

So the net rate is given by

$$\frac{df_{\mathbf{k}}}{dt} = \frac{2\pi}{\hbar}\sum_{\mathbf{q}}|\mathcal{K}_{\mathbf{k+q,q,k}}|^2$$
$$\times [(1+f_{\mathbf{k}})(1+f_{\mathbf{q}})f_{\mathbf{k+q}} - f_{\mathbf{k}}f_{\mathbf{q}}(1+f_{\mathbf{k+q}})]\delta(\mathcal{E}_{\mathbf{k+q}} - \mathcal{E}_{\mathbf{q}} - \mathcal{E}_{\mathbf{k}}). \qquad (3.71)$$

Note that in the equilibrium distribution, if $f_{\mathbf{k}}$ is given by the boson distribution function denoted by $f^0_{\mathbf{k}}$, this rate vanishes because $(1+f^0_{\mathbf{k}})(1+f^0_{\mathbf{q}})f^0_{\mathbf{k+q}} = f^0_{\mathbf{k}}f^0_{\mathbf{q}}(1+f^0_{\mathbf{k+q}})$ if $\mathcal{E}_{\mathbf{k+q}} = \mathcal{E}_{\mathbf{q}} + \mathcal{E}_{\mathbf{k}}$. This is known as the *detailed balance condition*. Hence, to study the quasi-particle lifetime, we consider the situation that $f_{\mathbf{k}}$ slightly deviates its equilibrium value as $f_{\mathbf{k}} = f^0_{\mathbf{k}} + \delta f_{\mathbf{k}}$, then we have

$$\frac{d\delta f_{\mathbf{k}}}{dt} = \frac{2\pi\delta f_{\mathbf{k}}}{\hbar}\sum_{\mathbf{q}}|\mathcal{K}_{\mathbf{k+q,q,k}}|^2(f^0_{\mathbf{k+q}} - f^0_{\mathbf{q}})\delta(\mathcal{E}_{\mathbf{k+q}} - \mathcal{E}_{\mathbf{q}} - \mathcal{E}_{\mathbf{k}}) \qquad (3.72)$$

and the damping rate

$$\frac{1}{\tau} = -\frac{2\pi}{\hbar} \sum_{\mathbf{q}} |\mathcal{K}_{\mathbf{k+q,q,k}}|^2 (f^0_{\mathbf{k+q}} - f^0_{\mathbf{q}}) \delta(\mathcal{E}_{\mathbf{k+q}} - \mathcal{E}_{\mathbf{q}} - \mathcal{E}_{\mathbf{k}}). \qquad (3.73)$$

Computation of this integration is quite involved. In the regime with $ck < k_B T < k_B T^*$, it gives [144, 177]

$$\frac{1}{\tau}_{\mathbf{k}} \propto \left(\frac{T}{T^*}\right)^4 ck, \qquad (3.74)$$

where T^* is defined as $k_B T^* = U n_0$. The Landau damping is less suppressed by the momentum factor compared with the Beliaev damping, because this damping process is through scattering with another thermally populated quasi-particle at high energy, and therefore the phase space restriction from the energy conservation is less important. But the Landau damping is still suppressed by the temperature factor because it requires thermal population of other quasi-particles. In any case, again, the quasi-particle is well defined at low temperature because $\hbar/\tau_{\mathbf{k}}$ is suppressed by a factor of $(T/T^*)^4$ compared with $\mathcal{E}_{\mathbf{k}}$, and therefore, the level broadening is smaller than the excitation energy itself. In Figure 3.7(b), we show the experimental measurement of how the damping rate of a low-lying excitation depends on temperature, which can be well explained by the Laudau damping mechanism.

The Bragg Spectroscopy. Spectroscopy refers to a probe by a time-periodical perturbation with varying frequency, from which one can investigate how the response of a system changes as the frequency varies. Here we will discuss the Bragg spectroscopy as an example of the spectroscopy measurement in ultracold atomic physics. This probe is a powerful experimental tool to measure the quasi-particle properties discussed above. For the Bragg spectroscopy, two lasers with same polarization, for instance, along $\hat{z}$, are applied to an ultracold atom system, and the two lasers have different momenta $\mathbf{k}_1$ and $\mathbf{k}_2$ and different frequencies ω_1 and ω_2, as shown in Figure 3.8(a). The electric field is then given by

$$\mathbf{E} = E_1 e^{i\mathbf{k}_1 \mathbf{r} - i\omega_1 t} \hat{z} + E_2 e^{i\mathbf{k}_2 \mathbf{r} - i\omega_2 t} \hat{z}, \qquad (3.75)$$

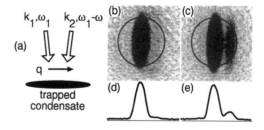

Figure 3.8   Experiment of the Bragg spectroscopy. (a) Schematic of the experimental scheme for the Bragg spectroscopy measurement. Two laser beams with different momenta and different frequencies are applied to the condensate. Typical absorption images (b, d) before and (c, e) after the Bragg pulse. (d) and (e) are integrated density from (b) and (c), respectively. A portion of atoms is transferred into momentum with finite $\mathbf{q} = \mathbf{k}_1 - \mathbf{k}_2$. Reprinted from Ref. [163].

where E_1 and E_2 are intensities of the two lasers. With the previous discussion of the scalar light shift in Section 1.3, these laser beams give a scalar light shift as

$$V(\mathbf{r}) \propto \cos(\mathbf{k} \cdot \mathbf{r} - \omega t), \tag{3.76}$$

where $\mathbf{k} = \mathbf{k}_1 - \mathbf{k}_2$ and $\omega = \omega_1 - \omega_2$. In the second quantized form, this perturbation $\hat{H}_{\text{perp}}$ acts as $\int d^3 \mathbf{r} V(\mathbf{r}) \hat{n}(\mathbf{r})$, and more explicitly,

$$\hat{H}_{\text{perp}} \propto \hat{\rho}_{\mathbf{k}} e^{-i\omega t} + \text{h.c.}, \tag{3.77}$$

where $\hat{\rho}_{\mathbf{k}} = \sum_{\mathbf{q}} \hat{a}^\dagger_{\mathbf{q}+\mathbf{k}} \hat{a}_{\mathbf{q}}$.

After applying the Bragg pulse, part of the atoms are transferred from zero to finite momentum state, as shown in Figure 3.8(b). Using the perturbation theory, one can show that the transition probability is proportional to $S_{\rho_{\mathbf{k}}}(\omega)$. Here, for a general operator $\hat{F}$, $S_F(\omega)$ is called the *dynamic structure factor* and is defined as

$$S_F(\omega) = \frac{1}{\mathcal{Z}} \sum_{m,n} e^{-E_m/(k_{\mathrm{B}}T)} |\langle n|\hat{F}|m\rangle|^2 \delta(\hbar\omega - (E_n - E_m)), \tag{3.78}$$

which reduces to

$$S_F(\omega) = \sum_n |\langle n|\hat{F}|0\rangle|^2 \delta(\hbar\omega - (E_n - E_0)) \tag{3.79}$$

at the zero temperature limit. $S_F(\omega)$ is always positive. It is easy to show that the dynamic structure factor defined in this way can also be written as temporal correlation function given by

$$S_F(\omega) = \int dt e^{i\omega t} \langle \hat{F}(t)\hat{F}(0)\rangle. \tag{3.80}$$

For each fixed momentum $\mathbf{k}$, one can vary ω and the resonant frequency of $S_{\rho_{\mathbf{k}}}(\omega)$ determines the excitation dispersion $\mathcal{E}(\mathbf{k})$, and the width can be related to the quasi-particle lifetime as $\hbar/\tau_{\mathbf{k}}$.

A typical measurement of the Bogoliubov spectrum $\mathcal{E}(\mathbf{k})$ of Eq. 3.56 by the Bragg spectroscopy is shown in Figure 3.9(a). In the zoom-in plot, it is shown that the dispersion is linear for the small momentum regime. When the momentum becomes larger than $1/\xi$, the spectrum becomes quadratic again and is parallel to $\hbar^2 k^2/(2m)$. As shown in Figure 3.6(b), $\mathcal{E}(k) - \hbar^2 k^2/(2m)$ indeed saturates to a constant value in large k. This constant value in fact is the chemical potential of the system.

Interaction Effect on a BEC. In the above two sections, we have used both the hydrodynamic theory and the Bogoliubov theory to address the interaction effects in a BEC. After these discussions, we now understand better why we emphasize the *macroscopic occupation* as the essential ingredient and the defining property for a BEC. This is because many other properties are actually different between a non-interacting BEC and an interacting one. We summarize these differences in Table 3.1. Before ending this section, we should also summarize that, in an interacting BEC, as the interaction strength increases, following effects occurs:

Table 3.1 Comparison of physical properties between a noninteracting BEC and a weakly interacting BEC in three dimensions

	Noninteracting BEC	Weakly interacting BEC
Chemical potential μ	$\mu = 0$	$\mu = U n_0$ (mean-field level)
Low energy dispersion	Quadratic	Linear
Superfluidity	No	Yes
Density in harmonic trap	Gaussian	Inverted parabola
Expansion dynamics	Ballistic	Hydrodynamic
Low-energy modes in trap	Equal spacing	Unequal spacing
Quasi-particle	Original bosons	Bogoliubov quasi-particle
Quantum depletion at $T = 0$	Zero	$\sim 1/\xi^3$
Quasi-particle lifetime	Infinite	Finite

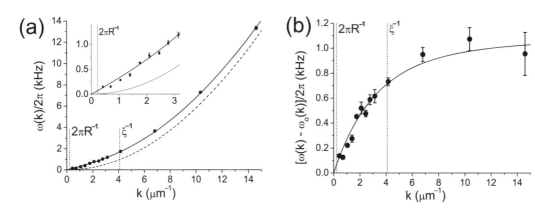

Figure 3.9 Measurement of the Bogoliubov dispersion. (a) The Bogoliubov dispersion $\omega(k)$ (i.e., $\mathcal{E}(\mathbf{k})$ in Eq. 3.56) measured by the Bragg spectroscopy. The inset shows the zoom-in plot of the small momentum regime. The dashed line shows $\omega_0(k) = \hbar^2 k^2/(2m)$. (b) $\omega(k) - \omega_0(k)$ as a function of k. Reprinted from Ref. [165].

- The mean-field interaction energy and the chemical potential increase.

- The phonon velocity and the superfluid critical velocity increase.

- The zero-point energy and the LHY correction increase.

- The healing length ξ becomes smaller.

- In the momentum space, the regime of collective excitation expands.

- The quantum depletion increases.

- The quasi-particle lifetime becomes shorter.

3.4 One-Dimensional Bosons

In this section, we will study interacting bosons in one dimension. In the literatures of ultracold atomic physics, "one dimension" sometimes means "real one dimension" and sometimes means "quasi one dimension." We clarify these two different cases in Box 3.1. Here we consider the situation of real one dimension. By the Bogoliubov theory, we can see from Eq. 3.53 that the occupation on single-particle mode with momentum $\mathbf{k}$ is always given by $n_{\mathbf{k}} = v_{\mathbf{k}}^2$, and in the small k regime, the momentum distribution behaves as

$$n_{\mathbf{k}} = v_{\mathbf{k}}^2 \to \frac{gn_0}{\hbar ck}. \tag{3.81}$$

This behavior holds for any dimension as long as that the Bogoliubov theory is valid. In three dimensions, the integration over the entire momentum converges, which gives a finite quantum depletion shown in Eq. 3.64. And as we have discussed in Section 3.3, this depletion is small compared with the total density, as long as the interaction is weak. However, in one dimension, it is easy to show that $\int_\Lambda v_k^2 dk$ diverges as $\log \Lambda$ when taking the infrared cutoff Λ to zero. As we discussed in Box 2.2, an infrared divergence means the low-energy and long-range physics is not treated correctly. In this case, it means that in one dimension, the low-energy quantum depletion diverges if one assumes a Bose condensate. Therefore, the Bogoliubov theory is no longer a self-consistent theory and fails in one dimension. This difference between one dimension and three dimensions is essentially due to their difference in the single-particle density-of-state. In three dimensions, the low-energy density-of-state vanishes as $\sqrt{\epsilon}$ when the energy ϵ approaches zero. In

Box 3.1	One Dimension and Quasi One Dimension

In the ultracold atom system, "one dimension" is achieved by applying a confinement potential in the transverse direction, and this confinement potential is usually a harmonic trap with large trapping frequency $\omega_\perp$. In the ultracold atom literature, the term "one dimension" sometimes means "real one dimension" and sometimes means "quasi one dimension." By "real one dimension," it means that the harmonic confinement energy $\hbar\omega_\perp$ is much larger than the chemical potential μ, i.e., $\hbar\omega_\perp \gg \mu$, such that the population in the single-particle excited state of the transverse direction is negligible. In this case, the wave function in the transverse direction is almost frozen to the ground state Gaussian wave function, as we discussed in the confinement-induced-resonance in Section 2.5. Such a strong confinement can usually be achieved either by using a strong two-dimensional optical lattice or by using atomic chip. As we discussed here, Bose condensation is absence in real one dimension. However, by "quasi one dimension," for bosons, it is still a condensate in three dimensions, but the Thomas–Fermi radius along the transverse direction is about the healing length ξ, i.e., $\sqrt{\mu/(m\omega_\perp^2)} \sim \xi$. Since $\hbar^2/(m\xi^2) \sim \mu$, this is equivalent to $\hbar\omega_\perp \sim \mu$. This situation can usually be satisfied by an anisotropic harmonic trap. In this case, the dynamics of the condensate is mostly along the longitudinal direction, and therefore the system is called quasi-one-dimensional. The soliton in a Bose condensate discussed in Section 4.1 belongs to this situation.

one dimension, the low-energy density-of-state diverges as $1/\sqrt{\epsilon}$ when ϵ approaches zero. Because of the large low-energy density-of-state in one dimension, Bose condensate is strongly depleted. Note that this argument is independent of the strength of interactions. That is to say, even for very weak interactions, the condensate will be destroyed inevitably. Hence, we need to introduce an alternative theory to capture the interacting effects in one dimension.

Bethe–Ansatz Solution. In Section 2.5 we have discussed that, unlike the three-dimensional case, the δ-function potential is well defined in one dimension. There, with a two-body problem, we have shown how to determine the interaction strength of the δ function potential from the three-dimensional scattering length. Here we will discuss that one-dimensional Bose gas interacting with the δ function potential is actually exactly solvable [111].

The reason that such one-dimensional models are exactly solvable is in fact due to a special feature of one dimension. Considering two atoms with position x_1 and x_2, suppose in the regime $x_1 < x_2$, their momenta are respectively k_1 and k_2, and the wave function is $e^{ik_1x_1+ik_2x_2}$. In the regime when $x_1 > x_2$, suppose their momenta are k_1' and k_2', the momentum conservation gives

$$k_1 + k_2 = k_1' + k_2'. \tag{3.82}$$

And outside the interaction range, the energy is purely kinetic energy and therefore, for the same eigenstate, it requires

$$k_1^2 + k_2^2 = k_1'^2 + k_2'^2. \tag{3.83}$$

In one dimension, it is easy to show that Eq. 3.82 and Eq. 3.83 together determine that either $(k_1, k_2) = (k_1', k_2')$ or $(k_1, k_2) = (k_2', k_1')$. That is to say, in the regime $x_2 > x_1$, the momenta of two atoms are either the same as, or a permutation of two momenta as in the regime $x_2 < x_1$. This property is unique for one dimension and does not hold in higher dimension.

The property can be straightforwardly generalized to a multiparticle situation. Considering N particles whose coordinates are labeled by $x_1, \ldots, x_N$, suppose that in the region $x_1 < x_2 < \cdots < x_N$, the momenta of each particle are $k_1, \ldots, k_N$ and the wave function is given by $e^{i\sum_i k_i x_i}$. In the same region and other spatial regions, if the many-body wave function also contains other components such as $e^{i\sum_i k_i' x_i}$, it can be shown that $\{k_1', \ldots, k_N'\}$ can only be a permutation of $\{k_1, \ldots, k_N\}$. Therefore, for a given set of $\{k_1, \ldots, k_N\}$, all the wave functions $e^{i\sum_i k_i' x_i}$ with $\{k_1', \ldots, k_N'\}$ being a permutation of $\{k_1, \ldots, k_N\}$ form a closed Hilbert space, whose dimensionality is always finite. In general, the many-body wave function is a superposition of all of them, and the coefficients in the superposition are determined by matching the boundary conditions when any two $x_i = x_j$. This will eventually reduce the Schrödinger equation to a set of algebraic equations. In this sense, this model is exactly solvable. The same method can be used to solve interacting spinful fermions in one dimension [184].

Note that in the weakly interacting limit, the interaction energy is estimated by $g_{1d}n_{1d}$, where g_{1d} is the one-dimensional interaction strength discussed in Section 2.5 and n_{1d} is the one-dimensional density. The typical kinetic energy is given by $\hbar^2 n_{1d}^2/(2m)$. Hence, the ratio of the interaction energy to the kinetic energy is given by

$$\gamma = \frac{g_{1d} n_{1d}}{\hbar^2 n_{1d}^2/(2m)} = \frac{2mg_{1d}}{\hbar^2 n_{1d}}. \tag{3.84}$$

It is interesting to note that the smaller the density, the larger the γ. The system is in a strongly interacting regime when $\gamma \gg 1$. This is in contrast to the three-dimensional case where the high density regime is strongly interacting. In practices, there are several different ways to achieve the $\gamma \gg 1$ regime. One can tune g_{1d} to be very large by utilizing the confinement-induced-resonance as discussed in Section 2.5. And one can also achieve the $\gamma \gg 1$ regime by reducing the density [90]. Moreover, an alternative tool is to increase the mass. The single-particle effective mass is defined through the single-particle dispersion, which can be tuned by applying an extra optical lattice along the longitudinal direction, as we will show in Section 7.1 [134].

In the limit $\gamma \gg 1$, the ground state should first try to minimize the interaction energy. To maximumly minimize the interaction energy of a δ-function repulsive potential, it requires that the wave function vanishes when the coordinates of any two particles coincide. If we write an antisymmetric wave function of two particles as

$$\Psi = e^{ik_1 x_1 + ik_2 x_2} - e^{ik_1 x_2 + ik_2 x_1}, \tag{3.85}$$

this wave function satisfies the requirement $\Psi = 0$ when $x_1 = x_2$. And it also requires $k_1 \neq k_2$ for otherwise the wave function vanishes everywhere. However, this wave function does not satisfy the Bose statistics. Hence, we should modify the wave function as

$$\Psi = |e^{ik_1 x_1 + ik_2 x_2} - e^{ik_1 x_2 + ik_2 x_1}| = \left| \sin\left((k_1 - k_2)\frac{x}{2}\right) \right|, \tag{3.86}$$

where $x = x_1 - x_2$. This consideration can be generalized to a many-body situation and the many-body wave function can be written as

$$\Psi = \text{Abs} \left[\text{Det} \begin{vmatrix} e^{ik_1 x_1} & e^{ik_1 x_2} & \dots & e^{ik_1 x_N} \\ e^{ik_2 x_1} & e^{ik_2 x_2} & \dots & e^{ik_2 x_N} \\ \dots & \dots & \dots & \dots \\ e^{ik_N x_1} & e^{ik_N x_2} & \dots & e^{ik_N x_N} \end{vmatrix} \right], \tag{3.87}$$

where "Det" denotes Slater Determinant and "Abs" denotes taking an absolute value. The Slater determinant is the many-body wave function of free fermions. However, because of taking the absolute value of the Slater determinant, the momentum distribution is different from that of free fermions and it does not display a sharp Fermi surface structure. Nevertheless, the momentum distribution is much more flat in the momentum space compared with that of the Bose condensation case. In fact, it can be shown that there is no macroscopic occupation in any modes, and the system is not a Bose condensate.

Since the interaction energy vanishes for this wave function, the total energy is given by the kinetic energy alone, and

$$\mathcal{E} = \sum_{i=1}^{N} \frac{\hbar^2 k_i^2}{2m}, \tag{3.88}$$

Box 3.2	Scale Invariant Quantum Gases

We consider the scaling transformation that the coordinates of all particles scale as $\mathbf{r}_i \rightarrow \lambda \mathbf{r}_i$. A quantum many-body system is called scale invariant if the total energy $\mathcal{E}$ scales as $\mathcal{E} \rightarrow \mathcal{E}/\lambda^2$ under the scaling transformation. Obviously noninteracting Bose and Fermi gases are scale invariant, since the Hamiltonian only contains the kinetic energy $\mathcal{T} = \sum_i \hbar^2 \nabla_i^2/(2m)$ that is scaled by $1/\lambda^2$ under the scaling transformation. For an interacting gas, if the total energy is proportional to the Fermi energy E_F only, such a system is also scale invariant because E_F is always scaled by $1/\lambda^2$ under the scaling transformation. Such examples include the one-dimensional Tonks–Girardeau gas discussed here and the three-dimensional unitary Fermi gas discussed in Chapter 6. Both are strongly interacting systems. There are also examples of weakly interacting systems that are scale invariant, for instance, a weakly interacting two-dimensional Bose gas. At the mean-field level, the interaction energy of weakly interacting bosons is always proportional to the two-dimensional density n_{2d}. In two dimensions, the density n_{2d} also scales as $1/\lambda^2$ under the scaling transformation, and therefore, both $\mathcal{T}$ and $\mathcal{V}$ scale as $1/\lambda^2$. Hence, the total energy also scales with $1/\lambda^2$. However, in this system, the scale invariance will be broken when quantum effects beyond mean field are included, which is known as an *anomaly* as in high-energy physics.

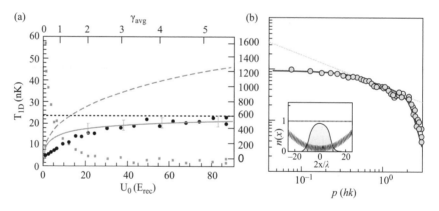

Figure 3.10   Experimental measurements of Tonks gases. (a) T_{1D} measures the total energy of the system and is plotted as a function of γ. The horizontal dashed line indicates the fermionized energy in the Tonks–Girardeau limit. The solid line is theory based on the Bethe–Ansatz solution. The deviation is attributed to the system not being in a pure one-dimensional regime, as indicated by the squares. (b) Momentum distribution with $\gamma \approx 200$. The solid line is theory based on the fermionized wave function in the Tonks–Girardeau limit. (a) is reprinted from Ref. [90], and (b) is reprinted from Ref. [134]. A color version of this figure can be found in the resources tab for this book at cambridge.org/zhai.

where $k_i = 2\pi n/L$ and L is the system size. Since all k_i should be different from each other, to minimize the total energy, k_i should be as small as possible. Thus, n should take all integers from $-N/2$ to $N/2$. This is reminiscent of filling the Fermi sea for a free Fermi gas. It also says that when $g_{1d} \rightarrow \infty$, the ground state energy of such a strongly interacting

Bose gas approaches the total kinetic energy of a free Fermi gas with same density. In this sense, we state that the system is fermionized, and such system is called the *Tonks–Girardeau gas* [174, 60]. This Tonks–Girardeau gas is one of the examples that are scale invariant. There are other systems in ultracold atomic gases that are scale invariant, and we summarize them in Box 3.2.

The first experimental evidences of the Tonks–Girardeau gas were reported in Ref. [90] and Ref. [134], where the $\gamma \gg 1$ regime is achieved either by reducing the density or by increasing the effective mass. The main experimental observations are presented in Figure 3.10. In Ref. [90], they observe that the energy of this one-dimensional system saturates to that of a free Fermi gas with the increasing of γ. In Ref. [134], they observe that the momentum distribution becomes more and more flat when γ increases. Both experimental measurements agree with the theoretical calculations using the Bethe–Ansatz method and agree with a fermionized wave function Eq. 3.87 in the large-γ limit.

3.5 Phase Coherence and Fragmentation

In previous sections, we mainly focus on the interaction effects on energy. In this section, we focus on the interaction effects on phase coherence. Here "phase" actually means the relative phase between different positions or different subsystems of the Bose condensate. To this end, we consider a simple "toy model" of interacting bosons with two spatial modes, and therefore, the "phase" simply means the relative phase between these two spatial modes.

Double-Well Model. Let us consider interacting bosons in a double-well potential, as shown in Figure 3.11. At each well, we only consider the lowest energy state, whose spatial wave functions are denoted by $\psi_1(\mathbf{r})$ and $\psi_2(\mathbf{r})$, respectively. When the barrier is sufficiently high and the tunneling is much weaker compared with the level separation in each

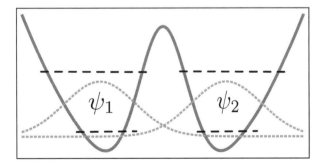

Figure 3.11 Illustration of the double-well model. The solid line is the double-well potential. Two dotted lines represent wave function $\psi_1(\mathbf{r})$ and $\psi_2(\mathbf{r})$ of the two lowest energy states in each well. The horizontal black dashed lines indicate the energy levels in absence of tunneling, which is much larger than the tunneling energy and the interaction energy considered here. A color version of this figure can be found in the resources tab for this book at cambridge.org/zhai.

well, tunneling can hardly couple these two low-lying states to other excited levels. Moreover, when we include interactions, we assume that the interaction energy is also much weaker compared with the level separations. Hence, we only need to consider the lowest energy mode in each well. Although in practical situations, these conditions are hard to be fully satisfied, this model can still serve as a sufficiently simple toy model to illustrate many key physical concepts.

Because of tunneling, the single-particle ground state $\psi_g(\mathbf{r})$ is given by

$$\psi_g(\mathbf{r}) = \frac{1}{\sqrt{2}} \left(\psi_1(\mathbf{r}) + \psi_2(\mathbf{r}) \right), \tag{3.89}$$

and the first excited state $\Psi_e(\mathbf{r})$ is given by

$$\psi_e(\mathbf{r}) = \frac{1}{\sqrt{2}} \left(\psi_1(\mathbf{r}) - \psi_2(\mathbf{r}) \right). \tag{3.90}$$

For a symmetric double-well, the Hamiltonian possesses the parity symmetry of $x \leftrightarrow -x$. Hence, both ψ_g and ψ_e respect this symmetry. And the symmetric superposition ψ_g is the ground state, following from the Feynman's argument that the ground state of a real Hamiltonian should be nodeless.

We can expand the field operator as

$$\hat{\Psi}(\mathbf{r}) = \psi_1(\mathbf{r})\hat{a}_1 + \psi_2(\mathbf{r})\hat{a}_2, \tag{3.91}$$

where $\hat{a}_1$ and $\hat{a}_2$ are boson annihilation operators in each well, and the second-quantized Hamiltonian for this system can be given by

$$\hat{H} = -J \left(\hat{a}_1^\dagger \hat{a}_2 + \hat{a}_2^\dagger \hat{a}_1 \right) + \frac{U}{2} \left(\hat{n}_1(\hat{n}_1 - 1) + \hat{n}_2(\hat{n}_2 - 1) \right), \tag{3.92}$$

where J denotes the single-particle tunneling rate between two wells, and U is the interaction strength between atoms in the same well.

There are two ways to solve this model. First, we can take the mean-field approximation as we discussed in Section 3.1. Assuming that all bosons are condensed in the same single-particle state, we can write down the mean-field wave function as

$$|\Psi_{\mathrm{MF}}\rangle = \frac{1}{\sqrt{N!}} \left(\cos\frac{\alpha}{2} e^{-i\theta/2} \hat{a}_1^\dagger + \sin\frac{\alpha}{2} e^{i\theta/2} \hat{a}_2^\dagger \right)^N |0\rangle, \tag{3.93}$$

where the state is taken as a general superposition of two modes. It is straightforward to show that for this wave function,

$$\langle \hat{a}_1^\dagger \hat{a}_1 \rangle = N \cos^2\frac{\alpha}{2}, \quad \langle \hat{a}_2^\dagger \hat{a}_2 \rangle = N \sin^2\frac{\alpha}{2}, \quad \langle \hat{a}_1^\dagger \hat{a}_2 \rangle = N \sin\frac{\alpha}{2} \cos\frac{\alpha}{2} e^{i\theta}, \tag{3.94}$$

and the energy of this mean-field state is given by

$$\mathcal{E} = \langle \Psi_{\mathrm{MF}} | \hat{H} | \Psi_{\mathrm{MF}} \rangle = -JN \sin\alpha \cos\theta + \frac{UN^2}{4} \cos^2\alpha. \tag{3.95}$$

Since $J > 0$, it is easy to see that the kinetic energy always favors $\theta = 0$. And for repulsive interaction $U > 0$, both kinetic and interaction term favor $\alpha = \pi/2$. Thus, as long as

$U > 0$, regardless of the value of U/J, the ground state determined by the mean-field ansatz is always

$$|\Psi_{\text{MF}}\rangle = \frac{1}{\sqrt{N!}} \left(\frac{\hat{a}_1^\dagger + \hat{a}_2^\dagger}{\sqrt{2}} \right)^N |0\rangle. \qquad (3.96)$$

This state means that all bosons occupy the same single-particle state, which is exactly the single-particle ground state Eq. 3.89.

Let us consider the situation that the interaction is attractive and $U < 0$.[5] When $2J > |U|N$, the minimum is still located at $\alpha = \pi/2$. When $2J < |U|N$, minimizing $\mathcal{E}$ with respect to α will yield two degenerate minima at $\alpha_0 = \arcsin(2J/(|U|N))$ and $\pi - \alpha_0$, respectively. In the limit $U \to -\infty$, $\alpha \to 0$, and these two degenerate solutions approach $\hat{a}_1^{\dagger N}|0\rangle$ and $\hat{a}_2^{\dagger N}|0\rangle$, respectively. It is easy to see that these two mean-field solutions break the parity symmetry between left and right wells.

Another way to solve this model is to write down the most general form of the wave function under the Fock bases as

$$|\Psi\rangle = \sum_l \Psi_l \left| \frac{N}{2} + l, \frac{N}{2} - l \right\rangle = \sum_l \Psi_l \frac{\hat{a}_1^{\dagger N_1} \hat{a}_2^{\dagger N_2}}{\sqrt{N_1! N_2!}} |0\rangle, \qquad (3.97)$$

where $N_1 = N/2 + l$ and $N_2 = N/2 - l$. Here since we take the two-mode approximation, the Hilbert space dimension increases linearly with the number of atoms, and we can easily solve this model up to a few thousands of atoms. Later we will compare the solution from these two approaches.

Josephson Effects. Let us first consider the $U > 0$ regime. According to the mean-field results, we have $\alpha = \pi/2$ and $\theta = 0$, thus $\langle \Delta \hat{N} \rangle = \langle \hat{N}_1 - \hat{N}_2 \rangle = 0$. That is to say, there are equal number of atoms residing in each well and the relative phase between them is also zero. Now we consider the situation when either α or $\langle \Delta \hat{N} \rangle$ deviates from this equilibrium situation and study their dynamics.

First, we have

$$\frac{d\langle \Delta \hat{N} \rangle}{dt} = -\frac{2J}{\hbar} \langle \hat{I} \rangle, \qquad (3.98)$$

and in this two-mode case, the current operator $\hat{I}$ is defined as $-i(\hat{a}_1^\dagger \hat{a}_2 - \hat{a}_2^\dagger \hat{a}_1)$. With Eq. 3.94, we have $-\langle \hat{I} \rangle = -N \sin\alpha \sin\theta$. We assume $2JN \sin\alpha$ is nearly a constant throughout the dynamics, denoted by E_J following the convention in literatures, and we will see that it is indeed the case in the two regimes that we will discuss below. Then, we have

$$\frac{d\langle \Delta \hat{N} \rangle}{dt} = -\frac{E_J}{\hbar} \sin\theta. \qquad (3.99)$$

In the weak tunneling regime, the phase dynamics of each mode is mainly governed by the local interaction energy. Therefore, on one site the phase evolves as $e^{-iUN_1(N_1-1)t/2}$ and on the other site the phase evolves as $e^{-iUN_2(N_2-1)t/2}$. Thus the time evolution of phase

---

[5] As a toy model with two discrete modes, let us disregard the issue that the condensate with attractive interactions is unstable because of the negative compressibility.

difference is proportional to ΔN. Also following the convention in literatures, we denote the proportional coefficient as E_c and therefore, we have

$$\frac{d\theta}{dt} = \frac{E_c}{\hbar}\langle \Delta \hat{N} \rangle. \tag{3.100}$$

From Eq. 3.99 and 3.100, one can already see that θ and ΔN are two quantities that form two coupled equations. Eq. 3.99 and 3.100 lead to

$$\frac{d^2\theta}{dt^2} = -\frac{E_c E_J}{\hbar^2}\sin\theta, \tag{3.101}$$

which coincides with the classical equation-of-motion of a simple pendulum. It is known that there are two solutions for a classical pendulum, depending on the initial velocity. Here these two solutions are respectively known as the *Josephson oscillation* and the *Self-trapping*. We shall remark that both two phenomena are consequences of the phase coherence and are absent in a normal gas.

- If the initial $\langle \Delta \hat{N} \rangle_0$ and θ are sufficiently small, it gives a simple harmonic oscillation with frequency

$$\omega_J = \frac{\sqrt{E_c E_J}}{\hbar}. \tag{3.102}$$

In this case, both the atom number imbalance, current and phase oscillate in time with this frequency. This is known as the Josephson oscillation. Here we should remark that, even initial $\langle \Delta \hat{N} \rangle_0 = 0$, as long as θ is nonzero, a finite current will be induced. This is counterintuitive and is strongly in contrast to a normal gas, where a macroscopic current is not possible if there is no bias voltage or particle imbalance applied between two sides.
- If initial $\langle \Delta \hat{N} \rangle_0$ is much larger than E_J/E_c, from Eq. 3.100, one finds that θ increases linearly in time as $tE_c\langle \Delta N \rangle_0/\hbar$. And if θ changes sufficiently fast, from Eq. 3.99, we can see that at the zeroth order the current averages out and $\langle \Delta \hat{N} \rangle$ remains as a constant $\langle \Delta \hat{N} \rangle_0$. This is also a counterintuitive result and also cannot happen in a normal gas. In a normal gas, when a strong density imbalance is applied between two sides, the system should gradually relax to the balanced situation. In contrast, here when the initial state strongly deviates from the ground state with balanced population, the system retains this far-from-equilibrium situation instead of relaxing back to the ground state. This regime is known as the self-trapping.

Therefore, as initial $\langle \Delta \hat{N} \rangle_0$ increases, the dynamics changes from the Josephson oscillation regime to the self-trapping regime. This has been demonstrated experimentally with a BEC in double-well potential [4]. Figure 3.12(a) shows the case with small atom number imbalance, where both the relative atom number and the relative phase oscillate around zero. Figure 3.12(b) shows the case with large atom number imbalance, where one can see that θ increases monotonically in time and the relative atom number is nearly a constant around its initial value.

Fragmentation and Phase Fluctuation. Now we return to another method to solve the Hamiltonian with the general wave function Eq. 3.97. First of all, in the noninteracting limit $U = 0$, it is easy to see that the ground state is given by

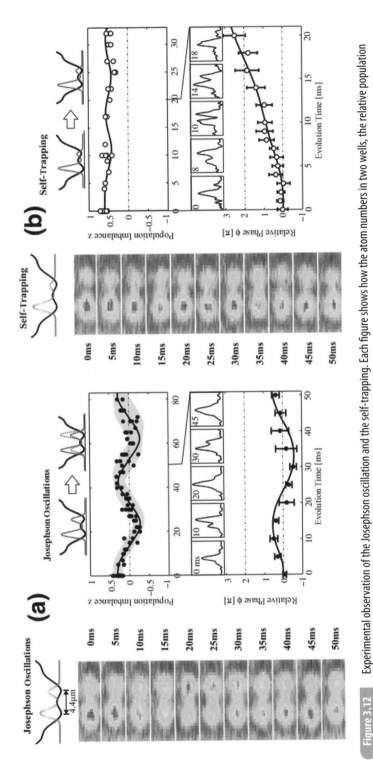

Figure 3.12 Experimental observation of the Josephson oscillation and the self-trapping. Each figure shows how the atom numbers in two wells, the relative population $\langle \Delta \hat{N} \rangle / N$, and the relative phase θ change as a function of time t. (a) Josephson oscillation. (b) Self-trapping. Reprinted from Ref. [4]. A color version of this figure can be found in the resources tab for this book at cambridge.org/zhai.

$$|\Psi\rangle = \frac{1}{\sqrt{N!}} \left(\frac{\hat{a}_1^\dagger + \hat{a}_2^\dagger}{\sqrt{2}} \right)^N |0\rangle, \tag{3.103}$$

and this wave function is also the mean-field state for all positive U/J. We expand the wave function Eq. 3.103 in terms of Eq. 3.97, and with the help of the Stirling's approximation $\log N! = N \log N - N$, we can obtain

$$\Psi_l = \sqrt{\frac{N!}{2^N \left(\frac{N}{2} + l \right)! \left(\frac{N}{2} - l \right)!}} \simeq \frac{e^{-2l^2/N}}{(\pi N/4)^{1/4}}. \tag{3.104}$$

That is to say, Ψ_l obeys a Gaussian distribution with the width $\sigma_{\Delta N}^2$ being proportional to N, and it is straightforward to show that $\langle (\Delta \hat{N})^2 \rangle \sim N$. This is actually a natural result of the central limit theorem.

Now when U is finite, we consider the Schrödinger equation [71]

$$E\Psi_l = -J_{l+1}\Psi_{l+1} - J_l\Psi_{l-1} + Ul^2\Psi_l, \tag{3.105}$$

where $J_l = J\sqrt{(N/2 + l)(N/2 - l + 1)}$. For sufficiently large and positive U, the wave function is localized around $l = 0$. For $l \ll N$, approximately we have $J_l \approx JN/2$ as a constant, and Eq. 3.105 is reduced to a discrete version of the single-particle Schrödinger equation in a harmonic trap. The ground state solutions gives

$$\Psi_l = \frac{1}{(\pi \sigma_{\Delta N}^2/2)^{1/4}} e^{-l^2/\sigma_{\Delta N}^2}, \quad \sigma_{\Delta N}^2 \propto \sqrt{\frac{JN}{U}}. \tag{3.106}$$

Thus, as U increases, the width of the Gaussian distribution $\sigma_{\Delta N}^2$ varies from being proportional to N to being proportional to $\sqrt{JN/U}$. The particle number fluctuation $\langle (\Delta \hat{N})^2 \rangle \sim \sigma_{\Delta N}^2$ is strongly suppressed as U increases. This is illustrated in Figure 3.13. In the limit $JN/U \ll 1$, this state approaches a Fock state

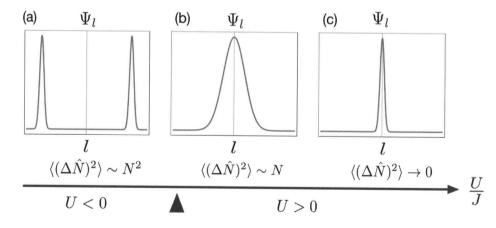

Figure 3.13 Schematic of atom number fluctuations. Here we plot Ψ_l as a function of $l = (N_1 - N_2)/2$ for different regimes of U/J. Here $U/J < 0$ and $|U| \gg J$ for (a), $U > 0$ but $U \ll J$ for (b), and $U > 0$ and $U \gg J$ for (c). A color version of this figure can be found in the resources tab for this book at cambridge.org/zhai.

$$|\Psi\rangle = \frac{\hat{a}_1^{\dagger N/2}\hat{a}_2^{\dagger N/2}}{(N/2)!}|0\rangle. \tag{3.107}$$

This state becomes an eigenstate of $\Delta\hat{N}$.

For noninteracting quantum state Eq. 3.103, the single-particle density matrix can be easily computed as

$$\rho = \begin{pmatrix} \langle\hat{a}_1^{\dagger}\hat{a}_1\rangle & \langle\hat{a}_1^{\dagger}\hat{a}_2\rangle \\ \langle\hat{a}_2^{\dagger}\hat{a}_1\rangle & \langle\hat{a}_2^{\dagger}\hat{a}_2\rangle \end{pmatrix} = \frac{N}{2}\begin{pmatrix} 1 & 1 \\ 1 & 1 \end{pmatrix}. \tag{3.108}$$

This density matrix can be diagonalized as

$$\rho = N\begin{pmatrix} 1 & 0 \\ 0 & 0 \end{pmatrix}. \tag{3.109}$$

That is to say, only the population on one of the modes is of the order N and the population on the other mode is zero. According to the definition given in Section 3.1, the system is a simple BEC. For strong repulsive interaction discussed above, because

$$\langle\hat{a}_1^{\dagger}\hat{a}_2\rangle = \sum_l \sqrt{(N/2-l)(N/2+l+1)}\Psi_{l+1}\Psi_l \approx (N/2)e^{-1/(2\sigma_{\Delta N}^2)}, \tag{3.110}$$

the single-particle density matrix is given by

$$\rho = \frac{N}{2}\begin{pmatrix} 1 & e^{-1/(2\sigma_{\Delta N}^2)} \\ e^{-1/(2\sigma_{\Delta N}^2)} & 1 \end{pmatrix}. \tag{3.111}$$

Two eigenvalues are $\lambda = (N/2)(1 \pm e^{-1/(2\sigma_{\Delta N}^2)})$, and both eigenvalues are of the order of N. In the limit $U \to +\infty$ and $\sigma_{\Delta N} \to 0$, the quantum state becomes the Fock state Eq. 3.113, and it is easy to see that the density matrix becomes

$$\rho = \frac{N}{2}\begin{pmatrix} 1 & 0 \\ 0 & 1 \end{pmatrix}. \tag{3.112}$$

According to the definition given in Section 3.1, the system is a fragmented condensate. Note that the Fock state can also be viewed as a single condensate in terms of two-particle density matrix, because this state can be written as

$$|\Psi\rangle = \frac{1}{(N/2)!}\left(\hat{a}_1^{\dagger}\hat{a}_2^{\dagger}\right)^{N/2}|0\rangle, \tag{3.113}$$

and $\hat{a}_1^{\dagger}\hat{a}_2^{\dagger}$ can be viewed as a boson pair operator.

To better understand what happens from the small U regime to the large U regime, we revisit the mean-field description. We can expand the mean-field energy Eq. 3.95 around its minimum $\alpha = \pi/2$ and $\theta = 0$, and it yields

$$\mathcal{E} = \frac{JN}{2}\theta^2 + \frac{UN^2}{4}(\Delta\alpha)^2, \tag{3.114}$$

where $\Delta\alpha = \alpha - \pi/2$. If we consider $\Delta\alpha$ and θ as a pair of conjugate variables like $\hat{x}$ and $\hat{p}$, Eq. 3.114 can be considered as a harmonic oscillator, and the width of the phase fluctuation is given by $\sigma_\theta^2 \propto \sqrt{UN/J}$. When $\sqrt{UN/J} \gg 2\pi$, we can consider that the phase can freely fluctuate between zero and 2π. Using the identity that

$$\int_{-\pi}^{\pi} d\theta \frac{1}{\sqrt{2^N N!}} (\hat{a}_1^\dagger e^{i\theta/2} + \hat{a}_2^\dagger e^{-i\theta/2})^N |0\rangle \propto \frac{\hat{a}_1^{\dagger N/2} \hat{a}_2^{\dagger N/2}}{(N/2)!} |0\rangle, \qquad (3.115)$$

we can state that a strong enough phase fluctuation renders the system from a single condensate to a fragmented condensate with fixed relative particle numbers. We shall revisit similar physics in Section 4.3 when we discuss spinor condensate.

It is also interesting to note that $\sigma_{\Delta N}^2 \sigma_\theta^2 \sim N$. It is actually better to state that the relative atom number difference $\langle \Delta \hat{N} \rangle$ and the relative phase θ can be viewed as a pair of conjugate variables. In the weakly interacting regime, there is a well-defined relative phase, meaning that the phase fluctuation is strongly suppressed, therefore, the relative atom number fluctuation is as large as $\langle (\Delta \hat{N})^2 \rangle \sim N$. As U increases, the relative atom number fluctuation is gradually suppressed and the phase fluctuation increases. Eventually, in the strongly repulsive interaction limit, the relative atom number fluctuation is strongly suppressed such that the system becomes a Fock state, but the phase fluctuation becomes very large such that the phase coherence between two sides is completely lost.

Quantum Measurement. When talking about interference in a quantum system, the double-slit experiment always the first one that occurs to one's mind. In order to see an interference pattern in the usual double-slit experiment, there are essentially two conditions. First, the waves passing through two slits have to have well-defined relative phase. Second, the particles have to be identical particles when they are detected. In other words, the detectors do not know which slit the particle comes from. If the particle are labeled by the path, the interference will disappear.

Decades ago, Anderson has asked the question that, if two superfluids have never met before, whether they can interfere when they meet. This question can be studied in ultracold atom experiments. The answer to this question has to involve the measurement process. Let us prepare a Bose condensate in a double-well potential. If the barrier is shallow, the system initially is a simple condensate with well-defined relative phase between two sides. Then, we turn off the potential and let the system expand. When atoms from two sides met and overlap, the density distribution will show interference pattern, similar as in the usual double-slit interference experiment. However, if we start with sufficiently high barrier, the system initially is a Fock state like Eq. 3.113. And as shown by Eq. 3.115, there is no well-defined relative phase between two sides. The question is whether we can still observe the interference pattern when the double-well potential is turned off and atoms overlap.

First of all, we consider an ensemble measurement, that is to say, we repeat the same measurements under identical conditions and take an average over them. In this case, the density is given by

$$n(\mathbf{r}) = \langle \hat{\Psi}^\dagger(\mathbf{r}) \hat{\Psi}(\mathbf{r}) \rangle = \langle (\Psi_1^*(\mathbf{r}) \hat{a}_1^\dagger + \Psi_2^*(\mathbf{r}) \hat{a}_2^\dagger)(\Psi_1(\mathbf{r}) \hat{a}_1 + \Psi_2(\mathbf{r}) \hat{a}_2) \rangle, \qquad (3.116)$$

where $\Psi_1(\mathbf{r})$ and $\Psi_2(\mathbf{r})$ label the wave function evolving from the left and the right sides, respectively. Note that $\langle \hat{a}_1^\dagger \hat{a}_2 \rangle = \langle \hat{a}_2^\dagger \hat{a}_1 \rangle = 0$ for the Fock state Eq. 3.113, Eq. 3.116 becomes

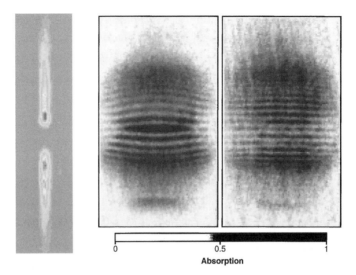

Figure 3.14 Experimental observation of interference pattern. A BEC is separated into two by a high barrier created by laser. When these two initially separated BECs are released from the trap, and when they overlap after expansion, an interference pattern can be observed in each run of measurements. Reprinted from Ref. [7]. A color version of this figure can be found in the resources tab for this book at cambridge.org/zhai.

$$n(\mathbf{r}) = \langle \hat{a}_1^\dagger \hat{a}_1 \rangle |\Psi_1(\mathbf{r})|^2 + \langle \hat{a}_2^\dagger \hat{a}_2 \rangle |\Psi_2(\mathbf{r})|^2. \tag{3.117}$$

It is obvious that no interference exists after the ensemble average.

However, if one actually performs such an experiment, one can observe the interference pattern in each run of measurement, as one can see in Figure 3.14. As we will explain below, this is a quantum measurement effect. Quantum measurement can be viewed as a projection. As we have shown in Eq. 3.115, a Fock state can be viewed as a superposition of single condensate with all possible relative phases. Thus, each measurement projects this state into one of the states in the superposition with fixed relative phase. To illustrate how this quantum measurement projects the Fock state into a state with fixed relative phase, we present following two-step arguments [28].

- First, it is important to emphasize again that the atoms are identical particles, and when one atom is detected, one cannot distinguish which side the atom comes from. When the first atom is detected, without loss of generality, we can assume that this atom has equal probability of coming from the left side or coming from the right side, with a relative phase ϕ. The many-body wave function after the first atom being detected $|\Phi_1\rangle$ is given by

$$|\Phi_1\rangle = \frac{\hat{a}_1 e^{-i\phi/2} + \hat{a}_2 e^{i\phi/2}}{\sqrt{2}} \left| \frac{N}{2}, \frac{N}{2} \right\rangle$$

$$= \frac{\sqrt{N}}{2} \left(e^{-i\phi/2} \left| \frac{N}{2} - 1, \frac{N}{2} \right\rangle + e^{i\phi/2} \left| \frac{N}{2}, \frac{N}{2} - 1 \right\rangle \right). \tag{3.118}$$

Now considering that the second atom is detected, this atom is also assumed to be in a superposition state with a relative phase ζ, then, the many-body wave function $|\Phi_2\rangle$ after the second atom being detected is given by

$$|\Phi_2\rangle = \frac{\hat{a}_1 e^{-i\zeta/2} + \hat{a}_2 e^{i\zeta/2}}{\sqrt{2}} |\Phi_1\rangle \tag{3.119}$$

$$\propto \sqrt{\frac{N}{2} - 1} \left(e^{-i(\phi+\zeta)} \left| \frac{N}{2} - 2, \frac{N}{2} \right\rangle + e^{i(\phi+\zeta)} \left| \frac{N}{2}, \frac{N}{2} - 2 \right\rangle \right)$$

$$+ 2\cos(\phi - \zeta)\sqrt{\frac{N}{2}} \left| \frac{N}{2} - 1, \frac{N}{2} - 1 \right\rangle, \tag{3.120}$$

and

$$\langle \Phi_2 | \Phi_2 \rangle = (N - 2) + 2N\cos^2(\phi - \zeta). \tag{3.121}$$

This shows that $\langle \Phi_2 | \Phi_2 \rangle$ has the largest amplitude when $\phi = \zeta$. That is to say, the possibility is the largest when successively two atoms with the same relative phase are detected.

- Second, if successively k-atoms are detected with same relative phase, we have

$$\left(\frac{\hat{a}_1 e^{-i\phi/2} + \hat{a}_2 e^{i\phi/2}}{\sqrt{2}} \right)^k \left| \frac{N}{2}, \frac{N}{2} \right\rangle$$

$$\propto \left(\frac{\hat{a}_1 e^{-i\phi/2} + \hat{a}_2 e^{i\phi/2}}{\sqrt{2}} \right)^k \int_{-\pi}^{\pi} d\theta \left(\frac{\hat{a}_1^\dagger e^{i\theta/2} + \hat{a}_2^\dagger e^{-i\theta/2}}{\sqrt{2}} \right)^N |0\rangle \tag{3.122}$$

$$\propto \int_{-\pi}^{\pi} d\theta (\cos(\theta/2 - \phi/2))^k \left(\frac{\hat{a}_1^\dagger e^{i\theta/2} + \hat{a}_2^\dagger e^{-i\theta/2}}{\sqrt{2}} \right)^{N-k} |0\rangle. \tag{3.123}$$

It is easy to see that $(\cos(\theta/2 - \phi/2))^k$ quickly approaches the delta function $\delta(\theta - \phi)$ as k increases, because for any number $x < 1$, x^k quickly approaches zero as k increases. Therefore, such a measurement picks up a coherent state with fixed relative phase, that is,

$$\left(\frac{\hat{a}_1 e^{-i\phi/2} + \hat{a}_2 e^{i\phi/2}}{\sqrt{2}} \right)^k \left| \frac{N}{2}, \frac{N}{2} \right\rangle \rightarrow \left(\frac{\hat{a}_1^\dagger e^{i\phi/2} + \hat{a}_2^\dagger e^{-i\phi/2}}{\sqrt{2}} \right)^{N-k} |0\rangle. \tag{3.124}$$

From above discussion, we see that quantum measurement can project a Fock state without phase coherence to a simple condense state with a well-defined relative phase. However, this relative phase is randomly picked up by the measurement process, and therefore, for different runs of experiment, the relative phases are different and one can see different interference pattern. Hence, after averaging over many measurements, the interference pattern is smeared out, which recovers the result of ensemble measurement discussed above.

The Hanbury–Brown–Twiss Effect. For a Fock state, above discussions have shown that although the ensemble measurement of density does not show interference pattern, the identical boson nature of atoms enables that each individual measurement does show interference. Here we can further show that, also due to the nature of identical bosons, the density-density correlation can also show an interference pattern even after the ensemble average. This is known as the *Hanbury–Brown–Twiss effect*. We summarize whether interference pattern can appear or not under different situations in the Table 3.2.

Following the same spirit of Eq. 3.116 and considering the density-density correlation function, we have

$$\langle \hat{\Psi}^\dagger(\mathbf{r})\hat{\Psi}(\mathbf{r})\hat{\Psi}^\dagger(\mathbf{r}')\hat{\Psi}(\mathbf{r}')\rangle$$
$$= n(\mathbf{r})n(\mathbf{r}') + \langle \hat{a}_1^\dagger \hat{a}_1\rangle\langle \hat{a}_2^\dagger \hat{a}_2\rangle(\Psi_1^*(\mathbf{r})\Psi_2^*(\mathbf{r}')\Psi_2(\mathbf{r})\Psi_1(\mathbf{r}') + \text{h.c.}), \qquad (3.125)$$

where $n(\mathbf{r})$ and $n(\mathbf{r}')$ are given by Eq. 3.117. Here we also take $\langle \hat{a}_1^\dagger \hat{a}_2\rangle = \langle \hat{a}_2^\dagger \hat{a}_1\rangle = 0$ because of the absence of phase coherence in the initial state. The interference appears in the second term of Eq. 3.125.

This effect has been first observed in ultracold atomic system by using the Mott insulator phase of bosons in optical lattices [56]. We will discuss the Mott insulator in Section 8.1. The basic idea is to use a deep optical lattice potential to localize atoms inside each well, such that there is no phase coherence between wells. Then, after turning off the lattice, the wave functions of atoms expand from different sites, and then they will overlap with each other, as shown in Figure 3.15(a). As shown in Figures 3.15(b) and (d), when measuring the density after the time-of-flight and averaging over a number of measurements, one cannot see any signal of interference pattern in the density profile. However, when one analyzes

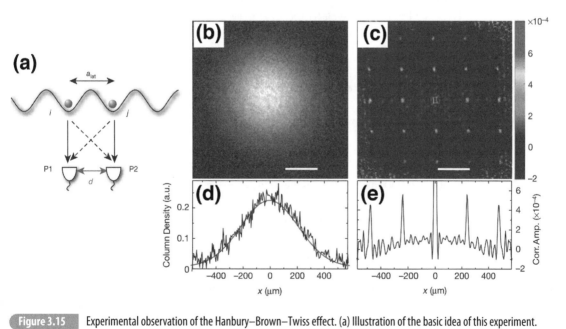

Figure 3.15 Experimental observation of the Hanbury–Brown–Twiss effect. (a) Illustration of the basic idea of this experiment. Atoms are initially prepared in a deep optical lattice, and density distributions are measured after the time of flight. (b) and (d) show the density after ensemble average, and (c) and (e) show the density-density correlation after the ensemble average. (d) and (e) are integrated results of (b) and (c), respectively. Reprinted from Ref. [56]. A color version of this figure can be found in the resources tab for this book at cambridge.org/zhai.

Table 3.2	On interference pattern of density and density-density correlation measurements	
	Single measurement	Ensemble measurement
$\langle n(\mathbf{r}) \rangle$	Yes	No
$\langle n(\mathbf{r})n(\mathbf{r}') \rangle$	Yes	Yes

Note: "Yes" and "No" denote the presence and the absence of interference pattern, respectively.

Box 3.3 | **Symmetry Breaking**

Let us first introduce two concepts of the symmetry group of the wave function and the symmetry group of the Hamiltonian. The symmetry group of the Hamiltonian is the group of operations that keep the Hamiltonian invariant, and the symmetry group of the wave function is the group of operations that keep the wave function invariant. When we say that a state breaks symmetry, it means that the symmetry group of the wave function is a subgroup of the symmetry group of the Hamiltonian. A direct consequence of symmetry breaking is degeneracy. For the single-particle case, we have emphasized that the ground state has no degeneracy, and therefore, it cannot break symmetry. However, symmetry breaking is a common phenomenon in the ground state of a many-body system, the reason for which will be explained in this section. In the Landau theory of phase transition, different phases are characterized by different symmetry groups of the wave function. In Section 4.5, we will discuss an example of a Bose system with various symmetry-breaking phases, and we shall discuss how symmetry breaking helps us determine the properties of phase transitions. In Section 8.2, we will discuss an example of a Fermi system with various symmetry-breaking phases.

---

the density-density correlation from these images, the interference peaks are clearly visible, as shown in Figures 3.15(c) and (e).

The Schrödinger Cat and Spontaneous Symmetry Breaking. Now we turn into the discussion of attractive interaction with $U < 0$. As mentioned above, when $|U|N > 2J$, the mean-field energy has two degenerate minima respectively located at $\alpha_0 = \arcsin(2J/(|U|N))$ and $\pi - \alpha_0$. In the limit $U \to -\infty$, these two solutions approach $\hat{a}_1^{\dagger N}|0\rangle$ and $\hat{a}_2^{\dagger N}|0\rangle$.

In Box 3.3, we have introduced the concept of the *symmetry group of the wave function* and the *symmetry group of the Hamiltonian*, and the important concept of *symmetry breaking*. Here the Hamiltonian has a Z_2 symmetry of exchanging $\hat{a}_1$ and $\hat{a}_2$. The mean-field states for very negative U do not respect this Z_2 symmetry, and consequently, the mean-field ground states are two-fold degenerate. It is noted in Box 3.3 that symmetry breaking is a common phenomenon in nature. However, we also note that, for a real Hamiltonian such as this case, the ground state should be unique and cannot has degeneracy. How should we resolve this paradox ?

Indeed, if we solve the Hamiltonian Eq. 3.105 with the general wave function Eq. 3.97, the system should have a unique ground state, and the corresponding Ψ_l respects the Z_2 symmetry and exhibits a double peak distribution

$$\Psi_l \propto e^{-(l-l_0/2)^2/(2\sigma^2)} + e^{-(l+l_0/2)^2/(2\sigma^2)}, \tag{3.126}$$

as shown in Figure 3.13. In the limit $U \to -\infty$, the ground state wave function, denoted by $|\Psi_+\rangle$, is a superposition as

$$\Psi_+ = \frac{1}{\sqrt{2}}\left(|N,0\rangle + |0,N\rangle\right), \tag{3.127}$$

which is also known as the *NOON state*. These two states $|N,0\rangle$ and $|0,N\rangle$ in the superposition are macroscopically distinct, because one corresponds to all atoms in the left well and the other corresponds to all atoms in the right well. Such a superposition of macroscopically distinct states is well known as the *Schrödinger Cat state*.

The Schrödinger Cat state has the same one-body density matrix as the Fock state discussed above, that is,

$$\rho = \frac{N}{2}\begin{pmatrix} 1 & 0 \\ 0 & 1 \end{pmatrix}, \tag{3.128}$$

therefore it is also a fragmented state. However, in contrast to the Fock state, its particle number fluctuation is $\langle(\Delta\hat{N})^2\rangle \propto N^2$. It also differs from the Fock state in terms of two-body density matrix. The Fock state can be viewed as a pair condensate in terms of two-body density matrix, but the Schrödinger Cat state cannot.

However, why is such a Schrödinger Cat state very rare in nature ? In fact, in this model, there exists the first excited state $|\Psi_-\rangle$ whose energy is very close to the ground state, and the wave function is given by

$$|\Psi_-\rangle = \frac{1}{\sqrt{2}}\left(|N,0\rangle - |0,N\rangle\right). \tag{3.129}$$

One can show that the energy difference between $|\Psi_+\rangle$ and $|\Psi_-\rangle$ exponentially decreases as the atom number N increases. Thus, even for N being of the order of ten, this energy separation is already extremely small. This is because, in order to couple $|N,0\rangle$ state to $|0,N\rangle$ state, there have to go through N-steps of single-particle tunneling as

$$|N,0\rangle \to |N-1,1\rangle \to |N-2,2\rangle \to \cdots \to |1,N-1\rangle \to |0,N\rangle, \tag{3.130}$$

and during this process, the interaction energy keeps increasing. Such a process is called *macroscopic quantum tunneling*. Therefore, the tunneling rate is strongly suppressed as N increases, and consequently, the energy splitting between $|\Psi_+\rangle$ and $|\Psi_-\rangle$ is strongly suppressed. In fact, this is not a specific feature of this particular model. It applies universally to all Schrödinger Cat states. We recall that the Schrödinger Cat state refers to superposition of macroscopically distinct states, and if two states are macroscopically distinct, it means that there are macroscopic number of degrees of freedom that are different between these two states. In nature, nearly all physical systems possess *locality*, which means each term can only change few number of degrees of freedom. Hence, in order to couple two macroscopically distinct states, one needs macroscopic number steps. Therefore, the coupling coefficient becomes extremely small. For instance, let us consider two low-lying ferromagnetic states whose spins point to different directions. In order to couple them, one needs to flip each spin from one direction to another direction, and the intermediate states are not ferromagnetic, which have higher energies. Hence, such a coupling will be strongly suppressed as the number of spins increases.

Because of this extremely weak coupling between macroscopically distinct states due to the macroscopic quantum tunneling, a Schrödinger Cat state is very unstable in nature. For instance, in this case, if we add infinitesimal small energy difference between the left and the right sides, that is, $\epsilon(\hat{a}_1^\dagger \hat{a}_1 - \hat{a}_2^\dagger \hat{a}_2)$, which is inevitable in reality, the energy difference between $|N,0\rangle$ and $|0,N\rangle$ is ϵN. This energy difference can easily overwhelm the exponential small coupling between $|N,0\rangle$ and $|0,N\rangle$. Hence, the actual ground state is either $|N,0\rangle$ or $|0,N\rangle$. Another way to view this problem is to consider a dynamical process. For instance, suppose that initially all atoms are condensed in one of the well because of this ϵ energy difference, when we turn off the energy difference such that the Hamiltonian restores the Z_2 symmetry, the system should oscillate between two wells to restore the symmetry. However, because of the extremely small tunneling rate, the oscillation period is extremely long, which can easily exceed the lifetime of our universe for system with large number of particles. In case of ferromagnetism, the presence of an infinitesimal magnetic field can pin the spins to certain direction and breaks the $SU(2)$ spin rotational symmetry. When the infinitesimal small magnetic field is turned off, it also takes extremely long time to rotate all spins from one to another direction and to restore the $SU(2)$ spin rotational symmetry.

In summary, we have shown that the instability of the Schrödinger Cat state and the stability of the symmetry-breaking phenomenon in nature are two sides of the same coin. From the discussion above, it is important to realize that symmetry breaking is a unique phenomenon in many-body system with large number of degrees of freedom.

Exercises

3.1   Discuss the Bose–Einstein condensation temperature for free bosons at one and two dimensions.

3.2   Calculate the two-body density matrix $\rho(\mathbf{r}, \mathbf{r}')$ for noninteracting bosons above and below the Bose–Einstein condensation temperature.

3.3   Show that if we define $\hat{\alpha}_\mathbf{k}$ and $\hat{\alpha}_{-\mathbf{k}}^\dagger$ as a unitary transformation of $\hat{a}_\mathbf{k}$ and $\hat{a}_{-\mathbf{k}}^\dagger$, $\hat{\alpha}_\mathbf{k}$ and $\hat{\alpha}_{-\mathbf{k}}^\dagger$ do not obey the boson commutative relation.

3.4   Show that the wave function Eq. 3.57 is the ground state wave function of the Bogoliubov Hamiltonian that satisfies $\langle G|\hat{a}_{\mathbf{k}=0}|G\rangle = \sqrt{N_0}$ and $\hat{\alpha}_\mathbf{k}|G\rangle = 0$.

3.5   Show that the detailed balance condition

$$(1+f_\mathbf{k})(1+f_\mathbf{q})f_{\mathbf{k}+\mathbf{q}} = f_\mathbf{k}f_\mathbf{q}(1+f_{\mathbf{k}+\mathbf{q}}) \tag{3.131}$$

can be satisfied if $f_\mathbf{k}$ satisfies the Bose distribution

$$f_\mathbf{k} = \frac{1}{e^{\beta \mathcal{E}_\mathbf{k}} - 1} \tag{3.132}$$

and $\mathcal{E}_\mathbf{k} + \mathcal{E}_\mathbf{q} = \mathcal{E}_{\mathbf{k}+\mathbf{q}}$. Also show that the detailed balance condition

$$(1-f_\mathbf{k})(1-f_\mathbf{q})f_{\mathbf{k}+\mathbf{q}} = f_\mathbf{k}f_\mathbf{q}(1-f_{\mathbf{k}+\mathbf{q}}) \tag{3.133}$$

can be satisfied if $f_{\mathbf{k}}$ satisfies the Fermi distribution

$$f_{\mathbf{k}} = \frac{1}{e^{\beta \mathcal{E}_{\mathbf{k}}} + 1} \tag{3.134}$$

and $\mathcal{E}_{\mathbf{k}} + \mathcal{E}_{\mathbf{q}} = \mathcal{E}_{\mathbf{k+q}}$.

3.6   Calculate $S_{\rho_{\mathbf{k}}}(\omega)$ with the Bogoliubov Hamiltonian and show the Feynman relation

$$S_{\rho_{\mathbf{k}}}(\omega) = \frac{\hbar^2 k^2/(2m)}{\mathcal{E}_{\mathbf{k}}}. \tag{3.135}$$

3.7   Considering two atoms in a one-dimensional ring with size L, compute the ground state energy as a function of g_{1d}.

3.8   Verify Eq. 3.94 for the mean-field state Eq. 3.96.

3.9   Numerically solve the Hamiltonian Eq. 3.92 for a finite number of particles, and discuss the wave function, density matrix, and relative particle fluctuations in three regimes (i) $U > 0$ and $U \lesssim J$; (ii) $U > 0$ and $U \gg J$; and (iii) $U < 0$ and $|U| \gg J$. In regime (iii), also discuss the excitation gap between the first excited state and the ground state as a function of total particle number N.

3.10   Discuss the Hanbury–Brown–Twiss effect for noninteracting fermions and compare its difference with noninteracting bosons.

4

Topology and Symmetry

Learning Objectives

- Introduce soliton in quasi-one-dimensional time-dependent Gross–Pitaevskii equation.
- Introduce the basic idea of topology and the homotopy groups.
- Introduce vortex in spinless condensate as a typical example of topological defect.
- Introduce the Berezinskii–Kosterlitz–Thouless transition as a topological defect-driven phase transition, and emphasize the topological and energy requirements for such a transition.
- Discuss the geometric configuration that minimizes the energy of a vortex lattice.
- Introduce the Majorana stellar representation as a useful tool to visualize the symmetry of a high-spin wave function.
- Introduce two different phases of spin-1 condensate.
- Introduce the relation between a mean-field state and the singlet pair condensate.
- Introduce the spin vortex and half vortex in a spinor condensate, and discuss various possibilities of a spin vortex core.
- Introduce two categories of topological excitations in a Bose condensate, and introduce monopole and skyrmion as the typical example of each category.
- Introduce the simulating Dirac monopole in a synthetic magnetic field.
- Discuss the symmetry of the Hamiltonian and the symmetry of various phases in a spin-orbit coupled BEC.
- Discuss the relation between symmetry and phase transitions, using spin-orbit coupled condensate as an example.
- Discuss the Galilean invariance and the superfluid critical velocity.

4.1 Soliton

Soliton is an excitation widely existed in many nonlinear systems in different branches of physics. In this section we will first discuss *soliton* as an exact solution of a type of excitations in one-dimensional GP equation, which describes a quasi-one-dimensional condensate. The difference between quasi one dimension and real one dimension has been clarified in Box 3.1. For quasi one dimension, we consider a Bose condensate in a very elongated trap whose transverse confinement is much stronger than the longitudinal one, and the size of condensate in the transverse direction is comparable with the healing length.

Thus, we can first consider that the motion in the transverse direction is frozen, and we shall also first ignore the harmonic confinement along the longitudinal direction. Therefore we start with a one-dimensional GP equation as

$$i\hbar \frac{\partial \psi}{\partial t} = -\frac{\hbar^2 \partial^2}{2m \partial^2 x}\psi + U|\psi|^2\psi. \tag{4.1}$$

Now let us consider a special wave function ansatz as

$$\psi(x) = \sqrt{n_0}\tanh\left(\frac{x}{\sqrt{2}\xi}\right), \tag{4.2}$$

where n_0 is the ground state density, and ξ is the healing length. It is easy to show that

$$-\frac{\hbar^2 \partial^2}{2m \partial^2 x}\psi + U|\psi|^2\psi$$
$$= \frac{\hbar^2}{2m\xi^2}\left(\mathrm{sech}^2\left(\frac{x}{\sqrt{2}\xi}\right) + \tanh^2\left(\frac{x}{\sqrt{2}\xi}\right)\right)\sqrt{n_0}\tanh\left(\frac{x}{\sqrt{2}\xi}\right)$$
$$= Un_0\psi, \tag{4.3}$$

where we have used the relations $d^2\tanh(y)/dy^2 = -2\mathrm{sech}^2(y)\tanh(y)$, $\mathrm{sech}^2(y) + \tanh^2(y) = 1$ and $\hbar^2/(2m\xi^2) = Un_0 = \mu$. This proves that the soliton wave function $e^{-i\mu t}\psi$ satisfies the time-dependent GP equation Eq. 4.1.

The soliton wave function has following properties that $\lim_{x\to\infty}\psi(x) = \sqrt{n_0}$ and $\lim_{x\to-\infty}\psi(x) = -\sqrt{n_0}$, that is to say, the wave function has a π phase different between two ends, as shown in Figure 4.1(b). This means that the global phase structure has been changed once such a soliton is presented. In other words, if one measures the phase of the condensate even far away from the center of the soliton, one can still notice the presence of the soliton. The phase jumps occurs around $x = 0$ where the density depletes to zero. And around the phase jump, there exists a regime with size ξ where the condensate density is strongly depleted, as shown in Figure 4.1(a). The rapid phase change creates a large velocity that locally destroys condensate. That is also the reason why ξ is called the healing length. The word "healing length" means the distance it takes for the condensate density

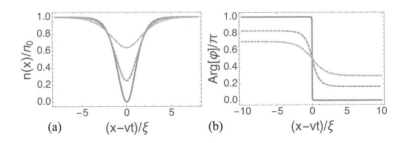

Figure 4.1   Soliton wave function. (a) The density $n(x) = |\psi(x)|^2$ (a) and (b) the phase arg $\psi(x)$ as a function of the coordinate in the comoving frame $(x - vt)/\xi$. Three different velocities $v/c = 0$ (solid line), $v/c = 0.5$ (dashed line), and $v/c = 0.8$ (dash-dotted line) are plotted. A color version of this figure can be found in the resources tab for this book at cambridge.org/zhai.

to recover its bulk value when the condensate density is locally destroyed. We will revisit this concept again when we discuss vortex core in Section 4.2. Here this soliton is called a *dark soliton*. In superconductors, such an excitation is also called the *phase slip*.

The soliton wave function Eq. 4.2 can be generalized to a moving soliton with velocity v, whose wave function is given by [93]

$$\psi(x,t) = \sqrt{n_0} \left[i\frac{v}{c} + \sqrt{1 - \frac{v^2}{c^2}} \tanh\left(\sqrt{1 - \frac{v^2}{c^2}} \frac{x - vt}{\sqrt{2}\xi}\right) \right], \tag{4.4}$$

where c is the sound velocity. With the help of the Galilean transformation, it is straightforward to show that in a comoving frame the GP equation is written as

$$-\frac{\hbar^2 \partial^2}{2m\partial^2 x'}\psi + i\hbar v\frac{\partial\psi}{\partial x'} + U|\psi|^2\psi = i\hbar\frac{\partial\psi}{\partial t}, \tag{4.5}$$

where $x' = x - vt$. It is straightforward to calculate that

$$-\frac{\hbar^2 \partial^2}{2m\partial^2 x'}\psi = \frac{\hbar^2\sqrt{n_0}}{2m\xi^2}s^3 \, \text{sech}^2\,(s\tilde{x})\tanh\,(s\tilde{x}), \tag{4.6}$$

$$U|\psi|^2\psi = \frac{\hbar^2\sqrt{n_0}}{2m\xi^2}\left[\frac{v^2}{c^2} + s^2\tanh^2\,(s\tilde{x})\right]\left[i\frac{v}{c} + s\tanh\,(s\tilde{x})\right], \tag{4.7}$$

$$i\hbar v\frac{\partial\psi}{\partial x'} = \frac{\hbar^2\sqrt{n_0}}{2m\xi^2}i\frac{v}{c}s^2\,\text{sech}^2\,(s\tilde{x}), \tag{4.8}$$

where $s = \sqrt{1 - v^2/c^2}$ and $\tilde{x} = (x - vt)/(\sqrt{2}\xi)$, and therefore

$$-\frac{\hbar^2 \partial^2}{2m\partial^2 x'}\psi + i\hbar v\frac{\partial\psi}{\partial x'} + U|\psi|^2\psi = Un_0\psi. \tag{4.9}$$

Here we have used the relation $\hbar c/(\sqrt{2}\xi) = Un_0$. Therefore $e^{-i\mu t}\psi(x,t)$ with $\psi(x,t)$ given by Eq. 4.4 is also a solution of time-dependent GP equation Eq. 4.1. In Figure 4.1 we also plot the density and the phase for solitons with finite velocity. The larger v/c, the smaller the density dip at the center of soliton, and the smaller the phase jump. We also notice that in the limit $v \to 0$, Eq. 4.4 recovers Eq. 4.2, and in the limit $v \to c$, Eq. 4.4 recovers a ground state wave function.[1] That v cannot be larger than the sound velocity c is also consistent with the fact that a soliton cannot move faster than the sound velocity.

Now regarding the excitation energy of these solutions, we consider

$$\mathcal{E} = \int dx \left(\frac{\hbar^2}{2m}\left|\frac{\partial\psi}{\partial x}\right|^2 + \frac{U}{2}|\psi|^4 - \mu|\psi|^2\right), \tag{4.10}$$

---

[1] In this regard, soliton is not a strictly defined topological excitation since it can be deformed back to the ground state continuously.

where $\mu = Un_0$. For a uniform ground state solution, $|\psi|^2 = n_0$ and the ground state energy $\mathcal{E}_g = -Un_0^2/2$. For the soliton solution, $\mathcal{E}_s$ can also be computed exactly with Eq. 4.4 and Eq. 4.10, and we obtain [93]

$$\mathcal{E}_s - \mathcal{E}_g = \frac{8}{3\sqrt{2}} \frac{\hbar^2 n_0}{2m\xi} \left(1 - \frac{v^2}{c^2}\right)^{3/2} = \frac{4\hbar m}{3U}(c^2 - v^2)^{3/2}. \tag{4.11}$$

Interestingly, the excitation energy of a soliton decreases with the increasing of the velocity v, and the excitation energy vanishes when v approaches c.

Now we include the effect of a smooth confinement potential $V(x)$ along the longitudinal direction. Since the contribution to the energy cost of a soliton mostly comes from the center regime of a soliton, with the help of the local density approximation, the energy for a soliton can be obtained by replacing c in Eq. 4.11 with local sound velocity $c(X)$ at X, where X is the center of soliton. Considering that the soliton energy is conserved during its motion, one obtains that [93]

$$\frac{4\hbar m}{3U} \left(c(X)^2 - v^2\right)^{3/2} = \text{constant}, \tag{4.12}$$

where $c(X)^2 = Un(X)/m = c_0^2 - V(X)/m$. With the soliton velocity v defined as dX/dt, we obtain

$$m\left(\frac{dX}{dt}\right)^2 + V(X) = \text{constant}. \tag{4.13}$$

This energy conservation implies that the oscillation of a soliton in a harmonic trap can be equivalent to an oscillation of a single particle with mass $2m$, which yields that the oscillation frequency of a soliton is $1/\sqrt{2}$ of the trap frequency [93].

Experimentally, soliton can be created by the so-called phase imprinting method. As shown in Figure 4.2(a), a laser beam is applied to the left-half of a BEC, which generates a constant potential difference between the left and the right regions, say, denoted by U_0. The laser is applied for a duration T_0 such that $U_0 T_0 = \pi/\hbar$ and then a π-phase difference is created between two ends. This will generate a dark soliton. If this dark soliton is not generated right at the center of the trap, it will start to oscillate. The oscillation of a dark soliton can be recorded by imaging the density of a BEC, as shown in Figures 4.2(b) and (c). In Figure 4.2(d), we show the center position of the dark soliton as a function of time, from which one can extract the oscillation frequency. With the consideration of unharmonicity and other experimental imperfections, the experimental result is consistent with a soliton mass equaling $2m$.

Before concluding this section, we shall also remark that soliton is a stable solution only in the quasi-one-dimensional regime when the healing length ξ is comparable to the transverse confinement. If the transverse size is much larger than ξ, the motion in the transverse direction becomes important. In that case, the dark soliton can also couple to other modes in the transverse direction. Actually, a pair of vortices in the transverse plane can also create a π phase difference between two different sides of the transverse plane. In fact, it has been observed experimental that a dark soliton can indeed decay into a vortex ring or a pair of vortices [50, 98]. We will discuss vortex in the next section.

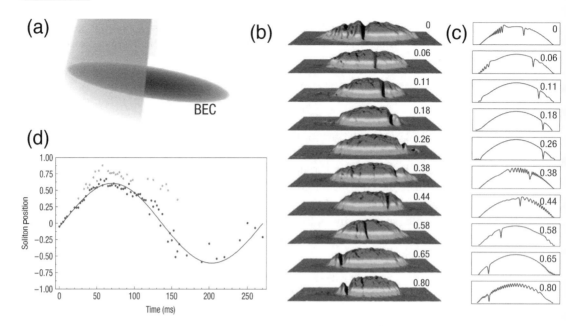

Figure 4.2 Experiment on soliton oscillation inside a harmonic trap. (a) Phase imprinting method for generating soliton. (b–c) Imaging for the motion of a soliton at different time slots after creation. (c) is the integrated density of (b). (d) Center position of a soliton as a function of time. Reprinted from Ref. [15]. A color version of this figure can be found in the resources tab for this book at cambridge.org/zhai.

4.2 Vortex

The terminology *topology* refers to global properties that are invariant under continuous deformation. Here, *global property* and *invariant under continuous deformation* are two key words. For example, considering a closed sphere with arbitrary shape, the famous Gauss-Bonnet theorem in topology says that integration the local Gauss curvature over the entire sphere always gives 4π. Here integration over the entire sphere means that this is a global property of the sphere. If one distorts the shape of the sphere, this value of 4π cannot be changed as long as one does not break the sphere. This means that this property is invariant under continuous deformation.

Homotopy Group. One of the useful tools to characterize topology is called the *homotopy group*. Here let us briefly discuss a few simple examples of the homotopy groups. One can find references such as Ref. [118] for more details. We consider a mapping from one-dimensional circle, denoted by S^1, to a target space $\mathcal{M}$, which can either be space of wave functions or be the space of order parameters. If one of such mapping, say f_0, can be continuously deformed to another mapping, say f_0', we say that these two mappings belong to one equivalent class, and we use $[f_0]$ to denote this equivalent class. All the equivalent classes form a group, which is called the first homotopy group, denoted by $\Pi_1(\mathcal{M})$. If a mapping maps all points in the preimage space to the same point in the image space, this is called a trivial map. And if a map can be continuously deformed into this trivial map, it belongs to the same class of trivial map. A homotopy group is called a trivial one if it only contains the trivial map class as its only element. In other words, all mappings can be continuously deformed into a trivial map. A homotopy group is called a nontrivial one if it contains equivalent classes other than the trivial map class. In other words, there exists some mappings that cannot be continuously deformed into the trivial map. Moreover, we can also consider a mapping from two-dimensional sphere S^2 to a target space $\mathcal{M}$. All the equivalent classes of such mappings form the second homotopy group, denoted by $\Pi_2(\mathcal{M})$. In general, we can consider the mapping from S^n to a target space $\mathcal{M}$, and the equivalent classes form the nth homotopy group, denoted by $\Pi_n(\mathcal{M})$. Below we discuss a few examples of homotopy groups that are used most often in physics problems.

- Let us consider that $\mathcal{M}$ is also S^1, this homotopy group is denoted by $\Pi_1(S^1)$. Now let us consider the following mapping, that is, when the preimage walks around the entire S^1 space clock-wisely, the image also winds around the entire S^1 space clock-wisely. One can see this mapping cannot be continuously deformed into the trivial mapping, and therefore, it belongs to another class. Moreover, when the preimage walks around the entire S^1 space clock-wisely once, the image can wind around the target S^1 space integer times, either clockwise or counterclockwise. This integer is call the *winding number*. Here clockwise winding and counterclockwise winding of image are characterized by positive and negative winding numbers, respectively. And notice that a clockwise winding can be untwisted by a counterclockwise winding such that the net winding number is zero. Hence, the net winding number is the integer that characterizes each equivalent class. Therefore, we reach the conclusion that $\Pi_1(S^1) = \mathcal{Z}$.
- Let us consider that $\mathcal{M}$ is S^2, this homotopy group is denoted by $\Pi_1(S^2)$. Since the image of an S^1 circle is always a circle in the S^2 space, it is easy to see that all the circles in the sphere can always be continuously shrink to one single point, no matter how many times it winds. That is to say, all mappings from S^1 to S^2 are smoothly connected to the trivial map. Therefore, $\Pi_1(S^2)$ is a trivial one.
- For the similar reason as $\Pi_1(S^1) = \mathcal{Z}$, it can be shown that $\Pi_2(S^2)$ is also the integer group $\mathcal{Z}$. That is to say, each equivalent class is also characterized by an integer, which is called the *Chern number*.[2] In this case, the Chern number describes that when the preimage covers the entire S^2 space, how many times the image covers the target S^2 space.
- For the similar reason that $\Pi_1(S^2)$ is trivial, $\Pi_2(S^3)$ is also a trivial one.
- Both $\Pi_3(S^3)$ and $\Pi_3(SO(3))$ are $\mathcal{Z}$.
- The mapping from S^3 to S^2 is called the *Hopf map*. $\Pi_3(S^2)$ also equals to $\mathcal{Z}$, and the topological number is called the *Hopf number*.

In this section and Section 4.4, we will discuss various topological objects in a BEC characterized by these homotopy groups.

Vortex. In Section 3.2 and Section 3.3, we have discussed the phonon excitation which only changes the density and phase of a Bose condensate locally. One cannot observe

---

[2] There is a more general definition of Chern number using the integration of local curvatures.

the physical effect of a phonon excitation at a distance far away from it. Here we will discuss vortex in a quasi-two-dimensional condensate. Strongly in contrast to the phonon excitation, it changes the global property of a condensate, namely, the presence of a vortex can be detected by walking around a loop enclosed the vortex, even when the loop is far away from the vortex center.

For simplicity, let us consider an isotropic vortex wave function written as

$$\psi(\mathbf{r}) = \sqrt{n_0} e^{i\kappa\varphi} f\left(\frac{r}{\xi}\right), \tag{4.14}$$

where we have taken the cylindrical coordinate. Here f is taken a real function, and ξ is the healing length. r measures the in-plane distance to the center of vortex and φ is the azimuthal angle in the two-dimensional plane. Here the phase θ of the condensate wave function is taken to be proportional to the azimuthal angle φ with the coefficient κ, which is required to be an integer to ensure the single-value condition of the wave function.

Using the relation that the velocity is defined as $\mathbf{v} = \hbar\nabla\theta/m$, this wave function gives a circulating velocity field

$$\mathbf{v}(\mathbf{r}) = \frac{\hbar\kappa}{mr} \hat{z} \times \hat{r}. \tag{4.15}$$

It is very important to note that the amplitude of the velocity field has a $1/r$ dependence on the radial coordinate. This has two important consequences:

- At small distance when $r \to 0$, there will be a singularity of the velocity field at the vortex center

$$\nabla \times \mathbf{v} = \frac{2\pi\hbar\kappa}{m} \delta^2(\mathbf{r})\hat{z}. \tag{4.16}$$

Generally speaking, the velocity field of any excitation can be decomposed into two components. One component has vanishing divergence everywhere, which corresponds to phonon excitation. The other component displays singularities in its curl, which corresponds to vortices. As we will show later, this singularity is the vortex center where density vanishes, and this feature is directly related to the fact that vortex is a topological excitation. We will explain this in detail below.

- At large distance, the $1/r$ dependence of the velocity field will lead to the logarithmical dependence of the vortex energy on the system size. As we will also explain below, this system size dependence has exact the same form as how entropy depends on system size, and this directly leads to the Kosterlitz–Thouless topological phase transition.

Below we will elaborate these two aspects in detail.

Vortex Core and Topological Defects. First of all, the energy of vortex wave function Eq. 4.14 is given by

$$\mathcal{E} = \int d^2\mathbf{r} \left\{ \frac{\hbar^2 n_0}{2m} \left[-\frac{f}{r} \frac{\partial}{\partial r} \left(r\frac{\partial f}{\partial r} \right) + \frac{\kappa^2 f^2}{r^2} \right] + \frac{U}{2} n_0^2 f^4 - \mu n_0 f^2 \right\}. \tag{4.17}$$

We now first consider the density distribution in the presence of a vortex. It is convenient to normalize r by the healing length ξ, and by using the relation $\hbar^2/(2m\xi^2) = Un_0 = \mu$, the energy minimization can yield an equation for f as a function of dimensionless quantity $\tilde{r} = r/\xi$ as

$$-\frac{1}{2\tilde{r}}\frac{\partial}{\partial\tilde{r}}\left(\tilde{r}\frac{\partial f}{\partial\tilde{r}}\right) + \frac{\kappa^2 f}{\tilde{r}^2} + f^3 = f, \tag{4.18}$$

which gives $f \to 1$ as $r/\xi \gg 1$, and $f \to 0$ as $\tilde{r} \to 0$. f deviates from unity and decreases toward zero when $\tilde{r}$ is of the order of unity. That is equivalent to saying, the density starts to deviate from n_0 and decreases toward zero when r is smaller than ξ, which determines that the size of the vortex core is of the order of ξ. This is similar to the size of soliton core, which is also of the order of ξ, as discussed in Section 4.1. Physically, this is because the amplitude of the local velocity becomes $\sim \hbar/(m\xi)$ at $r \sim \xi$, and it is easy to see that $\hbar/(m\xi) \sim c$. Hence, when $r < \xi$, the amplitude of the local velocity field becomes larger than the sound velocity. The condensate is depleted when the amplitude of the local velocity is greater than the critical velocity.

Now if we consider any physical region and we draw a boundary of this region. In this case, the physical region is a two-dimensional one and therefore the boundary is a one-dimensional loop. When the density of the wave function at the boundary is nonzero everywhere, the phase of the wave function is well defined, and it can take any value between zero and 2π. Because zero and 2π correspond to the same wave function, they are equivalent, and therefore the manifold of the phase of a wave function is also a one-dimensional loop. Thus, this defines a mapping from S_1 to S_1. As we have discussed, this mapping is characterized by the first order homotopy group $\Pi_1(S_1)$, and because $\pi_1(S_1) = \mathcal{Z}$, there are integer number of equivalent classes of mappings.

For instance, as shown in Figure 4.3, the four different cases correspond to four different mappings. In Figure 4.3(a) there is no vortex and the phase around the loop is a constant, that is to say, all points in the spatial loop S_1 (preimage) are mapped to one point in the manifold S_1 of the wave function phase (image). This is of course a topological

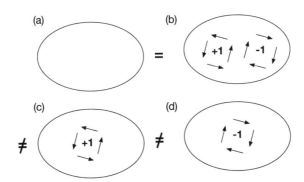

Figure 4.3 Four different vortex configurations. (a) No vortex. (b) Two vortices with winding number $+1$ and -1, respectively. (c) One vortex with winding number $+1$. (d) One vortex with winding number -1. (a) and (b) are topological equivalent, and they are not topologically equivalent to (c) and (d). (c) and (d) are also not topologically equivalent to each other.

trivial mapping. In Figure 4.3(b), there exists two vortices with opposite winding numbers. In this case, if one goes around the spatial loop, there is certainly some variation of the wave function phase. However, this mapping belongs to the same class as the trivial mapping shown in Figure 4.3(a), because one can continuously bring together these two vortices until they meet and annihilate, after which this mapping becomes the same as Figure 4.3(a). Figures 4.3(c) and (d) contain a single vortex with winding number +1 or −1, respectively. They are topologically distinct from Figures 4.3(a) and (b) and also they are not topologically equivalent to each other. In this case, the net winding number inside the loop is the topological invariant.

Since the change of the topological invariant must break the continuity in deforming the mapping, the topological nature of vortices directly connects to the fact that the vortex density vanishes at the vortex center. For instance, one can continuously move a vortex from outside to inside the loop, then the mapping of Figure 4.3(a) becomes that of Figure 4.3(c). Or vice versa, one can move the vortex from inside to outside of the loop, then the mapping of Figure 4.3(c) becomes that of Figure 4.3(a). Similarly, one can also move the antivortex outside the loop, then the mapping of Figure 4.3(b) becomes that of Figure 4.3(c). However, in all these cases, either a vortex or an antivortex has to cross the loop. Precisely because the density vanishes at the vortex center, when the vortex or the antivortex crosses the loop, the density in the loop vanishes at the crossing point, and consequently, the phase of the wave function is not well defined there. This breaks the continuity in deforming the mapping. Because the singularity plays an important role in defining the topology for vortices, this kind of topological object is also called *topological defects*.

Berezinskii–Kosterlitz–Thouless Transition. Second, we consider the excitation energy in the presence of a single vortex with $\kappa = 1$. Comparing the energy given in Eq. 4.17 to the ground state energy, only the contribution from $\kappa^2 f^2/r^2$ logarithmically depends on the system size, and the rest terms all give finite contributions. To estimate the logarithmic contribution, we approximate f as a step function with $f = 1$ for $r > \xi$ and $f = 0$ for $r < \xi$. This results in a simple expression of the energy of a single vortex in a disk with radius R as

$$E = \frac{\pi \hbar^2 n_0}{m} \ln\left(\frac{R}{\xi}\right), \tag{4.19}$$

where n_0 should be taken as the superfluid density. The error made by this step function approximation of f is also finite, and all these finite contributions can be ignored compared with the logarithmical energy dependence on system size for a large system.

Now let us turn to consider a finite temperature situation. The entropy of having a single free vortex in the disk is associated with the possible choices of the locations of the vortex, and all possible choices can be simply estimated as R^2/ξ^2. Thus, the entropy is given by

$$S = k_B \ln \frac{R^2}{\xi^2}. \tag{4.20}$$

It is an important observation that both energy and entropy scale with the system size R in the same way as $\log(R/\xi)$. With this observation, Berezinskii, and independently, Kosterlitz and Thouless proposed a transition at finite temperature due to the proliferation of vortices, which is now known as the *Berezinskii–Kosterlitz–Thouless (BKT) transition.*

The free-energy of a single free vortex at finite temperature is given by

$$\mathcal{F} = k_B T \left(\frac{\pi \hbar^2 n_0}{m k_B T} - 2 \right) \ln \frac{R}{\xi} = \frac{k_B T}{2} \left(n_0 \lambda^2 - 4 \right) \ln \frac{R}{\xi}, \tag{4.21}$$

where $\lambda = \sqrt{2\pi \hbar^2/(m k_B T)}$ is the thermal de Broglie wavelength. Thus, the BKT theory predicts that at low-temperature when $n_0 \lambda^2 > 4$, the presence of vortex causes positive free energy that logarithmically depends on the system size. Therefore, the thermal fluctuation of a single free vortex will be strongly suppressed. The low-energy thermal fluctuation is dominated by strongly confined vortex-anti-vortex pairs. This phase is also known as vortex confined phase. In this phase, a quasi-long-range order of phase coherence can be established and the system will be in a superfluid phase. As temperature increases, λ decreases, and when $n_0 \lambda < 4$, the entropy term dominates and the free energy $\mathcal{F}$ for a single vortex becomes negative. In this case, the thermal fluctuation of a single free vortex will be strongly enhanced. Therefore, the system contains lots of thermally excited free vortices or antivortices, which is called the proliferation of vortices. Once vortices are proliferated, it destroys the long-range phase coherence and the correlation function is exponentially decayed spatially. There will be no superfluidity at all, and the superfluid density n_0 vanishes. Thus, the BKT transition predicts a universal jump of superfluid density from $n_0 \lambda^2 = 4$ to zero at the transition point. This transition can also be viewed as the confinement-deconfinement transition of vortices, and is known as a *topological phase transition.* We discuss different meaning of the terminology of "topological phase transition" in Box 4.1 in different context.

The first experiment on the BKT physics in the ultracold atomic system was reported in Ref. [66]. Experimentally, they prepare two layers of two-dimensional (*xy* plane) ultracold Bose gases, as shown in Figure 4.4(a). Then, the confinement along the $\hat{z}$ direction is released. As discussed in Section 3.2, the gas expands mostly along the confinement

Box 4.1   **Topological Phase Transition**

Conventionally, different phases are distinguished by their different symmetry properties or different correlation properties. There are also cases that different phases are distinguished by different topological properties, such as by topological number of the band structure for band insulators, as we will discuss in Section 7.3, or by topological degeneracy, such as in the case of the fractional quantum Hall effect. In literature, the terminology "topological phase transition" means different contexts in different systems. In some cases, "topological phase transition" actually means topological defect-driven phase transition, such as the BKT transition discussed here. However, phases at different sides of the BKT transition are distinguished by their correlation properties, and in terms of topology, both are topologically trivial phases. In some other cases, "topological phase transition" can also mean transition between two different phases defined by distinct topological properties. Nevertheless, these transitions are usually not driven by topological defects.

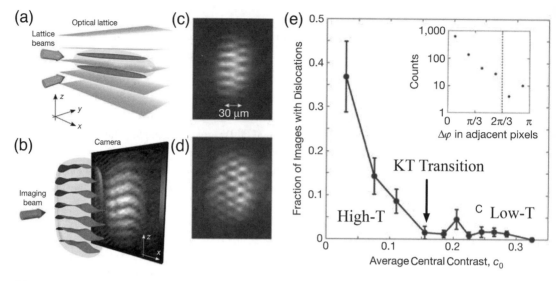

Figure 4.4 Experiment on the Kosterlitz–Thouless phase transition. (a-b) Schematic of experimental setup and measurement. (c–d) Two typical interference patterns with dislocations. (e) The number of free vortices changes as temperature varies. Reprinted from Ref. [66]. A color version of this figure can be found in the resources tab for this book at cambridge.org/zhai.

direction. Therefore, after expansion, two systems will overlap along the z direction and form an inference pattern, which can be measured by imagining beam perpendicular to the $\hat{z}$ direction, as shown in Figure 4.4(b). Then, if a free vortex is presented in one layer but not in another layer, it manifests itself as a dislocation in the interference pattern. Thus, the number of dislocations in the interference pattern reveals the number of free vortices. Figures 4.4(c) and (d) show examples where one (c) or several (d) free vortices are found. With this method, as shown in Figure 4.4(e), experimentally one finds that the vortex number suddenly increases as temperature increases beyond a critical temperature. Later, as shown in Figure 4.5, the critical velocity has also measured by stirring the Bose gas with a focused beam, and a jump of critical velocity is also found across a critical temperature [49], consistent with the universal jump of superfluid density predicted by the BKT theory.

From the discussion made above, it is clear that the existence of the BKT transition does not depend on the microscopic details of the system, but requires following two conditions:

- The first condition is the topological condition, which requires the existence of stable topological defects. This stable topological defect is ensured by a nontrivial $\Pi_{d-1}(\mathcal{M})$, where d is the spatial dimension. For instance, we can also consider the BKT transition in two-dimensional spin models with short-range interactions. In some models such as the XY model, the spins are restricted to the equator of the Bloch sphere, and therefore, the manifold of the spin is also $U(1)$. There are also vortices as topological defects and the same kind of BKT transition can occur. However, if the spin is free to choose any direction in the two-dimensional Bloch sphere, such as in the isotropic Heisenberg

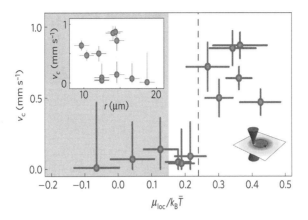

Figure 4.5 Experiment on the jump of superfluid density. The critical velocity as a function of $\mu/(k_B T)$. Reprinted from Ref. [49]. A color version of this figure can be found in the resources tab for this book at cambridge.org/zhai.

model, the manifold of the spin is S^2. Because $\Pi_1(S^2)$ is trivial, there is no topological defect in this model and therefore the BKT type transition does not exist in this model.

- The second condition is the energy condition, which requires the energy of the topological defect scales with system size in the same way as the entropy. Again, considering the isotropic spin models, as we will discuss in Section 4.4, there exists topological defects in three-dimensional system because of $\Pi_2(S^2) = \mathcal{Z}$, which is known as *monopole*. However, as we will discuss in Section 4.4, the energy of monopole does not scale the same way as entropy, and therefore, the BKT transition also cannot occur there.

Fast Rotation and the Vortex Lattices. Finally, let us turn into the case that the system is rotated along $\hat{z}$ with a fixed rotational frequency Ω. In the moving frame there exists an additional term $-\Omega \cdot L_z$ which is originated from the Coriolis force. This term gives an energy $\sim -\hbar\Omega\pi\kappa R^2 n_0$, which favors increasing the number of net vorticity. Thus, the ground state changes from a vortex free state to a state with vortices above a critical velocity, and this critical velocity is estimated by

$$\frac{\pi\hbar^2 n_0}{m} \ln\left(\frac{R}{\xi}\right) - \hbar\Omega_c \pi R^2 n_0 = 0, \tag{4.22}$$

which gives

$$\Omega_c \sim \frac{\hbar}{mR^2} \ln\left(\frac{R}{\xi}\right). \tag{4.23}$$

For typical parameters of ultracold atomic systems, Ω_c is smaller than the trapping frequency ω. In Figure 4.6 we show an experimental measurement of vortex number which changes from zero to nonzero above a critical rotational frequency, and increases with the increasing of the rotational frequency.

As the rotational frequency further increases, more and more vortices enter the condensate and the question is how these vortices arrange themselves into an energetically stable configuration. In fact, a natural expectation is that these vortices form a lattice. Then the next question is what is the lattice structure is the most stable one that minimizes the total

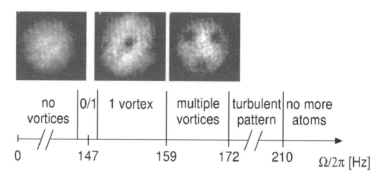

Figure 4.6   Experiment on vortex generation. Vortex number increases when the rotational frequency increases. Reprinted from
Ref. [116].

energy. This question is first answered by Abrikosov in 1957 in the context of type-II superconductor. He showed that the most stable lattice structure is a triangular lattice, using a
variational wave function approach. Therefore, this vortex lattice is also called *abrikosov
vortex lattice*. The similar variational wave function can also be applied to a fast rotating
Bose–Einstein condensate and one can reach the same conclusion that the triangular lattice is the most stable one [122]. Below we will briefly discuss the key idea behind this
variational wave function approach [122].

To this end, we first consider the single-particle Hamiltonian

$$\hat{H}_0 = -\frac{\hbar^2 \nabla^2}{2m} + \frac{1}{2}m\omega^2 r^2 - \Omega \hat{L}_z. \tag{4.24}$$

One can find the eigenspectrum is given by $E_{n_r,m} = n_r \hbar \omega - m\hbar\Omega$, where n_r and m are both
integers. In the fast rotating limit when $\Omega \to \omega$, all states with the same $n_r - m$ become
nearly degenerate, and the states with $n_r = m$ form the lowest Landau level. The reason
we call them the Landau levels is because Eq. 4.24 can be rewritten as

$$\hat{H}_0 = \frac{1}{2m}\left[(\hat{p}_x + m\Omega y)^2 + (\hat{p}_y - m\Omega x)^2\right] + \frac{1}{2}m(\omega^2 - \Omega^2)r^2. \tag{4.25}$$

Eq. 4.25 can be viewed as a particle in a magnetic field, with $A_x = -m\Omega y$ and $A_y = m\Omega x$.
Hence effectively, the magnetic field is given by $\mathbf{B}_{\text{eff}} = \nabla \times \mathbf{A} = 2m\Omega \hat{z}$. Therefore, the
vortex density is directly related to the flux density of the effective magnetic field. When
$\Omega \to \omega$, the residual trapping potential vanishes, and the Hamiltonian is purely a particle
in a magnetic field, whose eigenstates are the Landau levels. The wave function of states
in the Landau level are given by

$$\psi_m \propto z^m e^{-|z|^2/(2a^2)}, \tag{4.26}$$

where $a = \sqrt{\hbar/(m\Omega)}$ is the harmonic length. These wave functions are analytical function
of $z = x + iy$, aside from an exponential factor $e^{-|z|^2/(2a^2)}$. Therefore, any superposition of
the lowest Landau level wave functions can be expressed as

$$\psi = f(z)e^{-|z|^2/(2a^2)}, \tag{4.27}$$

and the zeros of $f(z)$ determine the location of vortices. Thus, if vortices form a lattice,
without loss of generality, we can choose the two lattice vectors as $\mathbf{B}_1 = b_1\hat{x}$ and $\mathbf{B}_2 =$

$b_1(u\hat{x} + v\hat{y})$. It means if z_0 is a zero of $f(z)$, $z_0 + b_1$ and $z_0 + b_2$ are also zeros of $f(z)$, where $b_2 = b_1(u + iv)$.

Therefore, what we should look for is an analytical function whose zeros form a regular lattice. In fact, the Jacobi Theta function is such an analytical function and therefore, we can use the Jacobi Theta function as a variational wave function to determine the lattice structure. We will not go into the details of the derivations here. In the end, using the mathematical properties of the Jacobi-theta function and the cylindrical symmetry, we can eventually obtain the density profile of such a vortex lattice state as [122]

$$n(\mathbf{r}) = \left[\frac{1}{v_c} \sum_{\mathbf{K}} n_{\mathbf{K}} e^{i\mathbf{K}\mathbf{r}} \right] e^{-\mathbf{r}^2/\sigma^2}. \tag{4.28}$$

Here $v_c = b_1^2 v$ is the area of a unit cell, and

$$\frac{1}{\sigma^2} = \frac{1}{a^2} - \frac{\pi}{v_c}. \tag{4.29}$$

Eq. 4.29 shows that

1. v_c is limited by πa^2. This is because the rotational frequency Ω has to be smaller than the trapping frequency ω, which imposes a constraint on the upper limit of the effective flux density.

2. In the fast rotation limit $\Omega \to \omega$, v_c approaches πa^2. In this case, σ^2 becomes much larger than πa^2 and the cloud size is significantly expanded.

In Eq. 4.28, $\mathbf{K} = l_1 \mathbf{K}_1 + l_2 \mathbf{K}_2$, and the two reciprocal vectors $\mathbf{K}_1$ and $\mathbf{K}_2$ are given by $\mathbf{K}_1 = 2\pi \mathbf{B}_2 \times \mathbf{z}/v_c = (2\pi/b_1)(\hat{x} - u\hat{y}/v)$ and $\mathbf{K}_2 = 2\pi \mathbf{B}_1 \times \mathbf{z}/v_c = -(2\pi/b_1)\hat{y}/v$. $n_{\mathbf{K}}$ is a function of $\mathbf{K}_1$ and $\mathbf{K}_2$ and the functional form can be obtained explicitly with the help of the Jacobi Theta function. Since we have made the lowest Landau level approximation at the beginning, the single-particle energy is already degenerate. Thus, the vortex lattice structure is eventually determined by minimizing the interaction energy given by

$$\mathcal{E}_{\text{int}} = \frac{U}{2} \int n^2(\mathbf{r}) d^2\mathbf{r}, \tag{4.30}$$

where U is the interaction strength. Using Eq. 4.28, minimizing the interaction energy can determine the vortex lattice structure. From this calculation one can find out that u and v of a triangular lattice gives the lowest energy. That is why vortex lattices are always triangular in superfluid and superconductor. In Figure 4.7 we show an experimental observation that vortices in a single component BEC always arrange themselves into a triangular lattice, when the vortex number varies from dozens to hundreds. The same variational wave function method can also be generalized to studying vortex lattices in other situations such as two-component Bose condensate [122], Bose condensate with dipolar interaction [188] or even Bose–Fermi superfluid mixture [83].

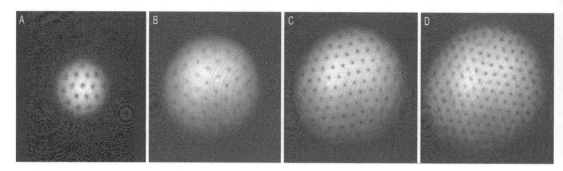

Figure 4.7 Experiment on triangular vortex lattice in BEC. The number of vortices is (a) 16, (b) 32, (c) 80, and (d) 130. In all these cases, the vortices are arranged into a triangular lattice. Reprinted from Ref. [2].

4.3 Spinor Condensate

So far we have focused on Bose–Einstein condensate of bosons without internal spin degree of freedom. However, as we discussed in Section 1.1, most atoms have spins. Taking alkali-metal atoms as an example, at the low-field limit the hyperfine spin F naturally gives the spin degree of freedom for an alkali-metal atom, which usually takes value either one or two. In Section 1.3, we have discussed that all these spin states can be simultaneously trapped in an optical trap. In Section 2.3, we have also discussed the spin-dependent two-body interaction between two spinful atoms, subjected to the total spin rotational symmetry. In this section, we will discuss the properties of Bose–Einstein condensate of these spinful atoms due to the spin-dependent interactions. We will discuss both the ground state property in this section and the topological excitations in these systems in next section.

Majorana Stellar Representation of High Spins. Before proceeding to discussing the property of a BEC, let us first address the question that what is the most intuitive way to represent a spin-F state. $F = 1/2$ is the situation that we are mostly familiar with, where we always represent a spin-1/2 as a unit vector in the unit sphere of the three-dimensional space, or a point in the Bloch sphere. This is a faithful representation of a spin-1/2. Because in the $|F, F_z\rangle$ bases, a spin-1/2 wave function can be written as a normalized two-component spinor $(\psi_{\frac{1}{2}}, \psi_{-\frac{1}{2}})$. Aside from a total phase, the wave function can be written as $(\cos(\theta/2), e^{i\phi} \sin(\theta/2))$, and the degree of freedom is the same as a point in the Bloch sphere. Nevertheless, for a spin-F wave function, it is represented by a normalized $2F + 1$-component spinor $(\psi_F, \psi_{F-1}, \ldots, \psi_{-F})$. Aside from the total phase, the degree of freedom are $4F$ real numbers. Thus, a single point in the Bloch sphere is certainly not a sufficient geometric representation of the state. By counting the degree of freedom, one may guess that such a spin state should be represented by $2F$ points in the Bloch sphere. In fact, such a representation indeed exists and it is known as the *Majorana Stellar Representation* of high spins [117, 10, 13, 164, 110].

To determine these $2F$ points for a given state, we first introduce the *Schwinger boson representation* of the spin operators. Let us introduce $\hat{a}$ and $\hat{b}$ as two bosonic operators, we denote

$$\hat{F}_x = \frac{1}{2}(\hat{a}^\dagger \hat{b} + \hat{b}^\dagger \hat{a}), \tag{4.31}$$

$$\hat{F}_y = \frac{1}{2i}(\hat{a}^\dagger \hat{b} - \hat{b}^\dagger \hat{a}), \tag{4.32}$$

$$\hat{F}_z = \frac{1}{2}(\hat{a}^\dagger \hat{a} - \hat{b}^\dagger \hat{b}). \tag{4.33}$$

It is easy to check that when $\hat{a}$ and $\hat{b}$ satisfy boson commutation relations, $\hat{F}_x$, $\hat{F}_y$ and $\hat{F}_z$ defined in this way satisfy the spin commutation relations. Using the Schwinger boson representation, the quantum state $|F, F_z\rangle$ can be represented by

$$|F, F_z\rangle = \frac{1}{\sqrt{(F+F_z)!\,(F-F_z)!}} \hat{a}^{\dagger F+F_z} \hat{b}^{\dagger F-F_z} |0\rangle, \tag{4.34}$$

where $|0\rangle$ represents the vacuum of both $\hat{a}$ and $\hat{b}$ bosons. In this way, for a given state described by a spinor $(\psi_F, \psi_{F-1}, \ldots, \psi_{-F})$, the wave function can be written as

$$|\psi\rangle = \sum_{F_z=-F}^{F} \psi_{F_z}|F, F_z\rangle = \sum_{F_z=-F}^{F} \psi_{F_z} \frac{1}{\sqrt{(F+F_z)!\,(F-F_z)!}} \hat{a}^{\dagger F+F_z} \hat{b}^{\dagger F-F_z} |0\rangle. \tag{4.35}$$

Eq. 4.35 can be viewed as a homogeneous polynomial of $\hat{a}^\dagger$ and $\hat{b}^\dagger$ of degree $2F$, therefore, without loss of generality, it can be factorized as

$$|\psi\rangle = \frac{1}{\mathcal{N}} \prod_{i=1}^{2F} (u_i \hat{a}^\dagger + v_i \hat{b}^\dagger)|0\rangle, \tag{4.36}$$

where $\{u_i, v_i\}$ are a pair of complex numbers, and $\mathcal{N}$ is a normalization factor to ensure that we can normalize all $\{u_i, v_i\}$ as $|u_i|^2 + |v_i|^2 = 1$. By absorbing an overall phase into $\mathcal{N}$, we can always write $u_i = \cos(\theta_i/2)$ and $v_i = e^{i\phi_i}\sin(\theta_i/2)$, such that each pair of $\{u_i, v_i\}$ can be represented by a unit vector $\hat{n}_i$ in the Bloch sphere with polar angle $\{\theta_i, \phi_i\}$. In this way, we determine $2F$ points in the Bloch sphere that form a complete description of a spin-F quantum state. This is the Majorana Stellar Representation.

The advantage of this representation is that it is very intuitive to visualize the symmetry operator acting on a state. Let $\{\hat{n}_i, i=1, \ldots, 2F\}$ be $2F$ Majorana points for state $|\psi\rangle$, one can show following two results [13, 110]

- If one applies a spin rotation $\hat{\mathcal{U}}$ to state $|\psi\rangle$, the Majorana stellar representation of $\hat{\mathcal{U}}|\psi\rangle$ corresponds to a Cartesian rotation of all $\hat{n}_i$ vectors simultaneously in the Euclidian space with the same rotational angles.
- If one applies a time-reversal operator $\hat{\mathcal{R}}$ to state $|\psi\rangle$, $\hat{\mathcal{R}}|\psi\rangle$ is described by $2F$ points of $-\hat{n}_i$ as a central reflection of the original vectors.

With these two results, the advantage of this representation becomes quite clear:

- It very intuitively illustrates the symmetry of a state. As discussed before, by "symmetry of a state," it means a group of symmetry operators and when these operators act on the state $|\psi\rangle$, it results in the same state up to a global phase. Thus, if the distribution of all these $2F$ points is invariant under certain kind of rotation or a reflection, it means that such a state has certain spin rotational symmetry or the time reversal symmetry.

- It can also help to easily identify whether two states are related by a spin rotation or the time-reversal symmetry. If the distribution of the $2F$ points of one state can coincide with the distribution of the $2F$ points of another state upon a rotation or reflection, that means these two states are related by a spin rotation or the time reversal symmetry. Furthermore, if the Hamiltonian is also invariant under the spin rotation or the time reversal symmetry, these two states should have same energy.

New we discuss a few examples:

- Let us consider a spin-1 state $(1, 0, 0)$, one can show that these two points sit together at the north pole, as shown in Figure 4.8(a). It is clear that this state is invariant under a $U(1)$ rotation along the $\hat{z}$ axes, but it is not invariant under rotation along other axes and is not invariant under the time reversal symmetry. Similarly, the state $(0, 0, 1)$ is represented by two points sitting together at the south pole, as shown in Figure 4.8(b). We can also conclude that states $(1, 0, 0)$ and $(0, 0, 1)$ can be related to each other either by a spin rotation or by the time reversal symmetry.

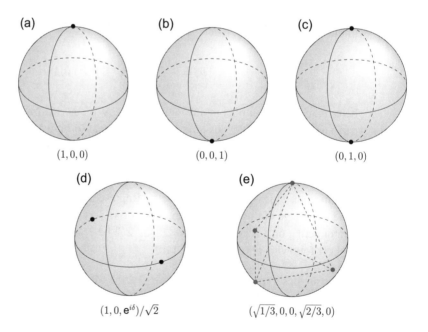

Figure 4.8 Examples of the Majorana stellar representation. (a–d) Four different spin-1 wave functions as shown below the figures. (e) A spin-2 wave function. A color version of this figure can be found in the resources tab for this book at cambridge.org/zhai.

- Considering another spin-1 state $(0,1,0)$, its Majorana stellar representation is given by one point sitting at the north pole and the other point sitting at the south pole, as shown in Figure 4.8(c). For this state, in addition to the rotational symmetry along the $\hat{z}$ axes, it is clear that this state is also invariant under the time reversal symmetry. Considering another state $(1,0,e^{i\delta})/\sqrt{2}$, its Majorana stellar representation are two points sitting at the equator and being opposite to each other, as displayed in Figure 4.8(d). Hence, states $(0,1,0)$ and $(1,0,e^{i\delta})/\sqrt{2}$ can also be related to each other by a spin rotation. But neither $(1,0,0)$ nor $(0,0,1)$ can be rotated to $(0,1,0)$ or $(1,0,e^{i\delta})/\sqrt{2}$ by spin rotation. Thus, for a spin rotational invariant Hamiltonian, the energies of $(1,0,0)$ and $(0,0,1)$ states are degenerate, and the energies of $(0,1,0)$ and $(1,0,e^{i\delta})/\sqrt{2}$ states are degenerate. However, the energies of $(1,0,0)$ and $(0,0,1)$ are not degenerate with the energies of $(0,1,0)$ and $(1,0,e^{i\delta})/\sqrt{2}$.

- Let us consider a spin-2 state as $(1/\sqrt{3},0,0,\sqrt{2/3},0)$, the four points form a tetrahedron in the Bloch sphere [13], as shown in Figure 4.8(e). Thus, this state has the full point group symmetry of a tetrahedron. This symmetry contains four different C_3 (rotation by $2\pi/3$) rotation along four different symmetry axes. This example illustrates the power of this representation, because, if not using the Majorana stellar representation, it is very difficulty to visualize that this spin state has such point group symmetry just from the five components spin wave function.

Ferromagnetic versus Polar Condensate. Now let us consider spin-1 bosons as an example. The many-body Hamiltonian is similar as Eq. 3.14, and the difference is that the interaction part is replaced by the form of Eq. 2.37 between two spin-1 bosons discussed in Section 2.3, that is,

$$\hat{\mathcal{H}} = \sum_{i=1}^{N}\left(\frac{\hat{\mathbf{p}}_i^2}{2m}+V(\mathbf{r}_i)\right) + \sum_{i<j}\frac{4\pi\hbar^2}{m}\left(a^{(n)}+a^{(s)}\mathbf{F}_i\cdot\mathbf{F}_j\right)\delta(\mathbf{r}_{ij})\frac{\partial}{\partial r_{ij}}r_{ij}. \quad (4.37)$$

Similar to the discussed made in Section 3.1, we consider a simple Bose condensate state where all bosons are condensed in the same state and the many-body wave function is given by

$$\Psi = \prod_{i=1}^{N}\left(\psi_{-1}(\mathbf{r}_i)|1,-1\rangle + \psi_0(\mathbf{r}_i)|1,0\rangle + \psi_1(\mathbf{r}_i)|1,1\rangle\right). \quad (4.38)$$

Here, without loss of generality, the condensate wave function can be represented by $\psi(\mathbf{r}) = (\psi_{-1}(\mathbf{r}),\psi_0(\mathbf{r}),\psi_1(\mathbf{r})) = \sqrt{n(\mathbf{r})}e^{i\theta}\zeta(\mathbf{r})$, where $\zeta(\mathbf{r})$ satisfies $\zeta(\mathbf{r})\zeta^\dagger(\mathbf{r})=1$. Using this as the variational wave function, and following similar discussion as in Section 3.2, and rescaling $\sqrt{N}\psi(\mathbf{r})\to\psi(\mathbf{r})$, we reach the energy function

$$\frac{\mathcal{E}}{N} = \int d^3\mathbf{r}\left[\psi^{*,T}(\mathbf{r})\left(-\frac{\hbar^2\nabla^2}{2m}+V(\mathbf{r})\right)\psi(\mathbf{r})+\frac{2\pi\hbar^2}{m}n^2(\mathbf{r})(a^{(n)}+a^{(s)}\langle\mathbf{F}\rangle^2)\right], \quad (4.39)$$

where $\langle \mathbf{F} \rangle^2$ explicitly means $(\zeta F_x \zeta^\dagger)^2 + (\zeta F_y \zeta^\dagger)^2 + (\zeta F_z \zeta^\dagger)^2$, and $F_{x,y,z}$ here denotes the spin-1 representation of the three generators for the $SO(3)$ group,[3] that is,

$$F_x = \begin{pmatrix} 0 & \frac{1}{\sqrt{2}} & 0 \\ \frac{1}{\sqrt{2}} & 0 & \frac{1}{\sqrt{2}} \\ 0 & \frac{1}{\sqrt{2}} & 0 \end{pmatrix}; \quad F_y = \begin{pmatrix} 0 & -\frac{i}{\sqrt{2}} & 0 \\ \frac{i}{\sqrt{2}} & 0 & -\frac{i}{\sqrt{2}} \\ 0 & \frac{i}{\sqrt{2}} & 0 \end{pmatrix};$$

$$F_z = \begin{pmatrix} 1 & 0 & 0 \\ 0 & 0 & 0 \\ 0 & 0 & -1 \end{pmatrix}. \tag{4.40}$$

We consider a repulsive density-density interaction with $a^{(n)} > 0$. As we discussed in Section 2.3, typically the absolute value of $a^{(s)}$ is much smaller than $a^{(n)}$, such that $a^{(n)} + a^{(s)}\langle \mathbf{F} \rangle^2$ is always positive that ensures the stability of the condensate. $a^{(s)}$ can be either positive, such as in the case of ^{23}Na, or negative such as in the case of ^{87}Rb. As we will see, the sign of $a^{(s)}$ is crucial for determining the spin structure of the condensate [70, 130, 164]. Here we first consider a uniform system with $V(\mathbf{r}) = 0$, where bosons are condensed in a zero-momentum state and $n(\mathbf{r})$ is a constant. Therefore, minimization of energy simply becomes minimizing $a^{(s)}\langle \mathbf{F} \rangle^2$.

Here we are interested in finding out the degenerate space of the ground state spin wave function. First of all, it is important to note that this Hamiltonian, as well as its mean-field version, is invariant under a $U(1)$ phase rotation $e^{i\delta}$ and a $SO(3)$ spin rotation $\mathcal{U} = e^{-i\hat{F}_z\alpha}e^{-i\hat{F}_y\beta}e^{-i\hat{F}_z\gamma}$, and we denote $\mathcal{G} = U(1) \times SO(3)$. That to say, a state $|\zeta\rangle$ has the same energy as the state $\mathcal{G}|\zeta\rangle$. Then, we should determine the symmetry group $\mathcal{H}$ under which the state is invariant. Hence, the degenerate space is given by $\mathcal{G}/\mathcal{H}$. We will see that the Majorana stellar representation is very useful for this purpose.

For $a^{(s)} < 0$, energy minimization favors maximizing $\langle \mathbf{F} \rangle^2$. This means that $|\langle \mathbf{F} \rangle|$ should be taken as unity and it can be satisfied by choosing $\zeta = (1,0,0)$. This is called the *ferromagnetic state*. Therefore, all elements in $\mathcal{G}$ acting on $\zeta = (1,0,0)$ generate all degenerate ferromagnetic states, that is,

$$\zeta_F^\dagger = e^{i\delta}\mathcal{U}\begin{pmatrix} 1 \\ 0 \\ 0 \end{pmatrix} = e^{i(\delta-\gamma)}\begin{pmatrix} e^{-i\alpha}\cos^2\frac{\beta}{2} \\ \frac{1}{\sqrt{2}}\sin\beta \\ e^{i\alpha}\sin^2\frac{\beta}{2} \end{pmatrix}. \tag{4.41}$$

As we have shown above using the Majorana stellar representation, the state $(1,0,0)$ is invariant under a $U(1)$ rotation along the $\hat{z}$ axes. Hence, $\mathcal{H} = U(1)$. Thus, the degenerate space of the ferromagnetic state is characterized by the coset

$$\mathcal{M} = U(1) \times SO(3)/U(1) = SO(3). \tag{4.42}$$

It can also be viewed intuitively from Eq. 4.41 that $\delta - \gamma$ can be combined as a single phase, and then the entire wave function can be viewed as a $SO(3)$ rotation applied to $\zeta = (1,0,0)$.

For $a^{(s)} > 0$, energy minimization favors minimizing $\langle \mathbf{F} \rangle^2$. This means that $|\langle \mathbf{F} \rangle|$ should be taken as zero and it can be satisfied by choosing $\zeta = (0,1,0)$. This is called the

---

[3] Note that $SO(3)$ group is isomorphic to $SU(2)$ group.

antiferromagnetic state or the *polar state*. Similarly, all degenerate polar states can be generated by applying all elements in $\mathcal{G}$ to $\zeta = (0, 1, 0)$, that is,

$$\zeta_P^\dagger = e^{i\delta}\mathcal{U}\begin{pmatrix} 0 \\ 1 \\ 0 \end{pmatrix} = e^{i\delta}\begin{pmatrix} -\frac{e^{-i\alpha}}{\sqrt{2}}\sin\beta \\ \cos\beta \\ \frac{e^{i\alpha}}{\sqrt{2}}\sin\beta \end{pmatrix}. \tag{4.43}$$

As we have also shown above using the Majorana stellar representation, the $(0, 1, 0)$ state is invariant under a $U(1)$ rotation along $\hat{z}$ axes and a Z_2 time-reversal symmetry. Hence, $\mathcal{H} = U(1) \times Z_2$. Thus, the degenerate space of the polar state is given by the coset

$$\mathcal{M} = \frac{U(1) \times SO(3)}{U(1) \times Z_2} = \frac{U(1) \times S^2}{Z_2}, \tag{4.44}$$

where S^2 denotes the space of a two-dimensional unit sphere. It is easily to show that $\langle \mathbf{F} \rangle = (0, 0, 0)$ for all polar states. Unlike the spin-1/2 case, here all three spin components can vanish simultaneously. In fact, the spin part of the wave function Eq. 4.43 can be represented by a nematic vector $\mathbf{n} = (\sin\beta\cos\alpha, \sin\beta\sin\alpha, \cos\beta)$, where α and β are respectively the azimuthal angle and the polar angle on the unit sphere. The wave function Eq. 4.43 can be rewritten as

$$\zeta_P^\dagger = e^{i\delta}\begin{pmatrix} -\frac{1}{\sqrt{2}}(\mathbf{n}_x - i\mathbf{n}_y) \\ \mathbf{n}_z \\ \frac{1}{\sqrt{2}}(\mathbf{n}_x + i\mathbf{n}_y) \end{pmatrix}. \tag{4.45}$$

Actually, this $\mathbf{n}$ vector is one of the points in the Majorana stellar representation of the polar state. Because the polar state is represented by a pair of two opposite points $\mathbf{n}$ and $-\mathbf{n}$, thus, $\mathbf{n}$ and $-\mathbf{n}$ actually represent the same state. That is why this vector is called the *nematic vector*, because nematic refers to an orientation without direction. From the wave function, one can also see that if (α, β) is changed to $(\alpha + \pi, \pi - \beta)$, that is, $\mathbf{n}$ is changed to $-\mathbf{n}$, the spinor part of the wave function only differs by a minus sign, which can be absorbed by changing $\delta \to \delta + \pi$. Thus, these two states become identical. By noticing this Z_2 redundancy, the actual degenerate space is $U(1) \times S^2/Z^2$.

Singlet Pair Condensate. Here we consider a simpler situation that the spatial wave function is fixed and the spatial fluctuation and dynamics of the wave function is frozen, and we only consider the spin wave function. This is also known as the single spatial mode approximation. Let $\hat{a}_{0,\pm1}$ be the annihilation operators for these three different F_z components on this single mode, the Hamiltonian Eq. 4.37 reduces to

$$\hat{H} = a^{(s)}\hat{\mathbf{F}}^2, \tag{4.46}$$

where $\hat{F}_x + i\hat{F}_y = \sqrt{2}(\hat{a}_1^\dagger\hat{a}_0 + \hat{a}_0^\dagger\hat{a}_{-1})$ and $\hat{F}_x - i\hat{F}_y = \sqrt{2}(\hat{a}_0^\dagger\hat{a}_1 + \hat{a}_{-1}^\dagger\hat{a}_0)$, and $\hat{F}_z = \hat{a}_1^\dagger\hat{a}_1 - \hat{a}_{-1}^\dagger\hat{a}_{-1}$. It can be shown that when $a^{(s)} > 0$, the exact ground state of this spin Hamiltonian is a total spin singlet given by [104]

$$|S\rangle \propto (-2\hat{a}_1^\dagger\hat{a}_{-1}^\dagger + \hat{a}_0^{\dagger 2})^{N/2}|0\rangle. \tag{4.47}$$

This state is a fragmented state in terms of one-body density matrix, similar as the Fock state that we have discussed in Section 3.5. In terms of two-body density matrix, it is a simple Bose condensation. Here the operator $-2\hat{a}_1^\dagger\hat{a}_{-1}^\dagger + \hat{a}_0^{\dagger 2}$ creates a pair of bosons that form a spin singlet of two $S = 1$ atoms. Intuitively, Eq. 4.47 means that two bosons form a singlet pair and these singlet pairs condense. Therefore, it is also called the *pair condensate*.

It is quite obvious that this exact ground state Eq. 4.47 is very different from a mean-field state. With the expression Eq. 4.45, the mean-field state can be written as

$$|\text{MF}\rangle = \frac{1}{\sqrt{N!}}\left(-\frac{\hat{a}_1^\dagger}{\sqrt{2}}(\mathbf{n}_x - i\mathbf{n}_y) + \mathbf{n}_z\hat{a}_0^\dagger + \frac{\hat{a}_{-1}^\dagger}{\sqrt{2}}(\mathbf{n}_x + i\mathbf{n}_y)\right)^N |0\rangle$$

$$= \frac{1}{\sqrt{N!}}\left(\mathbf{n}\cdot\hat{\mathbf{A}}^\dagger\right)^N |0\rangle, \qquad (4.48)$$

where $\hat{A}_x^\dagger = (\hat{a}_{-1}^\dagger - \hat{a}_1^\dagger)/\sqrt{2}$, $\hat{A}_y^\dagger = i(\hat{a}_{-1}^\dagger + \hat{a}_1^\dagger)/\sqrt{2}$ and $\hat{A}_z^\dagger = \hat{a}_0$. To show the relation between the total spin singlet state and the mean-field state, it can be shown that [123]

$$\int d\mathbf{n}\frac{1}{4\pi}(\mathbf{n}\cdot\mathbf{A}^\dagger)^N|0\rangle \propto (\mathbf{A}^\dagger\cdot\mathbf{A}^\dagger)^{N/2}|0\rangle. \qquad (4.49)$$

And it is easy to show that

$$\mathbf{A}^\dagger\cdot\mathbf{A}^\dagger = -2\hat{a}_1^\dagger\hat{a}_{-1}^\dagger + \hat{a}_0^{\dagger 2}. \qquad (4.50)$$

That is to say, if we start from a mean-field state and let the mean-field order strongly fluctuate and spread over the entire S^2 sphere, it yields a singlet pair condensate phase. This is actually the same physics as we have discussed in Section 3.5, where we have shown that a strong phase fluctuation can yield a fragmented condensate. In fact, it is generally true that strong fluctuation of mean-field order parameters can render a simple condensate into a boson pair condensate.

Quadratic Zeeman Effect. Finally, let us mention the role of a Zeeman field. As discussed in Section 1.2, Zeeman field creates both linear and quadratic Zeeman effects. Because the linear Zeeman effect is given by hF_z, and the total F_z is conserved during the spin-exchanging collision, the linear Zeeman field is therefore always a constant and does not play any role in determining the spin structure. Thus, we mainly focus on the quadratic Zeeman effect given by qF_z^2. q can be tuned to be either positive or negative. Here let us focus on the case with positive q. A positive q favors minimizing $\langle\hat{F}_z^2\rangle$. For a ferromagnetic state described by the wave function Eq. 4.41, $\langle\hat{F}_z^2\rangle = 1-(1/2)\sin^2\beta$. Thus, minimization of $\langle\hat{F}_z^2\rangle$ favors $\beta = \pi/2$, and the ferromagnetic wave function reduces to

$$\zeta)_{\text{F}}^\dagger = e^{i(\delta-\gamma)}\begin{pmatrix}\frac{1}{2}e^{-i\alpha}\\ \frac{1}{\sqrt{2}}\\ \frac{1}{2}e^{i\alpha}\end{pmatrix}. \qquad (4.51)$$

This corresponds to an in-plane ferromagnetic state, where $\langle\mathbf{F}\rangle = (\cos\alpha, \sin\alpha, 0)$. So the physical meaning of α is the azimuthal spin angle. The degenerate manifold is reduced to

$U(1) \times U(1)$. For a polar state described by Eq. 4.43, $\langle \hat{F}_z^2 \rangle = -\cos(2\beta)$. Thus, minimization of $\langle \hat{F}_z^2 \rangle$ favors $\beta = 0$, and the polar wave function reduces to

$$|\zeta_P^\dagger = e^{i\delta} \begin{pmatrix} 0 \\ 1 \\ 0 \end{pmatrix}. \tag{4.52}$$

This degenerate manifold becomes $U(1)$ alone.

4.4 Topological Excitations in Spinor Condensate

In Section 4.2 we have discussed that the topological defects can be classified by the homotopy group of the order parameter degenerate manifold. For a spinless condensate, the order parameter manifold is the $U(1)$ phase. In Section 4.3 we have shown that the order parameter manifold for a spinful condensate can possess much richer structure, therefore, it can host much richer topological excitations. We will discuss some typical ones in this session.

Spin Vortex in Ferromagnetic Condensate. Before we discuss more complicated topological structures, let us revisit vortex that we have discussed in Section 4.2. In Section 4.2 we have discussed vortex as topological excitation in a spinless condensate, which creates a point defect in the $U(1)$ phase of the order parameter. Now we revisit it for a spin vortex. Let us first consider the situation of the ferromagnetic condensate under a positive quadratic Zeeman field $q > 0$, whose ground state wave function is given by Eq. 4.51. It is quite clear that there are two $U(1)$ phases. The overall phase is $\delta - \gamma$, and a vortex in this phase is the same as a vortex in the spinless condensate, which manifests itself as a circulating density current. The new type of vortex is a vortex in the phase α, for instance, $\alpha = \kappa\varphi$, where κ is an integer and φ is the spatial azimuthal angle. This vortex does not generate circulating density current. Because as one can see from Eq. 4.51, this corresponds to that the $|1, 1\rangle$ component has a vortex with winding number $+\kappa$ and the $|1, -1\rangle$ component has a vortex with winding number $-\kappa$, therefore, the total density current cancels out. On the other hand, as discussed above, because the physical meaning of α is the azimuthal spin angle along the equator of the Bloch sphere, the geometric picture of this vortex is that the spin winds around the equator of the Bloch sphere for κ times. Hence, it is named as *spin vortex*.

A question here is what happens to the core of a spin vortex. Because both the $|1, 1\rangle$ and $|1, -1\rangle$ components contain vorticity that diverges at the vortex core, the densities of these two components must vanish at the vortex core for the same reason as we discussed in the spinless case. One can at least think about following three possibilities for the fate of the core of a spin vortex. As we will discuss below, the energy costs are different for different cases.

- *Empty Core Vortex:* The total density n vanishes and the core is completely empty, as shown in Figure 4.9(a). This costs the density-density interaction energy $a^{(n)}(n - \bar{n})^2$

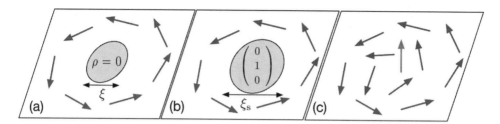

Different types of spin vortex. In a ferromagnetic condensate with positive quadratic Zeeman energy, there are three types of spin vortex with different core structures. (a) Empty core vortex where the total density vanishes inside the vortex core. (b) Polar core vortex where the spin wave function becomes a polar state inside the vortex core. (c) Mermin-Ho vortex where spins point out of the plane inside the vortex core. A color version of this figure can be found in the resources tab for this book at cambridge.org/zhai.

inside the vortex core, because the density deviates from the average density $\bar{n}$. In this case, the vortex core size is of the order of healing length ξ.

- *Polar Core Vortex:* Let us write a wave function ansatz as

$$e^{i(\delta-\gamma)} \begin{pmatrix} \frac{1}{2}f(r)e^{-i\kappa\theta} \\ \frac{1}{\sqrt{2}}\sqrt{2-f^2(r)} \\ \frac{1}{2}f(r)e^{i\kappa\theta} \end{pmatrix}. \tag{4.53}$$

where $f(r) \to 1$ for $r \gg \xi_s$ and $f(r) \to 0$ for $r \to 0$. When $f = 1$ the wave function recovers a ferromagnetic state Eq. 4.51. When $f = 0$, the wave function becomes a polar state Eq. 4.43. Thus, from $r \gg \xi_s$ to $r \to 0$, the density of the wave function remains a constant, and the spin wave function interpolates between a ferromagnetic state and a polar state. This is called *polar core vortex*, as shown in Figure 4.9(b). This vortex core causes the spin-dependent interaction energy $a^{(s)}\langle \mathbf{F} \rangle^2 n^2$. Hence, ξ_s should be determined by the balance between the spin-dependent interaction energy and the kinetic energy, such that ξ_s is called the *spin healing length* determined by $1/\sqrt{8\pi a^{(s)}n}$. Typically the spin healing length is larger than the healing length introduced in Section 3.3 because $a^{(n)}$ is usually much larger than $a^{(s)}$.

- *Mermin-Ho Vortex:* Let us consider a wave function ansatz

$$\begin{pmatrix} \cos^2 \frac{\beta(r)}{2} \\ e^{i\kappa\theta}\frac{1}{\sqrt{2}}\sin\beta(r) \\ e^{2i\kappa\theta}\sin^2\frac{\beta(r)}{2} \end{pmatrix} = e^{i\kappa\theta}\begin{pmatrix} e^{-i\kappa\theta}\cos^2\frac{\beta(r)}{2} \\ \frac{1}{\sqrt{2}}\sin\beta(r) \\ e^{i\kappa\theta}\sin^2\frac{\beta(r)}{2} \end{pmatrix}, \tag{4.54}$$

and $\beta(r) \to 0$ for $r \to 0$ and $\beta(r) \to \pi/2$ for $r \to \infty$. This is a combination of regular vortex with winding number κ in the total phase $\delta - \gamma$ and a spin vortex with the same winding number κ in α, as shown in Eq. 4.54. Consequently, the wave function shows no vortex in the $|1, 1\rangle$ component, a vortex with winding number κ in the $|1, 0\rangle$ component and a vortex with winding number 2κ in the $|1, -1\rangle$ component. Because $\beta(r) \to 0$ as $r \to 0$, the density in $|1, 0\rangle$ and $|1, -1\rangle$ components vanish as we expected. From the spin configuration point of view, the vortex in the total phase does not affect the spin configuration, and at long distance, spins still wind around the equator for κ time. But in short distance, when $\beta \to 0$, spins no longer lie in the equator but point to the north pole, as shown in Figure 4.9(c). This is known as the *Mermin-Ho vortex* [119]. Unlike the polar-core vortex, this wave function always stays in the ferromagnetic manifold in the entire space, and the lengths of the spins are always maximized. Thus, this state does not cause any interaction energy. However, because spins point out the equator, this state causes quadratic Zeeman energy for positive q.

In summary, the empty core vortex causes density interaction energy, the polar core vortex causes spin-dependent interaction energy and the Mermin-Ho vortex causes quadratic Zeeman energy. In reality, the energy competition determines which situation should take place.

Half-Vortex in Polar Condensate. Here we consider the polar condensate and return to consider the phase vortex. Usually for a phase vortex, the phase change around any closed loop should be integers times of 2π to ensure the uniqueness of the wave function. However, there exists such a phase vortex in polar condensate whose phase can only change by π for a closed loop. Such a vortex is called a *half vortex*.

Let us consider the wave function ansatz

$$e^{i\varphi/2} \begin{pmatrix} -\frac{e^{-i\alpha}}{\sqrt{2}} \sin(\varphi/2) \\ \cos(\varphi/2) \\ \frac{e^{i\alpha}}{\sqrt{2}} \sin(\varphi/2) \end{pmatrix}, \tag{4.55}$$

where φ is the spatial azimuthal angle, and around any closed loop, φ changes by 2π. Nevertheless, for this wave function, the total phase only changes by π, but the spin part also acquires a minus sign, such that the total wave function retains invariant. This property is because the spin wave function has a Z_2 symmetry which can be combined with the phase twist. In more sophisticated situation like spin-2 condensate with tetrahedral symmetry, one can combine the phase twist with the tetrahedral point group symmetry of the spin part to realize vortices with richer structure. Since the tetrahedral symmetry group is a non-abelian one, this realizes non-abelian topological defects [118].

Monopole versus Skyrmion. In Section 4.2, we have introduced vortex as a topological defect in two-dimensional space. We have discussed that the topological defects are classified by mapping from boundaries of a physical region to the order parameter space $\mathcal{M}$. With the help of the homotopy group, in two-dimensional space, such mappings are classified by $\Pi_1(\mathcal{M})$. When this mapping is nontrivial, for instance, when $\mathcal{M}$ is a $U(1)$ group, the boundary cannot be shrink to one point without crossing singularity of the mapping. Thus, this gives rise to a point topological defect called vortex. With similar spirit, we can consider a point defect in three dimensions. In this case, we consider a two-dimensional sphere as a boundary of three-dimensional physical region, and the mappings from this sphere to the order parameter manifold $\mathcal{M}$ are classified by $\Pi_2(\mathcal{M})$. If this mapping is nontrivial, one also cannot shrink the sphere to a point without crossing singularity. This gives rise to a point defect in three-dimensional space known as *monopole*. Thus, in order to find a system that can support a topological stable monopole excitation, one has to find a condensate whose order parameter manifold has nontrivial second homotopy group. The

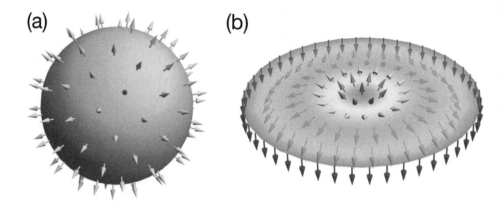

(a) (b)

Figure 4.10 Schematic of monopole versus skyrmion: (a) Monopole as a singular topological defect in three-dimensional space. (b) Skyrmion as a non=singular topological object in compacted two-dimensional space. Both are mathematically described by $\Pi_2(S^2) = \mathcal{Z}$. A color version of this figure can be found in the resources tab for this book at cambridge.org/zhai.

simplest nontrivial example of $\Pi_2(\mathcal{M})$ is that $\mathcal{M}$ is also a two-dimensional sphere S^2, such as the **n** space in the polor condensate discussed in Section 4.3. Hence, there is stable monopole excitation in a polar condensate. We shall note that because $\Pi_2(SO(3))$ is a trivial one, there does not exist stable topological monopole in the ferromagnetic condensate.

Let us now concretely construct a monopole excitation. We take $\mathbf{n}(\mathbf{r}) = \hat{\mathbf{r}}$, as shown in Figure 4.10(a). In this way, the order parameter **n** always fully covers the entire Bloch sphere for any surface enclosed $\mathbf{r} = 0$, and the **n** field displays a singularity when the sphere shrinks toward $\mathbf{r} = 0$. This gives rise to a monopole sitting at $\mathbf{r} = 0$ with Chern number being unity. It is not difficult to show that the kinetic energy of this monopole configuration is given by

$$\frac{\hbar^2}{2m} \sum_{i=x,y,z} (\partial_i \mathbf{n})^2 \propto \frac{\hbar^2}{2mr^2}. \tag{4.56}$$

Therefore, on one hand, similar as the vortex core, the density has to vanish when $r \to 0$, which gives rise to a monopole core of the order of healing length ξ. And on the other hand, at the long distance, the integration depends on the system size as

$$\frac{\hbar^2}{2m} \int_{\xi}^{R} r^2 dr \frac{1}{r^2} = \frac{\hbar^2}{2m} \frac{R}{\xi}. \tag{4.57}$$

In the case of vortex, the energy of a vortex logarithmically depends on system size, which scales the same way as the entropy. However, in this case, the energy of a monopole depends linearly on the system size. For sufficiently large system, the energy cost is always larger than the entropy gain. Therefore, at finite temperature the system is always in a monopole confinement phase. That is why the BKT transition does not exist in such a three-dimensional spin system.

Table 4.1 Different types of topological objects		
	Singular defects	Nonsingular objects
Mapping	From Boundary of a Region S_{d-1} to Order Parameter Space $\mathcal{M}$	From compactified space S_d to order parameter space $\mathcal{M}$
Homotopy group	$\Pi_{d-1}(\mathcal{M})$	$\Pi_d(\mathcal{M})$
Examples in 2D	$\Pi_1(S^1)$: Vortex in 2D (winding number)	$\Pi_2(S^2)$: Skyrmion in 2D (Chern number)
Examples in 3D	$\Pi_2(S^2)$: Monopole in 3D (Chern number)	$\Pi_3(S^2)$ Knots in 3D (linking number)

Unlike the topological defects, there also exists another type of nonsingular topological object. This topological object is defined through a mapping from the entire spatial space to the order parameter space. However, there is an obstacle if one wants to classify such mapping with homotopy group. Because the homotopy groups are defined by mapping from S^n to the order parameter space $\mathcal{M}$, and S^n is a closed manifold without boundary. However, a physical spatial space is an open manifold with boundary. Therefore, in order to classify such excitations with topology, one needs to first compactify the spatial space. This can be done by imposing a boundary condition that the wave function approaches the same value as $\mathbf{r}$ approaches infinity from different directions. When this boundary condition is imposed, as far as such mappings are concerned, all the infinite can be regarded as the same point, and a d-dimensional space is compacified as S^d. Hence, in d dimensions, such topological objects are classified by $\Pi_d(\mathcal{M})$.

For instance, considering two-dimensional space with order parameter space $\mathcal{M} = S^2$, an example of nontrivial mapping is shown in Figure 4.10(b) and such topological object is called *Skyrmion*. For example, we can write

$$\mathbf{n} = (\sin \beta(r) \sin \varphi, \sin \beta(r) \cos \varphi, \cos \beta(r)), \qquad (4.58)$$

where $\mathbf{r} = (r \sin \varphi, r \cos \varphi)$ and φ is the azimuthal angle. $\beta(r \to 0) \to 0$ and $\beta(r \to \infty) \to \pi$. In this configuration, spin points to the north pole at $\mathbf{r} = 0$, and all spins point to the south pole at $r = \infty$ regardless the azimuthal angle φ. Thus, it satisfies the boundary condition to compactify the two-dimensional space. It is easy to see that, for the entire two-dimensional space, the S^2 space is also fully covered once, and therefore, the topological invariant is unity. By continuously deforming the mapping, one can show that the topological charge cannot be changed as long as the boundary condition is not changed. From this example, it is also easy to show that this topological object is different from a topological defect, because the mapping is smooth everywhere and there is no singularity of the $\mathbf{n}(\mathbf{r})$ field in the entire space.

In Table 4.1 we compare two different types of topological objects as discussed above. As one can see, although both monopole in three dimensions and skyrmion in two dimensions are mathematically described by $\Pi_2(S^2)$, physically they are distinct topological excitations. In Sections 7.2 and 7.3, we will discuss band theory with nontrivial topology, where we will also use the homotopy group to characterize the mapping from the momentum space to the Bloch wave function space. Similarly, we will discuss two types of band structures. One is semimetal with singularities and the other is a band insulator without singularity. The semimetal has similar mathematical structure as topological defect, and the topological band insulator has similar mathematical structure as nonsingular topological object. As far as the mathematical structure is concerned, it is inspiring to compare Table 4.1 with Table 7.1. However, the physics contents are very different. Here, we are discussing the excitations in a condensate of bosons, and there, we will discuss classifying ground state with bands filled by noninteracting of fermions.

Dirac Monopole. In the Maxwell theory, *Dirac Monopole* refers to a singularity in the magnetic field, that is, for a single monopole, we have

$$\nabla \cdot \mathbf{B} = 4\pi \delta(\mathbf{r}). \tag{4.59}$$

Such a monopole so far has not been discovered in real electromagnetic field in nature. However, in Section 1.2 we have discussed the idea of simulating a synthetic gauge field $\mathbf{B}_{\text{syn}}$ coupled to the spatial motion of a neural atoms by spatially varying the spin direction of an atom. Thus, one may wonder whether we can create a spin configuration such that a Dirac magnetic monopole can exist in the resulting synthetic magnetic field $\mathbf{B}_{\text{syn}}$, that is,

$$\nabla \cdot \mathbf{B}_{\text{syn}} = 4\pi \delta(\mathbf{r}). \tag{4.60}$$

To this end we have to consider the ferromagnetic spinor condensate with wave function given by Eq. 4.41. Let us consider a spatial varying magnetic field $\mathbf{B}(\mathbf{r})$ and a simplified linear coupling $-h\mathbf{B} \cdot \mathbf{F}$ ($h > 0$) between magnetic field and spins. To be consistent with Eq.4.41, we denote $\mathbf{B} = (\sin \beta \cos \alpha, \sin \beta \sin \alpha, \cos \beta)$, such that Eq. 4.41 is the low-energy state polarized by this magnetic field. Here we consider that both α and β vary spatially. As one can show, this coupling can be diagonalized by a unitary matrix $\mathcal{U}(\alpha, \beta)$ as

$$\mathcal{U}^\dagger(\alpha, \beta)(-h\mathbf{B} \cdot \mathbf{F})\mathcal{U}(\alpha, \beta) = -h|\mathbf{B}| \begin{pmatrix} 1 & 0 & 0 \\ 0 & 0 & 0 \\ 0 & 0 & -1 \end{pmatrix}, \tag{4.61}$$

and the resulting synthetic gauge field is given by

$$\mathbf{A} = i\hbar\mathcal{U}^\dagger\nabla\mathcal{U} = -\hbar(I - \cos \beta F_z - \sin \beta F_x)\nabla\alpha. \tag{4.62}$$

For atoms staying in the lowest energy adiabatic spin bases, only the first diagonal component $\mathbf{A}_{11}$ matters, and $\mathbf{A}_{11} = -\hbar(1 - \cos \beta)\nabla\alpha$, which gives rise to a synthetic gauge field $\mathbf{B}_{\text{syn}} = \nabla \times A_{11}$.

Now let us consider the quadrupole trap with a uniform bias field, where $\mathbf{B} = B_0 x\hat{x} + B_0 y\hat{y} + (-2B_0 z + B_{\text{bias}})\hat{z}$. Let us make a coordinate transformation $x \to x$, $y \to y$ and $2z - B_{\text{bias}}/B_0 \to z$. This coordinate transformation corresponds to stretching and shifting the coordinate along the $\hat{z}$ direction and will not change the topology of the gauge field. In the new coordinate, one can find that this magnetic field configuration corresponds to

$\beta = \pi - \theta \equiv \theta'$ and $\alpha = \varphi$, where θ and φ are polar and azimuthal angles in the spatial spherical coordinate, and using Eq. 4.62, we can obtain [147]

$$\mathbf{A}_{11} = -\hbar \frac{1 + \cos\theta}{r\sin\theta} \hat{\varphi} \tag{4.63}$$

$$\mathbf{B}_{\mathrm{syn}} = \frac{\hbar}{r^2} \hat{r}. \tag{4.64}$$

From Eq. 4.64, it is clear that the synthetic magnetic field corresponds to a magnetic monopole sitting at the zero of the new coordinate.

The wave functions of this Dirac monopole can always be written as

$$\begin{pmatrix} e^{-2i\varphi} \cos^2 \frac{\theta'}{2} \\ \frac{1}{\sqrt{2}} e^{-i\varphi} \sin\theta' \\ \sin^2 \frac{\theta'}{2} \end{pmatrix}, \tag{4.65}$$

which can be viewed as a doubly quantized vortex in $m_z = 1$ component and a singly quantize vortex in $m_z = 0$ component. Hence, there is a line singularity with $\theta' = 0$, which is also the line singularity in Eq. 4.63 of A_{11}. This corresponds to the so-called *Dirac string*. Note that in the Maxwell theory, the location of the Dirac string depends on the gauge choice. Here the location of the Dirac string also depends on the gauge choice. Note that the $-h\mathbf{B} \cdot \mathbf{F}$ coupling only fixes the spin direction, we can choose another wave function written as

$$\begin{pmatrix} \cos^2 \frac{\theta'}{2} \\ \frac{1}{\sqrt{2}} e^{i\varphi} \sin\theta' \\ e^{i2\varphi} \sin^2 \frac{\theta'}{2} \end{pmatrix}, \tag{4.66}$$

and this wave function does not change the spin direction compared with the wave function Eq. 4.65. However, the singly quantized and doubly quantized vortices move to the $m_z = 0$ and $m_z = -1$ components, respectively, and the line singularity now appears at $\theta' = \pi$. The wave function Eq. 4.65 and Eq. 4.66 differ by a total phase $e^{i2\varphi}$, and therefore, their gauge fields differ by $2\hbar\hat{\varphi}/(r\sin\theta)$. This change of gauge field also moves the line singularity from $\theta' = 0$ to $\theta' = \pi$.

Experimentally, one can control the bias field B_{bias} to move the monopole position in and out of the condensate. When the monopole is below the condensate, the synthetic magnetic field at the condensate point to nearly the same direction, which results in a classical Hall effect from the Lorentz force [36]. When the monopole is gradually moved into the condensate, the singularity of the Dirac string can be observed by directly imaging the condensate [147].[4]

Finally, we should emphasize that the Dirac monopole is *not* the topological excitations we discussed above. Above, we have classified topological excitation by considering the mappings from the real space to the wave function space, but here the Dirac monopole refers to singularity in the synthetic gauge field. Since here we consider a ferromagnetic

---

[4] Here we should note that although the location of the Dirac string depends on the gauge choice, it can still be observed experimentally. This is because nearby the singularity, the spins rotate so fast that the adiabatic approximation breaks down.

condensate with $SO(3)$ degenerate manifold, and because $\Pi_2(SO(3))$ is trivial, this Dirac monopole is in fact not a topological stable object in terms of our definition of topological excitation discussed above. To see this, taking the wave function Eq. 4.65 as an example, we can construct a path $\beta = (1 - t)\theta' + t\pi$. When $t = 0$, the wave function is Eq. 4.65, but when $t = 1$, the wave function becomes $(0, 0, 1)^T$, which corresponds to a uniform wave function. Hence, the wave function for Dirac monopole can be continuously deformed into a topological trivial state.

4.5 Spin-Orbit-Coupled Condensate

In the spinor condensate discussed above, the kinetic energy is always the same for all spin components. In Section 1.3, we have discussed a scheme based on atom-light interaction to couple the spin degree of freedom with the spatial motion, which realizes the spin-orbit coupling effect as given by Eq. 1.51. Spin-orbit coupling has been extensively studied in the electronic systems in condensed matter physics. Realizing spin-orbit coupling in ultracold atomic systems, for the first time, introduces this effect into a Bose gas. In this section, we will focus on the effects of the spin-orbit coupling on Bose–Einstein condensate and superfluidity.

In Box 3.3, we have introduced the concepts of symmetry of wave function $\mathcal{G}^W$ and the symmetry of the Hamiltonian $\mathcal{G}^H$. We have introduced the concept of symmetry breaking for the situations that $\mathcal{G}^W$ is a subgroup of $\mathcal{G}^H$. In the Landau theory of phase transitions, two states, labeled by a and b, belong to different phases if their symmetry groups $\mathcal{G}^{W,a}$ and $\mathcal{G}^{W,b}$ are different. The order of phase transition between them is largely determined by their symmetry properties. The spin-orbit coupled BEC exhibits various symmetry-breaking phases and phase transitions. Below we will discuss the phase diagram of this system, and we will emphasize the role of symmetry in determining these phase transitions. Therefore, we will first discuss the symmetry of the Hamiltonian and the symmetry of possible phases.

Symmetry of the Hamiltonian. We first consider the symmetry of the single-particle Hamiltonian as we have described in Eq. 1.51:

$$\hat{H} = \frac{\hbar^2}{2m}(k_x - k_0\sigma_z)^2 + \frac{\hbar^2 \mathbf{k}_\perp^2}{2m} + \frac{\delta}{2}\sigma_z + \Omega\sigma_x. \tag{4.67}$$

The symmetry properties of this Hamiltonian include following aspects:

- This Hamiltonian possesses the spatial translational symmetry, and the momentum is a good quantum number.
- Because spin and momentum are locked, the Hamiltonian loses the spin rotational symmetry and also loses the spatial reflection symmetry $\hat{x} \rightarrow -\hat{x}$. But when $\delta = 0$, the system possesses a Z_2 symmetry of simultaneously reflecting both spin and space together, that is, $\hat{x} \rightarrow -\hat{x}$, $\hat{k} \rightarrow -\hat{k}$ and $\sigma_z \rightarrow -\sigma_z$. If $\delta \neq 0$, this symmetry is also explicitly broken.

- The system loses the Galilean invariance. When performing the Galilean transformation $\mathbf{r} \to \mathbf{r} - \mathbf{v}t$, the system acquires an additional term $-\hbar\mathbf{v} \cdot \mathbf{k}$. Normally this term can be further absorbed by performing a gauge transformation and the Hamiltonian is left invariant. But in this case, when applying the same procedure, it yields an extra term $\hbar k_0 v_x \sigma_z$, where v_x is the x-component of $\mathbf{v}$. Therefore, the Hamiltonian is no longer invariant. Physically, this is because of the spin-momentum locking, atoms in different frame have different momentum, and consequently, have different spin states.

Therefore, the single-particle Hamiltonian possesses the translational symmetry and a Z_2 symmetry when $\delta = 0$, and only translational symmetry for $\delta \neq 0$. Now we include the interaction between atoms, given by

$$\hat{V} = \int d^3\mathbf{r}[U_{\uparrow\uparrow}n_{\uparrow}^2(\mathbf{r}) + 2U_{\uparrow\downarrow}n_{\uparrow}(\mathbf{r})n_{\downarrow}(\mathbf{r}) + U_{\downarrow\downarrow}n_{\downarrow}^2(\mathbf{r})], \qquad (4.68)$$

where $U_{\uparrow\uparrow}$, $U_{\downarrow\downarrow}$ and $U_{\uparrow\downarrow}$ are the intracomponent and intercomponent interaction strengths, respectively. The interaction part is always invariant under spatial translation. However, since the "spin" here is in fact the pseudo-spin, and as we have discussed in Box 2.3, there is no spin rotational symmetry for interaction between different pseudo-spins. Hence, generally $U_{\uparrow\uparrow}$, $U_{\uparrow\downarrow}$ and $U_{\downarrow\downarrow}$ are all different. When we choose $U_{\uparrow\uparrow} = U_{\downarrow\downarrow} = U$, the interaction is also invariant under the Z_2 symmetry introduced above. Below we will always focus on this situation.

Symmetry of the Wave Function. Considering the spin-orbit coupled single-particle Hamiltonian Eq. 4.67, and the single-particle spectrum for $\delta = 0$ is shown in Figure 4.11(a). Here we mainly focus on the following three phases:

- When $\delta = 0$ and $\Omega < 4E_r$, where E_r denotes the recoil energy $\hbar^2 k_0^2/(2m)$, the single-particle ground state is doubly degenerate, and the two minima of the single-particle spectrum are located at $\pm k_{\min}$ ($k_{\min} \neq 0$). The wave functions of these two degenerate single-particle ground states are given by

$$|\psi_+\rangle = e^{ik_{\min}x}\begin{pmatrix} \sin\alpha \\ \cos\alpha \end{pmatrix}; \quad |\psi_-\rangle = e^{-ik_{\min}x}\begin{pmatrix} \cos\alpha \\ \sin\alpha \end{pmatrix}, \qquad (4.69)$$

where the parameter α depends on Ω/E_r. These two states have opposite momenta, and due to the spin-momentum locking, the spins of these two states are also polarized to opposite directions along $\hat{z}$. $|\psi_+\rangle$ and $|\psi_-\rangle$ are invariant under the spatial translation $x \to x+a$, under which the wave function only acquires a phase factor $e^{\pm ik_{\min}a}$. However, this wave function does not possess the Z_2 symmetry, because the Z_2 operation transfers $k_{\min}$ to $-k_{\min}$ and inverts the spin direction along $\hat{z}$. If bosons all condense into one of the minima, it is called a *plane wave state* (PW). This phase breaks the Z_2 symmetry but respects the spatial translational symmetry. When $\delta \neq 0$, the Hamiltonian also breaks the Z_2 symmetry, and these two states are no longer degenerate.

- This single-particle degeneracy can significantly modify the scenario of Bose–Einstein condensation, because any superposition of these two degenerate ground states has the same single-particle energy. Thus, it has to rely on the interaction energy to pick up a unique ground state. Now Let us consider a superposition as $\sin\beta e^{i\theta/2}|\psi_+\rangle +$

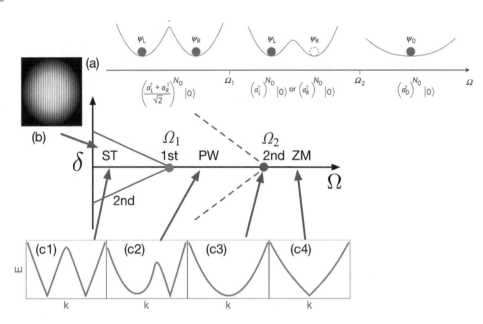

Figure 4.11 Schematic of the phase diagram of spin-orbit coupled bosons. Three different phases are plotted in terms of δ and Ω for the case $U_{\uparrow\downarrow}^2 < U^2$. "ST" stands for the stripe phase, "PW" stands for the finite momentum plane wave phase, and "ZM" stands for the zero-momentum phase. The solid line and the point at $\delta = 0$ and $\Omega = \Omega_2$ represent the second-order phase transition, and the points at $\delta = 0$ and $\Omega = \Omega_1$ represent a first-order phase transition. The dashed lines stand for crossover instead of phase transitions. Inset: (a) Schematically shows single-particle dispersion and three phases. (b) Density wave of the stripe phase. (c) Quasi-particle spectrums in the ST phase (c1), the PW phase (c2), at the transition point of $\delta = 0$ and $\Omega = \Omega_2$ (c3), and at the ZM phase (c4). Inset (a) is reprinted from Ref. [81], and inset (b) is reprinted from Ref. [74]. A color version of this figure can be found in the resources tab for this book at cambridge.org/zhai.

$\cos\beta e^{-i\theta/2}|\psi_-\rangle$ with $\sin\beta \neq 0,1$. The superposition of two states with different momenta leads to a spatial density modulation, and such a density modulation breaks the spatial translational symmetry, as shown in Figure 4.11(b). A Bose condensation into such a phase is first discovered in the Rashba spin-orbit coupled Bose gas and is named as the *stripe phase* (ST)[180]. When $\delta = 0$, $\sin\theta = \cos\theta = 1/\sqrt{2}$, this stripe phase respects the Z_2 symmetry. When $\delta \neq 0$, such a superposition state can still persist in certain parameter regime, but the Z_2 symmetry is broken.

- When $\delta = 0$ and Ω approaches $4E_r$ from below, $k_{\min}$ approaches zero. When $\Omega > 4E_r$, it becomes a single minimum at $k = 0$ without degeneracy. When bosons all condense into this state, we call it the *zero-momentum phase*(ZM). This phase respects the translational symmetry, and it respects the Z_2 symmetry. When $\delta \neq 0$, there is also only one single minimum at large Ω, but the location of the minimum is not at the zero momentum and the state does not obey the Z_2 symmetry if $\delta \neq 0$.

Symmetry and Phase Transitions. Now we discuss the basic structure of the phase diagram of this spin-orbit coupled Bose condensate [74, 109, 186]. The transitions between these phases can be either a first-order one or a second-order one. The order of phase transitions can be determined by symmetry and the general principles are as follows. Considering two phases and the symmetry group of these two phases denoted by $\mathcal{G}_1^W$ and $\mathcal{G}_2^W$, respectively, if $\mathcal{G}_1^W$ is a subgroup of $\mathcal{G}_2^W$, then generically, the transition between them is a second-order phase transition. If $\mathcal{G}_1^W$ is not a subgroup of $\mathcal{G}_2^W$, and if there is still a direct phase transition between them, then the transition must be a first order one. Such a first order transition is enforced by symmetry and is stable against perturbations. A phase diagram of spin-orbit coupled BEC is schematically shown in Figure 4.11, which can be precisely determined by minimizing the total energy. With this symmetry principle, we can determine the first order and the second order of phase boundaries as discussed below, and the orders of phase transitions are also marked in Figure 4.11.

- For $\delta = 0$, because the ZM phase has both spatial translational symmetry and Z_2 symmetry, and the PW phase only has spatial translational symmetry. Thus $\mathcal{G}_{PW}^W$ is a subgroup of $\mathcal{G}_{ZM}^W$, and therefore, the transition between them can be a second-order phase transition.

- For $\delta = 0$, because the ST phase preserves the Z_2 symmetry and breaks the spatial translational symmetry, and the PW phase preserves the spatial translational symmetry and breaks the Z_2 symmetry. Thus, $\mathcal{G}_{ST}^W$ and $\mathcal{G}_{PW}^W$ are not mutually subgroup of each other, and therefore, when there is a direct transition between them, it has to be a first-order one.

- For $\delta \neq 0$, the Hamiltonian does not have the Z_2 symmetry, and consequently, both the PW states and the ZM states do not have the Z_2 symmetry. Thus, there is no symmetry distinction between the PW state and the ZM states, and therefore, there is no phase transition between them.

- For $\delta \neq 0$, both $\mathcal{G}_{ST}^W$ and $\mathcal{G}_{PW}^W$ loss the Z_2 symmetry. Hence, $\mathcal{G}_{ST}^W$ has no symmetry and $\mathcal{G}_{PW}^W$ still has spatial translational symmetry. Thus, $\mathcal{G}_{ST}^W$ is a subgroup of $\mathcal{G}_{PW}^W$, and therefore, the transition between them can be a second-order one.

This phase diagram has also been observed in the first spin-orbit coupled Bose gas experiment using ^{87}Rb atoms [114], as shown in Figures 4.12(a)–(c). In this experiment, they show that in the ST phase regime, two spin components are mixed, and in the PW phase regime, two spin components tend to separated [114]. This result is shown in Figure 4.12(c). Another evidence of the stripe phase is provided by magnetization histogram [81], as shown in the inset of Figure 4.12(d). For the PW phase, bosons either condense into $|\psi_+\rangle$ or condense into $|\psi_-\rangle$, thus, the magnetization histogram shows two peaks at opposite magnetizations. Instead, for the ST phase, the magnetization histogram shows a Gaussian distribution centered at zero. This feature has also been observed and has been used to determine the phase diagram at finite temperature [81], as shown in Figure 4.12(d). Finally a direct evidence of the ST phase is observing the density wave structure that breaks spatial translational symmetry. In a solid, the density order of a crystal can be observed by the Bragg spectroscopy, and similar Bragg spectroscopy experiment can also be performed in ultracold atom experiments [107]. This provides a direct evidence of the ST phase, as shown in Figure 4.12(e).

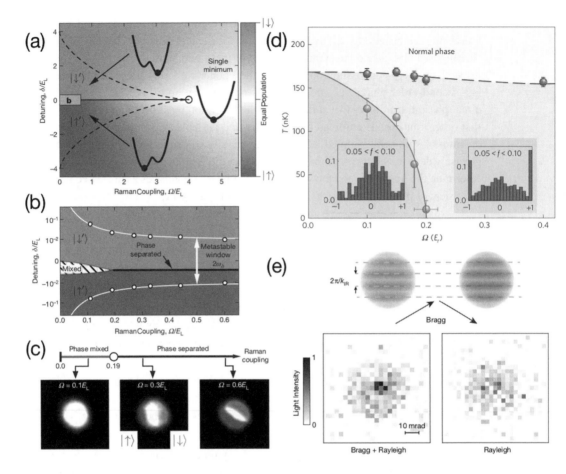

Figure 4.12 Experiments on spin-orbit coupled Bose condensate. (a) Observed phase diagram of the spin-orbit coupled condensate in terms of δ/E_r and Ω/E_r. (b) A zoom-in of (a) for small δ/E_r and small Ω/E_r. (c) The evidence from mixed two-spin components to separated two-spin components, as Raman coupling Ω increases. (d) The finite temperature phase diagram of the spin-orbit coupled bosons in terms of temperature and Raman coupling. The inset shows two typical histograms that are used for distinguishing ST and PW phases. (e) Bragg spectroscopy shows a peak in the ST phase, which is strong evidence of breaking translational symmetry. (a–c) are reprinted from Ref. [114], (d) is reprinted from Ref. [81], and (e) is reprinted from Ref. [107]. A color version of this figure can be found in the resources tab for this book at cambridge.org/zhai.

Symmetry not only can determine the order of phase transitions, but also can impose strong constraints on the low-energy excitation. The low-energy gapless modes are usually associated with the fact that the ground state breaks continuous symmetry. Usually a Bose condensate breaks the $U(1)$ phase symmetry, and therefore, there is one linear gapless mode that is the Bogoliubov phonon mode discussed in Section 3.3. For instance, the quasi-particle spectrum in the ZM phase displays such a linear gapless mode as shown in (c4) of Figure 4.11. In the ST phase, the Bose condensate not only breaks the $U(1)$ phase symmetry, but also breaks the spatial translational symmetry. Spatial translational symmetry is another continuous symmetry. Breaking spatial translational symmetry leads to another gapless mode. This is essentially the same as the gapless acoustic phonon mode in a crystal, which is also caused by breaking the spatial translational symmetry. More explicitly, when we write the wave function of a ST phase as $\sin \beta e^{i\theta/2}|\psi_+\rangle + \cos \beta e^{-i\theta/2}|\psi_-\rangle$, varying θ corresponds to sliding the density wave spatially, which does not cause energy as long as θ remains spatially uniform. Therefore, the ST phase exhibits two gapless linear modes, as shown in (c1) of Figure 4.11.

For the PW phase, say, if bosons are condensed in $|\psi_+\rangle$, then there exists a mode with linear dispersion and being gapless at k_{min}. Meanwhile it exhibits a local minimum at $-k_{min}$ with a finite gap, as shown in Figure 4.11(c2) [195]. This is known as the *roton* structure in the excitation spectrum. Similarly, roton structures have also been found in the liquid helium, and in dipolar Bose condensate and Bose condensate in cavity. There is a common physics behind the roton structure in different systems. Although the state itself is not a crystal and does not break the spatial translational symmetry, there is a strong crystallization tendency in this phase. Therefore, the excitation at the corresponding crystal momentum displays a minimum. When Ω decreases and the system approaches the transition to the ST phase, the gap at $-k_{min}$ becomes smaller and smaller. At the transition, the roton gap vanishes, indicating an instability of the PW phase toward forming the density wave order. The roton structure and vanishing roton gap at the transition have been successfully observed experimentally with the help of the Bragg spectroscopy [82]. Thus, at the zero-temperature phase transition between these two phases, the low-energy excitation displays two linear gapless modes on the ST phase side, and displays one linear mode and a quadratic mode with vanishing small gap on the PW phase side. Comparing these two low-lying excitations, the latter has larger density-of-state. Therefore, at low temperature, the PW phase acquires more entropy and is more favorable. This explains the experimental measurement shown in Figure 4.12(d) that the ST phase is suppressed and the PW phase is expanded as the temperature increases [81]. This gives an example that symmetry consideration can also help in determining the finite temperature phase diagram.

Finally, let us comment on a physical consequence of the absence of the Galilean invariance. In Section 3.2, we have discussed the critical velocity for superfluidity. We consider an impurity moving in the superfluid with velocity $\mathbf{v}$, and we have discussed that the Landau criterion of the critical velocity is determined by whether this moving impurity can excite quasi-particles in a static superfluid. Alternatively, another way to discuss the critical velocity is to consider a superfluid moving in the presence of static impurities. With the Galilean invariance, these two critical velocities are equivalent. For this spin-orbit coupled system without the Galilean invariance, these two critical velocities are no longer equal [195]. The difference is most significant at the transition point from the PW phase to the ZM phase, where the two degenerate minima at finite momenta merge into a single minimum at $k = 0$ [195]. At this critical point, the single-particle dispersion behaves as k_x^4 around its minimum, and according to Eq. 3.55, the quasi-particle dispersion behaves as k_x^2, as shown in Figure 4.11(c3). Then, according to the Landau criterion Eq. 3.31, the Landau critical velocity vanishes. This critical velocity is for a moving impurity in a static superfluid, however, the critical velocity for a moving superfluid is different [195]. For a moving superfluid, the Hamiltonian acquires an extra term $\hbar k_0 v_x \sigma_z$ as discussed above in the comoving frame, because of which the k_x^4 dispersion no longer exists. Thus, in that case, the critical velocity is always finite.

Exercises

4.1 Considering a quasi-one-dimensional condensate along $\hat{z}$, show that the following two cases give the same phase difference between two ends. In one case, there is a soliton located at $z = 0$ plane, and in the other case, there is a vortex and antivortex pair in the xy plane with $z = 0$. Estimate and compare their energies.

4.2 Let $\{\hat{\mathbf{n}}_i, i = 1, \ldots, 2F\}$ be $2F$ Majorana stellar points for state $|\psi\rangle$. Show that (i) if one applies a spin rotation $\hat{\mathcal{U}}$ to state $|\psi\rangle$, the Majorana stellar representation of $\hat{\mathcal{U}}|\psi\rangle$ corresponds to a $SO(3)$ rotation of all $\hat{\mathbf{n}}_i$ vectors simultaneously, and (ii) if one applies a time-reversal operator $\hat{\mathcal{R}}$ to state $|\psi\rangle$, $\hat{\mathcal{R}}|\psi\rangle$ are described by $2F$ points of $-\hat{\mathbf{n}}_i$ as a central reflection of the original vectors.

4.3 Calculate the Majorana stellar representation for states (i) $(1, 0, 0)$; (ii) $(0, 0, 1)$; (iii) $(0, 1, 0)$; (iv) $\frac{1}{\sqrt{2}}(1, 0, e^{i\phi})$; (v) $(1/\sqrt{3}, 0, 0, \sqrt{2/3}, 0)$, and (vi) $(\frac{1}{2}, 0, -\frac{i}{\sqrt{2}}, 0, \frac{1}{2})$.

4.4 Show that $\langle \zeta|\mathbf{F}|\zeta\rangle^2$ is invariant under a $SU(2)$ rotation of $|\zeta\rangle$.

4.5 Consider the Hamiltonian $\hat{H}_s = -h\mathbf{B} \cdot \mathbf{F}$, where $\mathbf{F}$ is the Pauli matrix for a spin-1 atom and $\mathbf{B} = (\sin\beta\cos\alpha, \sin\beta\sin\alpha, \cos\beta)$. Show (1) the explicit forms of $\mathcal{U}(\alpha, \beta)$ that diagonalize $\hat{H}_s$; (2) that the gauge field from this $U(\alpha, \beta)$ is given by Eq. 4.62; (3) Eq. 4.63 and Eq. 4.64 for $\alpha = \varphi$ and $\beta = \pi - \theta$ (θ and φ are the polar and the azimuthal angles in the spherical coordinate).

4.6 Considering the spin-orbit coupled Hamiltonian Eq. 4.67 with $\delta = 0$, compute the single-particle dispersion and the spin polarization at different momentums. Determine the critical coupling Ω_c at which the single-particle ground state changes from doubly degenerate to single minimum.

4.7 Considering the spin-orbit coupled Hamiltonian Eq. 4.67 with $\delta = 0$, in the regime when the single-particle ground states are doubly degenerate as given by Eq. 4.69, compare the interaction energy between ψ_+ or ψ_- and $(\psi_+ + \psi_-)/\sqrt{2}$, with the interaction part of the Hamiltonian given by Eq. 4.68 and $U_{\uparrow\uparrow} = U_{\downarrow\downarrow} = U$.

Part III

Degenerate Fermi Gases

5

The Fermi Liquid

Learning Objectives

- Introduce density distribution and momentum distribution of noninteracting Fermi gas.
- Discuss the transport property of a Fermi gas through a quantum point contact.
- Discuss the Fermi surface as a key of a noninteracting Fermi gas and its generalization to a Fermi liquid.
- Introduce the effective mass and the quasi-particle residue as two key parameters of the Fermi liquid.
- Introduce the self-energy and the universal relations between the effective mass, the quasi-particle residue, and the self-energy.
- Discuss the Fermi polaron as an example of the Fermi liquid, and show these universal relations with this example.
- Discuss the divergence of the effective mass and vanishing of quasi-particle residue when Fermi liquid description fails.
- Introduce radio-frequency spectroscopy and how it measures the polaron energy and the quasi-particle residue.
- Summarize different spectroscopy measurements in ultracold atomic systems.

5.1 Free Fermions

Momentum and Density Distributions. For a noninteracting Fermi gas, each plane wave state with wave vector $\mathbf{k}$ is a single-particle eigenstate, and the many-body eigenstate can be labeled by the occupation $n(\mathbf{k})$ on each of these plane wave states, with $0 \leqslant n(\mathbf{k}) \leqslant 1$. For the ground state, all fermions fill the Fermi sea up to the Fermi energy, and the many-body wave function is given by

$$|\text{FS}\rangle = \prod_{k<k_{\text{F}}} c_{\mathbf{k}}^{\dagger}|0\rangle, \tag{5.1}$$

where $\hat{c}_{\mathbf{k}}^{\dagger}$ and $\hat{c}_{\mathbf{k}}$ are creation and annihilation operators of fermions at momentum $\mathbf{k}$, and $\hbar k_{\text{F}}$ is the Fermi momentum. For this ground state, $n_0(\mathbf{k}) = 1$ for $k < k_{\text{F}}$ and $n_0(\mathbf{k}) = 0$ for $k > k_{\text{F}}$, as shown in Figure 5.1(a).

Considering excitations above the Fermi sea, one can either add a particle above the Fermi sea or take a particle away below the Fermi sea, which are called the *particle excitation* and the *hole excitation*, respectively. The wave function for the particle excitation is

$\hat{c}_{k>k_F}^\dagger|FS\rangle$, whose excitation energy is given by $\mathcal{E}_k = \epsilon_k - \mu$, where $\epsilon_k = \hbar^2 k^2/(2m)$ and $\mu = \hbar^2 k_F^2/(2m)$. The wave function for the hole excitation is $\hat{c}_{k<k_F}|FS\rangle$, whose $E_k = \mu - \epsilon_k$. Hence, the excitation energy is always positive and can be given by

$$\mathcal{E}_k = |\epsilon_k - \mu|. \tag{5.2}$$

For a general deviation from the ground state, the excitation energy is measured by the deviation from the equilibrium momentum distribution $n_0(\mathbf{k})$, that is, $\delta n(\mathbf{k}) = n(\mathbf{k}) - n_0(\mathbf{k})$, and the change of the total energy is given by

$$\delta E = \sum_{\mathbf{k}} (\epsilon_{\mathbf{k}} - \mu)\, \delta n(\mathbf{k}). \tag{5.3}$$

δE is therefore always positive.

At finite temperature and in the presence of the harmonic trap $V(\mathbf{r})$, it can be shown that when the total fermion number N is very large, the semiclassical approximation can be safely applied and the distribution is given by

$$f(\mathbf{r}, \mathbf{k}) = \frac{1}{e^{\beta(\epsilon_{\mathbf{k}} + V(\mathbf{r}) - \mu)} + 1}, \tag{5.4}$$

where $\beta = 1/(k_B T)$. And the local density $n(\mathbf{r})$ is given by

$$n(\mathbf{r}) = \frac{1}{(2\pi)^3} \int d^3 k f(\mathbf{r}, \mathbf{k}). \tag{5.5}$$

This is nothing but the local density approximation we have discussed in Section 3.2, by which we can replace μ as $\mu - V(\mathbf{r})$ in the equation-of-state for a uniform system. The chemical potential can be determined by the conservation of the total number of atoms as

$$N = \int d^3 r n(\mathbf{r}). \tag{5.6}$$

Here are a few remarks about this distribution:

- At zero temperature, the density will become

$$n(\mathbf{r}) \propto (\mu - V(\mathbf{r}))^{3/2} \tag{5.7}$$

for $\mu > V(\mathbf{r})$ and $n(\mathbf{r}) = 0$ for $\mu < V(\mathbf{r})$. This density distribution actually is similar as the density distribution of interacting bosons, which is proportional to $\mu - V(\mathbf{r})$.

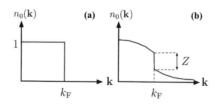

Figure 5.1   Schematic of the momentum distribution. The momentum distribution for (a) the noninteracting Fermi gas and (b) the Fermi liquid. Here k_F labels the Fermi wave vector, and Z labels the quasi-particle residue.

However, the physical difference is that, in that case of interacting bosons, the density distribution is determined by the balance between the interaction energy and the potential energy. And in this case of noninteracting fermions, the density distribution is determined by the balance between the kinetic energy and the potential energy.

- At zero temperature, we can determine $\mu = (6N)^{1/3}\hbar\bar{\omega}$, where $\bar{\omega} = (\omega_x\omega_y\omega_z)^{1/3}$ and ω_i ($i = x, y, z$) is the trapping frequency along direction-i. This dependence can be understood intuitively in the following way. For chemical potential μ, the radius of the cloud along direction-i will be $R_i = \sqrt{2\mu/(m\omega_i^2)}$. Therefore, roughly speaking, the density will be

$$n \propto \frac{N}{R_x R_y R_z} \propto \frac{Nm^{3/2}\bar{\omega}^3}{\mu^{3/2}}, \tag{5.8}$$

hence,

$$\mu \propto \frac{\hbar^2}{m}n^{2/3} \propto \frac{\hbar^2\bar{\omega}^2 N^{2/3}}{\mu}. \tag{5.9}$$

Solving Eq. 5.9, it leads to $\mu \propto N^{1/3}\hbar\bar{\omega}$.

- In an anisotropic trap, although the density distribution is anisotropic, the momentum distribution obtained by

$$n(\mathbf{k}) = \int d^3\mathbf{r} f(\mathbf{r}, \mathbf{k}) \tag{5.10}$$

is always isotropic. This is similar as the thermal gas of bosons above Bose condensation temperature. Thus, in the time-of-flight measurement, the aspect ratio of the cloud should finally approach unity after sufficiently long expansion time. As we have discussed in Section 3.2, hydrodynamical expansion leads to an anisotropic expansion with inverted aspect ratio. Therefore, the time-of-flight expansion can distinguish a free Fermi gas from a Fermi system with hydrodynamic behavior, and the latter can be either due to superfluidity or due to strong interactions.

- With the distribution $n(\mathbf{k})$ one can also extract the kinetic energy per particle E_K from the time-of-flight measurement as

$$E_K = \frac{V}{N}\frac{1}{(2\pi)^3}\int d^3\mathbf{k}\frac{\hbar^2\mathbf{k}^2}{2m}n(\mathbf{k}). \tag{5.11}$$

At high temperature, E_K is proportional to $k_B T$, and at low temperature, it approaches a constant due to the Fermi statistics. Thus that $E_K/(k_B T)$ deviates from a constant is taken as an evidence for the onset of the Fermi degeneracy, which has been observed experimentally [47] soon after the Fermi gas of atoms has been cooled to quantum degeneracy in 1999 [46]. The experimental results are shown in Figure 5.2.

Quantized Conductance through a Quantum Point Contact. Transport measurements play an important role in studying Fermi gases. In condensed matter physics, most of the states are characterized by and named after the transport properties of electrons, such as conductor, insulator and superconductor. Various kind of transport experiments have also been carried out in ultracold atom experiments, such as hydrodynamics expansion

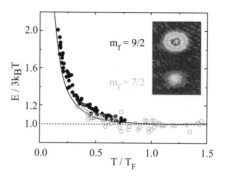

m_f = 9/2

Figure 5.2 Kinetic energy versus temperature. The kinetic energy of a noninteracting Fermi gas is measured at different temperatures. Here the kinetic energy is measured by releasing atoms from the trap and performing the time-of-flight measurement. Reprinted from Ref. [47]. A color version of this figure can be found in the resources tab for this book at cambridge.org/zhai.

dynamics discussed in Section 3.2, Bloch oscillation driven by an external forces to be discussed in Section 7.1, as well as diffusion dynamics and loss induced transport. In general, these transport dynamics can also be classified into two categories. One is near-equilibrium dynamics driven by small deviation from the equilibrium state, which usually probes the equilibrium properties. The other is far-from-equilibrium dynamics, which sometimes can also reveal properties in equilibrium, such as the quench dynamics of a topological band that we will discuss in Section 7.3. And in many other situations, the far-from-equilibrium dynamics can also reveal novel physics with no correspondence at equilibrium.

Here we focus on a near-equilibrium transport and consider a typical setup in condensed matter systems, called the *two-terminal transport*. This kind of transport measurement now can also be realized in ultracold atom systems [96]. In this setup, two reservoirs, denoted by the left and the right reservoirs, are connected by a one-dimensional tunneling channels, as shown in Figure 5.3. In this case, each reservoir is a noninteracting three-dimensional Fermi gas. Here two reservoirs have different chemical potentials, denoted by μ_L and μ_R, respectively. Without loss of generality, we assume $\mu_L > \mu_R$. Hence, driven by this potential difference, particles can flow from the left reservoir to the right reservoir, and the current is given by the Landauer–Büttiker formalism as [96]

$$I = \sum_n \int dE v(E) g(E) \mathcal{T}_n(E) [f(E, \mu_L) - f(E, \mu_R)]. \tag{5.12}$$

Here the index n labels the tunneling channels, and $\mathcal{T}_n(E)$ denote the transmission coefficients for each channel. Generally, $0 < \mathcal{T}_n(E) < 1$. In ideal situation, for ballistic transmission, we have $\mathcal{T}_n(E) = 1$ for an open channel.[1] Because of the Fermi statistics, in order to contribute to the current from the left to the right reservoirs, it has to satisfy two conditions: i) the states with energy E has to be occupied in the left reservoir, and ii) the states with energy E has to be empty in the right reservoir. The energy window that satisfy

---

[1] Here the meanings of open channel and closed channel are different from those used in the discussion of Feshbach resonances.

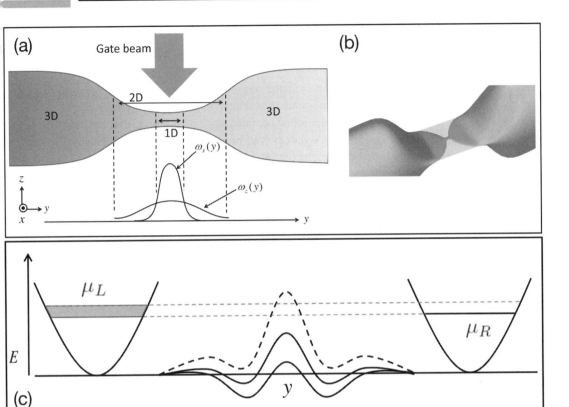

Figure 5.3   Schematic of two-terminal transport. (a–b) Schematic of quantum point contact. In this case, a quantum point
contact is made of a confinement potential along $\hat{x}$, a confinement potential along $\hat{z}$, and a gate potential. (c)
Schematic of the Landauer–Büttiker formalism for transport of fermions between the left and the right reservoirs. A
color version of this figure can be found in the resources tab for this book at cambridge.org/zhai.

these two conditions is marked by the shaded area in Figure 5.3(c), and at low temperature
with $\Delta\mu \gg k_B T$, it is counted by the factor $f(E, \mu_L) - f(E, \mu_R)$ in Eq. 5.12. Here, $v(E)$
denotes the group velocity and $g(E)$ denotes the density-of-state. In one dimension, since
$v(E) = (1/\hbar)\partial E/\partial k$ and $g(E) = 1/(2\pi|\partial E/\partial k|)$, we obtain $v(E)g(E) = 1/h$, and Eq. 5.12
is written as

$$I = \frac{1}{h}\sum_n \int dE\, T_n(E)[f(E, \mu_L) - f(E, \mu_R)]. \qquad (5.13)$$

In general, the transmission coefficient has to be determined by solving the Schrödinger
equation for each tunneling channel. Here we are interested in a specifically designed tunneling channel called *quantum point contact* (QPC), where $T_n(E)$ can reach nearly unity
for all open channels. In ultracold atom systems, such a QPC is made of three laser beams,
and the potential is given by [96, 95]

$$V_{QPC} = \frac{1}{2}m\omega_z^2(y)z^2 + \frac{1}{2}m\omega_x^2(y)x^2 - V_g f_g(y). \qquad (5.14)$$

Here we have set $\hat{y}$ direction as the transport direction. The first term in Eq. 5.14 creates a confinement potential along $\hat{z}$, such that the system becomes a two-dimensional one. The confinement is only applied to the central regime around $y = 0$, and $\omega_z = \omega_z^0 e^{-y^2/l_z^2}$. The second term in Eq. 5.14 creates another confinement potential along $\hat{x}$, and $\omega_x = \omega_x^0 e^{-y^2/l_x^2}$. With the first two terms, a one-dimensional tunneling channel is created. Here $l_z > l_x$. That is to say, the tunneling channel is first confined into a two-dimensional one, and then further confined into a one-dimensional one, as shown in Figures 5.3(a)–(b). Finally, the last term in Eq. 5.14 is called the gate potential, which can tune the relative energy between the tunneling channels and the reservoirs. It is also a focused Gaussian potential around $y = 0$ with $f_g = e^{-y^2/l_g^2}$ and $l_g \approx l_z$.

With the semiclassical approximation along $\hat{y}$, the potential Eq. 5.14 gives rise to a set of discrete tunneling channel and the potential curves along $\hat{y}$ are given by

$$V_{mn}(y) = \left(m + \frac{1}{2}\right)\hbar\omega_z^0 e^{-y^2/l_z^2} + \left(n + \frac{1}{2}\right)\hbar\omega_x^0 e^{-y^2/l_x^2} - V_g e^{-y^2/l_g^2}, \qquad (5.15)$$

which exhibit maximum at $y = 0$ as schematically shown in the central regime of Figure 5.3(c). The design of the QPC potential as described above is to ensure following three conditions:

- The tunneling channel is one-dimensional. This ensures $v(E)g(E) = 1/h$ as discussed above.
- The tunneling channel is discrete. Here the transverse modes have discrete energy levels, and the level separation has to be larger than temperature, for otherwise the discreteness is smeared out by temperature effect.
- The backward scattering is forbidden. This requires that the potential curve along $\hat{y}$ should be smooth enough and

$$\frac{\left\langle\left(\frac{\partial V_{mn}^2(y)}{\partial y^2}\right)\right\rangle}{\langle V_{mn}(y)\rangle} \ll k_F^2. \qquad (5.16)$$

As we discussed above, the incoming fermions are mainly from the Fermi surface. Thus, when Eq. 5.16 is satisfied, the potential can hardly scatter the incoming fermions to the backward direction. Therefore, we can show that the transmission coefficient $T(E) \approx 1$ if $E > V_{mn}(0)$, and these tunneling channels are called open channel. And $T(E) \approx 0$ if $E < V_{mn}(0)$, and these tunneling channels are called closed channels.

With these three conditions satisfied, Eq. 5.13 becomes

$$I = \frac{1}{h}\sum_{mn}\int dE\,\Theta(E - V_{mn}(0))[f(E, \mu_L) - f(E, \mu_R)]. \qquad (5.17)$$

where $\Theta(x)$ is the Theta-function with $\Theta(x) = 1$ when $x > 0$ and $\Theta(x) = 0$ when $x < 0$. In the zero-temperature limit, $f(E, \mu)$ approaches $\Theta(\mu - E)$, and therefore $\int dE[f(E, \mu_L) - f(E, \mu_R)] = \Delta\mu$, where $\Delta\mu = \mu_L - \mu_R$ is the bias voltage. Hence, we reach the quantization of conductance

$$G = \frac{I}{\Delta\mu} = \frac{1}{h}N_{open}, \qquad (5.18)$$

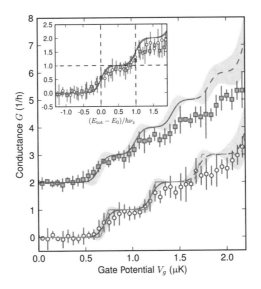

Experiment on quantized conductance. The conductance through a quantum point contact as a function of gate voltage. The gate voltage tunes the number of tunneling channels. The observed conductance is quantized in units of $1/h$. Reprinted from Ref. [95]. A color version of this figure can be found in the resources tab for this book at cambridge.org/zhai.

where N_{open} denotes the number of open channels and N_{open} can be tuned by varying gate voltage V_{g}. The experimental results are shown in Figure 5.4, where the quantization of conductance in unit of $1/h$ has been observed. Hence, we have demonstrated that when these three conditions are satisfied, the conductance is quantized. In Section 7.3, we will show that these three conditions can also be satisfied in a two-dimensional topological band insulator, which lead to quantized Hall conductance known as the *Quantum Anomalous Hall Effect*.

Concepts of the Fermi Liquid Theory. In Section 3.1, in order to generalize the concept of BEC from noninteracting to interacting boson system, we first discuss BEC in a non-interacting Bose gas, from which we identify the macroscopic occupation as the essential defining property of a BEC that can be generalized to an interacting system. Here, we also first study the noninteracting Fermi gas, and the purpose is also to identify a defining property that can be generalized to interacting cases. Here the defining property is the *Fermi Surface*, and the generalization is called the *Landau's Fermi liquid*. In a uniform Fermi liquid, the quasi-particles are fermionic with a well-defined wave vector **k**. The notion of Fermi surface remains, and it is defined through two features: i) The momentum distribution $n_0(\mathbf{k})$ of the ground state displays a discontinuity at the Fermi surface, as shown in Figure 5.1(b), although the jump of $n_0(\mathbf{k})$ is not unity. The discontinuity in $n_0(\mathbf{k})$ is called the *quasi-particle residue* Z, and $0 < Z < 1$ for the Fermi liquid. ii) The quasi-particle excitations are gapless at the Fermi surface. We emphasize that although the momentum distribution and the excitation spectrum appear as two different quantities, the discontinuity of momentum distribution and the gapless excitation spectrum occur simultaneously at the Fermi surface. These two are defining properties of a Fermi liquid. If we alter $n_0(\mathbf{k})$ by $\delta n(\mathbf{k})$, the change of total energy, to the first order of $\delta n(\mathbf{k})$, is given by

$$\delta E = \sum_{\mathbf{k}} \mathcal{E}_{\mathbf{k}} \delta n(\mathbf{k}), \tag{5.19}$$

where $\mathcal{E}_{\mathbf{k}} = \delta E/\delta n(\mathbf{k})$ is introduced as energy of the quasi-particle with wave vector $\mathbf{k}$. The quantized conductance through a QPC also holds when the reservoirs are Fermi liquids instead of noninteracting fermions.

Similar as the noninteracting fermions, in an isotropic system, we can define the velocity of the quasi-particle as

$$v_k = \frac{1}{\hbar} \frac{\partial \mathcal{E}_{\mathbf{k}}}{\partial k}, \tag{5.20}$$

and we can further define an *effective mass* m^* as

$$v_k = \frac{\hbar k}{m^*}. \tag{5.21}$$

In an interacting system, in general, $\mathcal{E}_{\mathbf{k}}$ is not a simple parabolic function, thus, m^* should depend on k. At finite density, we shall be particularly interested in the m^* nearby the Fermi momentum. If the system is anisotropic, m^* is in general a tensor. The effective mass m^* and the quasi-particle residue Z are two of the most basic parameters for describing a Fermi liquid, which are known as the *Fermi liquid parameters*. These Fermi liquid parameters largely determine the low-energy properties of a Fermi liquid. For instance, since the density-of-state nearby the Fermi surface can be straightforwardly calculated as

$$\mathcal{D}(E_{\mathrm{F}}) = \frac{V k_{\mathrm{F}} m^*}{2\pi^2 \hbar^2}, \tag{5.22}$$

one can obtain that the low-temperature specific heat only depends on effective mass as

$$C_{\mathrm{v}} = \frac{V k_{\mathrm{B}}^2 k_{\mathrm{F}} T m^*}{3\hbar^2}. \tag{5.23}$$

Expanding δE to the next order of $\delta n(\mathbf{k})$, it gives

$$\delta E = \sum_{\mathbf{k}} \mathcal{E}_{\mathbf{k}} \delta n(\mathbf{k}) + \frac{1}{2V} \sum_{\mathbf{k}\mathbf{k}'} f(\mathbf{k}, \mathbf{k}') \delta n(\mathbf{k}) \delta n(\mathbf{k}'), \tag{5.24}$$

where the second term is interpreted as interaction between quasi-particles. There are other Fermi liquid parameters that describe interaction between quasi-particles. In the next Sec 5.2, we will not discuss these parameters in detail and focus mainly on m^* and Z.

Finally, we should note that not all interacting Fermi systems satisfy the condition of a Fermi liquid. That is to say, there are other possible phases for an interacting Fermi system, as we summarized in Box 5.1. When the system approaches a transition at which the Fermi liquid description fails, m^* diverges and Z vanishes.

Here we summarize several possible phases of interacting Fermi gases at low temperature. In this section we have discussed one possibility of being a Fermi liquid, which is characterized by the existence of the Fermi surface defined by discontinuity in momentum distribution and gapless excitation. Another possibility is developing orders and breaking certain symmetry. In Chapter 6, we will introduce BCS pairing order as one such example. In the discussion of the Fermi–Hubbard model in Section 8.2, we will introduce other orders like the spin-density-wave and the charge-density-wave. At last, there is the third possibility, that is, the system neither satisfies the conditions of a Fermi liquid nor breaks any symmetry. This situation is called a non-Fermi liquid. Usually there is an absence of well-defined quasi-particle description in the non-Fermi liquid phase. Possible non-Fermi liquid states include interacting fermions in one dimension, a unitary Fermi gas above superfluid transition discussed in Section 6.2, and the Fermi–Hubbard model slightly away from half-filling discussed in Section 8.2.

5.2 Fermi Polaron

Here we consider a concrete example of the Fermi liquid. We will introduce an important concept called the *self-energy*, and the self-energy is a function of momentum $\mathbf{p}$ and energy ω, denoted by $\Sigma(\mathbf{p}, \omega)$. For a Fermi liquid, both the effective mass and the quasi-particle residue Z are determined from Σ as

$$\frac{m^*}{m} = \frac{1 - \frac{\partial \Sigma}{\partial \omega}}{1 + \frac{\partial \Sigma}{\partial \mathbf{p}^2/(2m)}} \tag{5.25}$$

and

$$\frac{1}{Z} = \left(1 - \frac{\partial}{\partial \omega} \Sigma(\mathbf{p}, \omega)\right). \tag{5.26}$$

It can be proved that these two relations hold for all Fermi liquid states. Here we will not discuss the general proof, instead, and we will show them in a specific example of the Fermi polaron.

Let us first consider a fully spin polarized Fermi gas in the thermodynamic limit with all atoms in the spin-up state. Only considering the *s*-wave interaction between fermions, the spin polarized Fermi gas is a noninteracting one. Now let us add one spin-down into this fully polarized Fermi sea, this spin-down atom can interact with all other spin-up atoms by the *s*-wave interaction and it will form a dressed state due to this interaction. This dressed state is called the *Fermi polaron*. In condensed matter physics, polaron is a concept first proposed by Landau, which describes a fermionic quasi-particles of an electron dressed up by the cloud of phonons. Here it is also a fermionic quasi-particle, but it describes fermions of spin-down fermion dressed up by the spin-up fermions. Hence it is called Fermi polaron to emphasize the difference that the constitution of the dressing cloud are fermions instead of phonons.

The Hamiltonian of this single impurity problem is written as

$$\hat{H} = \sum_{\mathbf{k}\sigma} \epsilon_{\mathbf{k}\sigma} \hat{c}^{\dagger}_{\mathbf{k}\sigma} \hat{c}_{\mathbf{k}\sigma} + \frac{g}{V} \sum_{\mathbf{k},\mathbf{k}',\mathbf{q}} \hat{c}^{\dagger}_{\mathbf{k}+\mathbf{q}\uparrow} \hat{c}^{\dagger}_{-\mathbf{k}+\mathbf{q}\downarrow} c_{-\mathbf{k}'+\mathbf{q}\downarrow} c_{\mathbf{k}'+\mathbf{q}\uparrow}, \tag{5.27}$$

where $\hat{c}^{\dagger}_{\mathbf{k}\sigma}$ and $\hat{c}_{\mathbf{k}\sigma}$ are the creation and annihilation operators of fermions with wave vector $\mathbf{k}$ and spin σ, and g obeys the renormalization condition that

$$\frac{1}{g} + \frac{1}{V} \sum_{\mathbf{k}} \frac{1}{2\epsilon_{\mathbf{k}}} = \frac{m}{4\pi \hbar^2 a_{\mathrm{s}}}. \tag{5.28}$$

Here, we use $\epsilon_{\mathbf{k}} = \hbar^2 k^2/(2m)$ to denote the free particle dispersion. Since the spin-up is in the thermodynamic limit and the spin-down is a single atom, hence, we distinguish them as $\epsilon_{\mathbf{k}\uparrow} = \epsilon_{\mathbf{k}} - \mu$, where $\mu = \hbar^2 k_{\mathrm{F}}^2/(2m)$, and $\epsilon_{\mathbf{k}\downarrow} = \epsilon_{\mathbf{k}}$.

Variational Wave Function. First we consider that the total momentum of this single impurity system is zero. The polaron wave function consists of the single spin-down atom and the particle-hole pairs on top of the Fermi sea of the spin-up atoms. Here we use a trial wave function [33]

$$|\Psi\rangle_{\mathbf{p}=0} = \psi_0 \hat{c}^{\dagger}_{\mathbf{p}=0\downarrow} |\mathrm{FS}\rangle + \sum_{|\mathbf{k}|>k_{\mathrm{F}},|\mathbf{q}|<k_{\mathrm{F}}} \psi_{\mathbf{k},\mathbf{q}} \hat{c}^{\dagger}_{\mathbf{q}-\mathbf{k}\downarrow} \hat{c}^{\dagger}_{\mathbf{k}\uparrow} \hat{c}_{\mathbf{q}\uparrow} |\mathrm{FS}\rangle, \tag{5.29}$$

with normalization condition $|\psi_0|^2 + \sum_{\mathbf{kq}} |\psi_{\mathbf{kq}}|^2 = 1$. Here the first term describes simply adding a spin-down atom without disturbing spin-up atoms, and the second term describes that the spin-down atom excites one particle-hole pair on top of the Fermi sea. In principle, this spin-down atom can also excite more particle-hole pairs in the Fermi sea. Here, for simplicity, we truncate the wave function by only including one particle-hole pair, which has been justified as a good approximation [38].

The total energy of this variational wave function is given by

$$\mathcal{E} = \sum_{\mathbf{kq}} |\psi_{\mathbf{kq}}|^2 (\epsilon_{\mathbf{k},\uparrow} + \epsilon_{\mathbf{q}-\mathbf{k},\downarrow} - \epsilon_{\mathbf{q},\uparrow})$$

$$+ \frac{g}{V} \left(\sum_{\mathbf{q}} |\psi_0|^2 + \sum_{\mathbf{kk'q}} \psi_{\mathbf{k'q}} \psi^*_{\mathbf{kq}} + \sum_{\mathbf{kq}} (\psi^*_0 \psi_{\mathbf{kq}} + \psi_0 \psi^*_{\mathbf{kq}}) \right), \tag{5.30}$$

where we have ignored the term $\sum_{\mathbf{kqq'}} \psi_{\mathbf{kq}} \psi^*_{\mathbf{kq'}}$ because this summation converges and therefore it vanishes after applying the renormalization condition for g. Then, by minimizing Eq. 5.30 with respect to $\psi_{\mathbf{kq}}$ and ψ_0, it yields

$$E_{\mathbf{kq}} \psi_{\mathbf{kq}} + \frac{g}{V} \chi_{\mathbf{q}} = \mathcal{E} \psi_{\mathbf{kq}} \tag{5.31}$$

$$\frac{g}{V} \sum_{\mathbf{q}} \chi_{\mathbf{q}} = \mathcal{E} \psi_0, \tag{5.32}$$

where $E_{\mathbf{kq}}$ denotes $\epsilon_{\mathbf{k},\uparrow} + \epsilon_{\mathbf{q}-\mathbf{k},\downarrow} - \epsilon_{\mathbf{q},\uparrow}$, and

$$\chi_{\mathbf{q}} = \sum_{\mathbf{k}} \psi_{\mathbf{kq}} + \psi_0, \tag{5.33}$$

and the Lagrange multiplier $\mathcal{E}$ is the polaron energy. From Eq. 5.31 one can obtain

$$\psi_{\mathbf{kq}} = \frac{g}{V} \frac{\chi_{\mathbf{q}}}{\mathcal{E} - E_{\mathbf{kq}}}. \tag{5.34}$$

Then, by substituting Eq. 5.34 into Eq. 5.33, we obtain

$$\chi_{\mathbf{q}} = \frac{\psi_0}{1 - \frac{g}{V} \sum_{|\mathbf{k}| > k_F} \frac{1}{\mathcal{E} - E_{\mathbf{kq}}}}. \tag{5.35}$$

Further substituting Eq. 5.35 into Eq. 5.32, and eliminating a nonzero ψ_0 from both sides, and using the renormalization condition for g, it yields a self-consistent equation for eigenenergy $\mathcal{E}$ as

$$\mathcal{E} = \frac{1}{V} \sum_{|\mathbf{q}| < k_F} \frac{1}{\frac{m}{4\pi \hbar^2 a_s} - \frac{1}{V} \sum_{\mathbf{k}} \frac{1}{2\epsilon_{\mathbf{k}}} - \frac{1}{V} \sum_{|\mathbf{k}| > k_F} \frac{1}{\mathcal{E} - E_{\mathbf{kq}}}}. \tag{5.36}$$

Solving Eq. 5.36 yields the zero-momentum polaron energy $\mathcal{E}$ as a function of $-1/(k_F a_s)$, as shown in Figure 5.5(a). Here the horizontal axis $-1/(k_F a_s)$ is a useful convention in studying strongly interacting gases across a scattering resonance, and we will always use this notation in this and next chapter. As we have shown in Section 2.1, for a square well potential, as the attraction monotonically increases, the scattering length first becomes more and more negative until it diverges to negative infinity, and then decrease from positive infinity to a finite positive value. This corresponds to $-1/(k_F a_s)$ monotonically decreases from large positive value to very negative value.

Eq. 5.36 has two solutions. The negative energy one is called the *attractive polaron* [33] and the positive energy one is called the *repulsive polaron* [40]. This is reminiscent of the upper and the lower branches of the two-body problem, as we discussed in Section 2.7. The attractive polaron can be used to determine the critical Zeeman field for stabilizing a fully polarized a Fermi gas in attractive interacting Fermi gas [33], and the repulsive

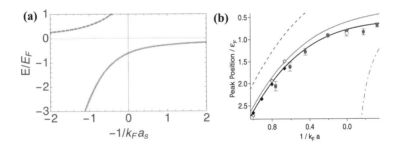

Figure 5.5   Polaron energy. (a) The energy of attractive polaron (negative energy solution) and repulsive polaron (positive energy solution) in units of E_F as a function of $-1/(k_F a_s)$. (b) The energy of the attractive polarons measured from the peak energy of the rf spectroscopy. The dashed and dotted lines are calculated from the variational wave function calculation without (dashed) and with (dotted) the final state correction. The solid line is obtained from the Monte Carlo calculation. (b) is reprinted from Ref. [153]. A color version of this figure can be found in the resources tab for this book at cambridge.org/zhai.

polaron can be used for determining the critical interaction strength for stabilizing a fully polarized ferromagnetism in repulsive interacting Fermi gas [40], which is also related to the ferromagnetism issue to be discussed again in Section 8.2. At unitarity with $1/(k_F a_s) = 0$, $\mathcal{E}/E_F$ becomes a universal constant, meaning that the interaction energy becomes the same order as the kinetic energy. For the attractive polarons, we find $\mathcal{E} \approx -0.6 E_F$ at the unitarity.

The Effective Mass Relation. This calculation can be straightforwardly generalized to polaron with finite momentum $\mathbf{p}$, with wave function

$$|\Psi\rangle_{\mathbf{p}} = \psi_0 \hat{c}^{\dagger}_{\mathbf{p}\downarrow} |FS\rangle + \sum_{|\mathbf{k}|>k_F, |\mathbf{q}|<k_F} \psi_{\mathbf{k},\mathbf{q}} \hat{c}^{\dagger}_{\mathbf{p}+\mathbf{q}-\mathbf{k}\downarrow} \hat{c}^{\dagger}_{\mathbf{k}\uparrow} \hat{c}_{\mathbf{q}\uparrow} |FS\rangle, \qquad (5.37)$$

and the self-consistent equation becomes

$$\mathcal{E}_{\mathbf{p}} = \epsilon_{\mathbf{p}} + \frac{1}{V} \sum_{|\mathbf{q}|<k_F} \frac{1}{\frac{m}{4\pi\hbar^2 a_s} - \frac{1}{V}\sum_{\mathbf{k}} \frac{1}{2\epsilon_{\mathbf{k}}} - \frac{1}{V}\sum_{|\mathbf{k}|>k_F} \frac{1}{\mathcal{E}_{\mathbf{p}} - E_{\mathbf{kq}}(\mathbf{p})}}, \qquad (5.38)$$

where $E_{\mathbf{kq}}(\mathbf{p}) = \epsilon_{\mathbf{k},\uparrow} + \epsilon_{\mathbf{p}+\mathbf{q}-\mathbf{k},\downarrow} - \epsilon_{\mathbf{q},\uparrow}$. By introducing the *self-energy* $\Sigma(\mathbf{p}, \omega)$ as

$$\Sigma(\mathbf{p}, \omega) = \frac{1}{V} \sum_{|\mathbf{q}|<k_F} \frac{1}{\frac{m}{4\pi\hbar^2 a_s} - \frac{1}{V}\sum_{\mathbf{k}} \frac{1}{2\epsilon_{\mathbf{k}}} - \frac{1}{V}\sum_{|\mathbf{k}|>k_F} \frac{1}{\omega - E_{\mathbf{kq}}(\mathbf{p})}}. \qquad (5.39)$$

$\mathcal{E}_{\mathbf{p}}$ is a solution of the equation

$$\omega - \epsilon_{\mathbf{p}} - \Sigma(\mathbf{p}, \omega) = 0. \qquad (5.40)$$

The self-energy includes all the interaction effects in a Fermi liquid. Eq. 5.39 is a specific form of the self-energy of this problem under this variational wave function approximation. However, Eq. 5.40 is a general equation and actually, one of the essential equations of the Fermi liquid. It shows how the self-energy determines quasi-particle dispersion. In the dilute limit, the effective mass m^* can be defined as $\partial\omega/\partial\mathbf{p}^2 = 1/(2m^*)$, thus Eq. 5.40 gives

$$\frac{\partial\omega}{\partial\mathbf{p}^2} - \frac{1}{2m} - \frac{\partial\Sigma}{\partial\mathbf{p}^2} - \frac{\partial\Sigma}{\partial\omega}\frac{\partial\omega}{\partial\mathbf{p}^2} = 0, \qquad (5.41)$$

from which the effective mass m^* can be obtained as

$$\frac{m^*}{m} = \frac{1 - \frac{\partial\Sigma}{\partial\omega}}{1 + \frac{\partial\Sigma}{\partial\mathbf{p}^2/(2m)}}, \qquad (5.42)$$

where the derivatives are taken at $\mathbf{p} = 0$ and $\omega = \mathcal{E}_{\mathbf{p}=0}$. Hence, we have reached Eq. 5.25, which in fact is a general relation for all Fermi liquids.

For the attractive polaron, at unitary, one finds $m^*/m = 1.17$. In fact, no matter the interaction is repulsive or attractive, the general expectation is that $m^*/m > 1$, because dressing up always makes the quasi-particle heavier than the bare one. On the other hand, it also shows that even the interaction energy is already as large as the Fermi energy, the change of effective mass can still be small. However, when attractive interaction further increases, the spin-down atom turns to form a bound state with one spin-up atom, and this bound

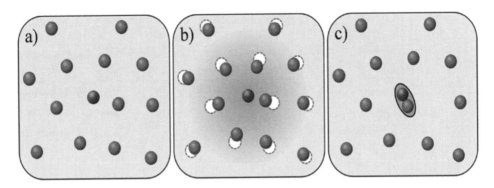

Figure 5.6   Cartoon of the polaron to molecule transition. (a) The noninteracting case. (b) The polaron state where the majority atomic cloud is polarized by the minority atom. (c) The molecular state where the minority atom bounds together with one majority atom and forms diatomic molecules. Reprinted from Ref. [153]. A color version of this figure can be found in the resources tab for this book at cambridge.org/zhai.

state is bosonic, which is different from the fermionic polaron. This leads to a polaron to molecule transition. The different scenarios of polaron and molecule are schematically shown in Figure 5.6. After the transition, with finite density of spin-down atoms, these molecules can condense at the low-temperature and the system becomes a mixture of the molecule BEC and the excess majority fermions. When this happens, the Fermi liquid picture breaks down.

The Quasi-Particle Residue Relation. Returning to the wave function Eq. 5.29, the first term does not disturb the Fermi surface, and therefore, the jump of the momentum distribution retains. For the second term in Eq. 5.29, the particle-hole excitation smears out the jump in the momentum distribution. Since the entire wave function is normalized, the jump of the momentum distribution at the Fermi surface is therefore proportional to the coefficient of the first term, that is, $Z = |\psi_0|^2$. Defining

$$\Pi_{\mathbf{q}} = \frac{1}{\frac{m}{4\pi\hbar^2 a_s} - \frac{1}{V}\sum_{\mathbf{k}}\frac{1}{2\epsilon_{\mathbf{k}}} - \frac{1}{V}\sum_{|\mathbf{k}|>k_F}\frac{1}{\omega - E_{\mathbf{kq}}(\mathbf{p})}} \tag{5.43}$$

and

$$\Sigma(\mathbf{p}, \omega) = \frac{1}{V}\sum_{|\mathbf{q}|<k_F}\Pi_{\mathbf{q}}(\mathbf{p}, \omega), \tag{5.44}$$

the finite $\mathbf{q}$ version of Eq. 5.34 and Eq. 5.35 can be rewritten as

$$\psi_{\mathbf{kq}} = \frac{1}{V}\frac{\Pi_{\mathbf{q}}}{\omega - E_{\mathbf{kq}}(\mathbf{p})}\psi_0, \tag{5.45}$$

and using the normalization condition, we obtain

$$\frac{1}{|\psi_0|^2} = \frac{1}{Z} = \left(1 + \frac{1}{V^2}\sum_{|\mathbf{k}|>k_F, |\mathbf{q}|<k_F}\frac{\Pi_{\mathbf{q}}^2}{(\omega - E_{\mathbf{kq}}(\mathbf{p}))^2}\right). \tag{5.46}$$

Table 5.1 Comparison of different types of spectroscopy measurements, including the Bragg spectroscopy, the Raman spectroscopy, and the radio-frequency spectroscopy

	Operator	Realization
Bragg	$\sum_{\mathbf{q}} \hat{a}^{\dagger}_{\mathbf{q}+\mathbf{k}} \hat{a}_{\mathbf{q}} e^{-i\omega t} + \text{h.c.}$	Two lasers with same polarization
Raman	$\sum_{\mathbf{q}} \hat{c}^{\dagger}_{\mathbf{q}+\mathbf{k}\uparrow} \hat{c}_{\mathbf{q}\downarrow} e^{-i\omega t} + \text{h.c.}$	Two lasers with different polarization
Radio-frequency	$\sum_{\mathbf{k}} \hat{c}^{\dagger}_{\mathbf{k}3} \hat{c}_{\mathbf{k}\downarrow} e^{-i\omega t} + \text{h.c.}.$	A radio-frequency laser

Noting that

$$\frac{\partial \Pi_{\mathbf{q}}^{-1}}{\partial \omega} = -\frac{1}{\Pi_{\mathbf{q}}^{2}} \frac{\partial \Pi_{\mathbf{q}}}{\partial \omega} = \frac{1}{V} \sum_{|\mathbf{k}|>k_{\mathrm{F}}} \frac{1}{(\omega - E_{\mathbf{k}\mathbf{q}}(\mathbf{p}))^{2}}, \tag{5.47}$$

we have

$$\frac{1}{V^2} \sum_{|\mathbf{k}|>k_{\mathrm{F}}, |\mathbf{q}|<k_{\mathrm{F}}} \frac{\Pi_{\mathbf{q}}^{2}}{(\omega - E_{\mathbf{k}\mathbf{q}}(\mathbf{p}))^{2}} = -\frac{1}{V} \sum_{|\mathbf{q}|<k_{\mathrm{F}}} \frac{\partial \Pi_{\mathbf{q}}}{\partial \omega} = -\frac{\partial \Sigma(\mathbf{p},\omega)}{\partial \omega}, \tag{5.48}$$

and we obtain

$$\frac{1}{Z} = \left(1 - \frac{\partial}{\partial \omega} \Sigma(\mathbf{p},\omega) \right). \tag{5.49}$$

Hence, we reach the quasi-particle residue relation Eq. 5.26.

Radio-Frequency Spectroscopy. In Table 5.1 we have listed and compared several different types of spectroscopy measurements in ultracold atomic systems. Both the Bragg and the Raman spectroscopy can change the momentum of atoms, and the radio-frequency spectroscopy cannot. Both the Raman and the radio-frequency spectroscopy change internal states of atoms, and the Bragg spectroscopy does not.

Here we focus on the radio-frequency spectroscopy detection of the properties of the polarons. There are two ways to perform the radio-frequency spectroscopy measurement. The first way is that the radio-frequency wave brings a spin-down particle to a third internal state $|3\rangle$, and the third state $|3\rangle$ is out of the two-component Fermi gas system. The second way is that the atom is initially prepared in the $|3\rangle$ state, and the radio-frequency wave injects the atom to the spin-down state. The latter is also known as the *spin-injection spectroscopy*. Here we focus on the former. Note that the wave length of the radio-frequency wave is much larger than the size of the system, and therefore, the momentum transfer during the transition is negligible, and the atom in the final state $|3\rangle$ has the same momentum as the atom in the initial spin-down state. In performing the radio-frequency spectroscopy, either we can collect atoms in all momenta in the third state, which corresponds to the integrated radio-frequency spectroscopy, or we can also record the momentum of atoms in the third state, which corresponds to the momentum-resolved radio-frequency spectroscopy. The latter is equivalent to the angle-resolved photoemission spectroscopy (ARPES) in condensed matter experiments. In ARPES, X-ray laser ejects electrons outside of a solid, and one records both the momentum of electrons detected and energy of the X-ray laser. Similar to the radio-frequency spectroscopy, the photon momentum is relatively small compared with the electron momentum, and the momentum transfer can be ignored during the transition. Hence, the electron detected outside solid also has the same momentum as the initial electron inside solid.

Thus, the radio-frequency spectroscopy measurement can be modeled as

$$\hat{V} = \hbar\Omega_R \sum_{\mathbf{k}} \hat{c}_{\mathbf{k}3}^{\dagger} \hat{c}_{\mathbf{k}\downarrow} e^{-i\omega t} + \text{h.c.} \tag{5.50}$$

Similar as the Bragg spectroscopy discussed in Section 3.3, the radio-frequency transition rate from an initial state $|i\rangle$ to a final state $|f\rangle$ can also be calculated by the Fermi's golden rule as

$$S(\omega) = \frac{2\pi}{\hbar} \sum_{f} |\langle f|\hat{V}|i\rangle|^2 \delta(\hbar\omega - (\mathcal{E}_f - \mathcal{E}_i)), \tag{5.51}$$

where $\mathcal{E}_i$ and $\mathcal{E}_f$ are energy of the initial and the final states, respectively. Here for the case of the Fermi polaron, the initial state wave function is taken as Eq. 5.37, and the final state can be either (i) a particle in the third state with momentum $\mathbf{p}$ embedded in a Fermi sea of spin-up atoms, denoted by $c_{\mathbf{p},3}^{\dagger}|\text{FS}\rangle$, or (ii) a particle in the third state with momentum $\mathbf{p} + \mathbf{q} - \mathbf{k}$ embedded in a Fermi sea with a particle-hole excitation, denoted by $\hat{c}_{\mathbf{p}+\mathbf{q}-\mathbf{k},3}^{\dagger}\hat{c}_{\mathbf{k}\uparrow}^{\dagger}\hat{c}_{\mathbf{q}\uparrow}|\text{FS}\rangle$. Therefore we obtain the spectrum as

$$S(\omega) \propto \left(Z\delta(\hbar\omega - (\epsilon_{\mathbf{p}} - \mathcal{E}_{\mathbf{p}})) + \sum_{\mathbf{k},\mathbf{q},|\mathbf{k}|>k_F,|\mathbf{q}|<k_F} |\psi_{\mathbf{k},\mathbf{q}}|^2 \delta(\hbar\omega - (E_{\mathbf{kq}}(\mathbf{p}) - \mathcal{E}_{\mathbf{p}})) \right). \tag{5.52}$$

The first part is a sharp peak at $\omega = \epsilon_{\mathbf{p}} - \mathcal{E}_{\mathbf{p}}$ with weight $|\psi_0|^2$, which comes from the contribution of the (i)-type of the final state. This part is named as the coherent peak. It is also this part of the initial state wave function that contributes to the discontinuity of the momentum distribution, and therefore, its coefficient is proportional to the quasi-particle residue Z. The second part is a broad distribution for $\omega > \epsilon_{\mathbf{p}} - \mathcal{E}_{\mathbf{p}}$, which comes from the contribution of the (ii)-type of the final state. This part is called the incoherent broadening. The typical radio-frequency spectroscopy for the spin-down atoms are shown by the red lines in Figures 5.7(b)–(d). From the location of the coherent peak in the frequency domain, we can extract the polaron energy. From the weight of the coherent peak, we can determine the quasi-particle residue Z. As shown in Figure 5.5(b), the polaron energy is plotted as a function of $1/(k_F a_s)$. Eq. 5.52 does not include the final state interaction effect between the $|3\rangle$ state and the spin-up atoms. In practice, the strength of this final state interaction effect depends on the choices of atomic species and the internal states. When this final state interaction effect is not negligible, it can also be taken into account in the Fermi's golden rule calculation, yielding certain corrections to the spectroscopy. By taking the final state interaction effect into account, the theoretical results from this variational wave function approach agrees reasonably well with the experimental data, as shown in Figure 5.5(b). This quasi-particle residue Z is plotted in Figure 5.8 in terms of the impurity density and the

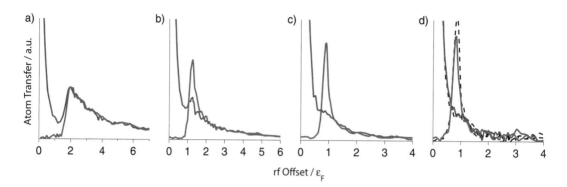

Figure 5.7   Radio-frequency spectroscopy measurement. The rf spectroscopy for the spin-up majority atoms (the solid line always peaked at zero frequency) and for spin-down minority atoms (another solid line), for $1/(k_F a_s) = $ (a) 0.76, (b) 0.43, (c) 0.20, and (d) 0. Reprinted from Ref. [153]. A color version of this figure can be found in the resources tab for this book at cambridge.org/zhai.

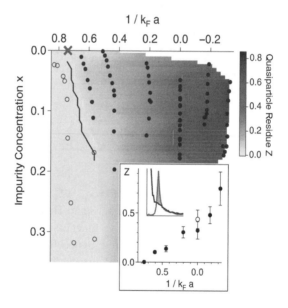

Figure 5.8   The quasi-particle residue. Z obtained from the radio-frequency spectroscopy measurement and plotted in terms of impurity density and $1/(k_F a_s)$. Inset shows that Z vanishes at the polaron to molecule transition. The shaded area is the area that is under the minority peak and is not included by the majority atom's response, and the weight of the shaded area is counted as Z. Reprinted from Ref. [153]. A color version of this figure can be found in the resources tab for this book at cambridge.org/zhai.

interaction parameter $1/(k_F a_s)$, and indeed it shows that Z vanishes at a critical interaction strength, at which the Fermi liquid undergoes a transition to the molecular BEC.

Similarly, one can also perform the radio-frequency spectroscopy on the majority spin-up atoms. In the polaron state, the change to the Fermi sea is negligible when annihilating one majority atom, and therefore, the polaronic state is not affected. Hence, the radio-frequency spectroscopy will be the same as that for free atoms, as shown in Figures 5.7(b)–(d). When entering the molecular regime, each spin-down atom should pair up with a spin-up atom. If the radio-frequency transition is acted on a spin-up atom in the pair, it requires the same amount energy to break the pair as the radio-frequency wave acted on the spin-down atom. Hence, the spectroscopy displays the same line-shape for the spin-up atoms and for the spin-down atoms in that energy regime. If the radio-frequency transition is acted on the excess spin-up atoms, it displays the same spectroscopy as the free atoms. This double peak structure is shown in Figure 5.7(a), which is taken in the molecular regime.

Exercises

5.1 Considering single component noninteracting fermions in a three-dimensional isotropic trap $V(\mathbf{r}) = m\omega^2 r^2/2$ with total number N, compute the density profile with two methods: (1) by filling the single-particle eigenstates from the lower energy to the higher energy; (2) by the local density approximation. Compare the density distribution obtained by these two methods when N is large enough.

5.2 Considering a two-component Fermi gas in three dimensions, compute the total $S_z = N_\uparrow - N_\downarrow$ when the chemical potential $\mu_\uparrow \neq \mu_\downarrow$. Defining $h = \mu_\uparrow - \mu_\downarrow$, when h is small, compute the spin susceptibility given by S_z/h in the limit of $h \to 0$.

5.3 Solve Eq. 5.36 perturbatively when a_s is a small positive value or a small negative value, and show that the results of polaron energy are consistent with the perturbation theory.

5.4 Considering a fully polarized Fermi gas with spin-up atoms only, now let us flip one spin-up fermion into a spin-down fermion, and this spin-down fermion forms a repulsive polaron with the rest of the spin-up fermions. Estimate when this spin flip process is energetically favorable and the fully polarized ferromagnetic state is not energetically stable.

5.5 Considering a Fermi gas fully polarized by a Zeeman field $-h(N_\uparrow - N_\downarrow)$, now let us flip one spin-up fermion into a spin-down fermion, and this spin-down fermion forms an attractive polaron with the rest of the spin-up fermions. Estimate the critical Zeeman field h_c above which the spin flip process is *not* energetically favorable and the fully polarized spin-up Fermi gas is energetically stable.

6

The Fermi Superfluid

Learning Objectives

- Compare the difference between the two-body problem in a vacuum and on top of the Fermi sea.
- Introduce the BCS mean-field theory and the BCS ground state wave function and its excitation.
- Introduce the physical picture of BCS pairing.
- Introduce the concept of contact using the BCS state as an example.
- Discuss how the BCS state responds to a Zeeman field.
- Discuss different pairing symmetries.
- Discuss how the BCS-BEC crossover mean-field equation recovers features in both the BCS and the BEC limit.
- Discuss different behaviors of the excitation spectrum in the BCS and the BEC regimes and determine superfluid critical velocity from the excitation spectrum.
- Discuss how the superfluid transition temperature changes from the BCS regime to the BEC regime.
- Introduce a list of experimental observations on the BCS-BEC crossover.
- Discuss the challenging issues in studying the unitary Fermi gas.
- Introduce the concept of holographic duality and the prediction of the η/s bound.

6.1 BCS Pairing

In the previous Section 5.2, we have considered interacting spin-1/2 fermions, but we only consider the situation that few minority spin-down fermions are embedded in the majority spin-up fermions. In this chapter, we consider spin-1/2 fermions, and in most cases, with equal density of spin-up and spin-down fermions. We will still use the renomalizable contact potential introduced in Section 2.2 and the Hamiltonian is given by

$$\hat{H} = \sum_{\mathbf{k}}(\epsilon_{\mathbf{k}} - \mu)\hat{c}_{\mathbf{k}\sigma}^\dagger \hat{c}_{\mathbf{k}\sigma} + \frac{g}{V}\sum_{\mathbf{k}\mathbf{k}'\mathbf{q}} \hat{c}_{\mathbf{k}+\frac{\mathbf{q}}{2}\uparrow}^\dagger \hat{c}_{-\mathbf{k}+\frac{\mathbf{q}}{2}\downarrow}^\dagger \hat{c}_{-\mathbf{k}'+\frac{\mathbf{q}}{2}\downarrow} \hat{c}_{\mathbf{k}'+\frac{\mathbf{q}}{2}\uparrow}, \tag{6.1}$$

where $\epsilon_{\mathbf{k}} = \hbar^2\mathbf{k}^2/(2m)$. Here the density of each spin component is fixed as $n = k_{\mathrm{F}}^3/(6\pi^2)$. $E_{\mathrm{F}} = \hbar^2 k_{\mathrm{F}}^2/(2m)$ and $T_{\mathrm{F}} = E_{\mathrm{F}}/k_{\mathrm{B}}$ are the Fermi energy and the Fermi temperature for noninteracting Fermi gas, respectively. They will be taken as energy units in following discussions.

Pairing Instability of Fermi Surface. Now we first revisit the two-body problem we have studied in Section 2.1, but with a different approach. Considering two-particle with opposite spins in vacuum with zero center-of-mass momentum, the wave function can be generally written as

$$|\Psi\rangle = \sum_{\mathbf{k}} \psi_{\mathbf{k}} \hat{c}^{\dagger}_{\mathbf{k}\uparrow} \hat{c}^{\dagger}_{-\mathbf{k}\downarrow} |0\rangle. \tag{6.2}$$

Below we will compare the two-body problem in vacuum and the two-body problem in the presence of a Fermi sea. And as an important difference between these two cases, the chemical potential is set as zero for the former case but is nonzero for the latter. Hence, the Schrödinger equation for the former case becomes

$$2\epsilon_{\mathbf{k}} \psi_{\mathbf{k}} + \frac{g}{V} \sum_{\mathbf{k}'} \psi_{\mathbf{k}'} = E \psi_{\mathbf{k}}, \tag{6.3}$$

which leads to a self-consistent equation

$$\frac{1}{g} = \frac{1}{V} \sum_{\mathbf{k}} \frac{1}{E - 2\epsilon_{\mathbf{k}}}. \tag{6.4}$$

Using the renormalization condition for g, we have

$$\frac{m}{4\pi\hbar^2 a_s} = \frac{1}{V} \sum_{\mathbf{k}} \left(\frac{1}{E - 2\epsilon_{\mathbf{k}}} + \frac{1}{2\epsilon_{\mathbf{k}}} \right). \tag{6.5}$$

Next we consider a two-body problem on top of a Fermi sea. Such a problem is first analyzed by Cooper and is known as the Cooper problem [39]. For this problem the wave function is modified as

$$|\Psi\rangle = \sum_{|\mathbf{k}|>k_F} \psi_{\mathbf{k}} \hat{c}^{\dagger}_{\mathbf{k}\uparrow} \hat{c}^{\dagger}_{-\mathbf{k}\downarrow} |\text{FS}\rangle, \tag{6.6}$$

where we have replaced $|0\rangle$ in Eq. 6.6 by the Fermi sea $|\text{FS}\rangle$, and $|\text{FS}\rangle$ is given by

$$|\text{FS}\rangle = \prod_{|\mathbf{k}|<k_F} \hat{c}^{\dagger}_{\mathbf{k}\uparrow} \hat{c}^{\dagger}_{-\mathbf{k}\downarrow} |0\rangle. \tag{6.7}$$

The momentum summation of these two extra particles is restricted to $|\mathbf{k}| > k_F$ because of the Pauli exclusion principle. Here we have ignored the distortion of the Fermi sea due to the presence of two extra particles, and μ is taken as E_F. Correspondingly, Eq. 6.5 is modified as

$$\frac{m}{4\pi\hbar^2 a_s} = \frac{1}{V} \left[\sum_{|\mathbf{k}|>k_F} \frac{1}{E - 2(\epsilon_{\mathbf{k}} - \mu)} + \sum_{\mathbf{k}} \frac{1}{2\epsilon_{\mathbf{k}}} \right]. \tag{6.8}$$

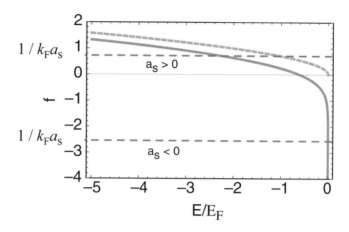

Figure 6.1 Solution of two-body problem. This plot shows the f-function of Eq. 6.9 for $E < 0$. The dashed line is the f-function for the two-body problem in vacuum, and the solid line is the f-function for the two-body problem in the presence of a Fermi sea. The solution of Eq. 6.9 can be obtained by looking at the intersection of a constant $1/(k_F a_s)$ with the f-functions. A color version of this figure can be found in the resources tab for this book at cambridge.org/zhai.

Using E_F as the energy unit and $1/k_F$ as the length unit, both Eq. 6.5 and Eq. 6.8 can be casted into

$$\frac{1}{k_F a_s} = f\left(\frac{E}{E_F}\right). \tag{6.9}$$

The function-f for Eq. 6.5 and Eq. 6.8 are shown in Figure 6.1, and the intersection between a constant $1/(k_F a_s)$ and the f-function gives rise to the solution to these equations. Here we would like to make two remarks regarding these solutions.

- For negative a_s, because the f function given by Eq. 6.5 is always positive for $E < 0$ and it terminates at $f = 0$ when $E = 0$, Eq. 6.5 does not have a $E < 0$ solution for negative a_s. But the f function given by Eq. 6.8 approaches $-\infty$ as $E \to 0$. Hence, it always has a bound state with energy $E < 0$ for any a_s. That is to say, for negative a_s, two particles cannot form a bound state if they are alone, but they can form a bound state with the presence of a Fermi sea. In other words, pairing is a collective effect in this regime. The existence of the bound state solution means that the two-particle excitation of a free Fermi sea always has negative excitation energy with any attractive interaction, which in fact indicates the Fermi sea is unstable toward fermion pairing. Therefore, one has to reconstruct a many-body state that includes this pairing effect, which leads to the famous Bardeen-Cooper-Schrieffer theory for superconductors, well known as the BCS theory [12].

- For positive a_s, pairing occurs even for a two-body system in vacuum. It is easy to show that Eq. 6.5 has a bound state solution with $E = -\hbar^2/(ma_s^2)$ when $a_s > 0$, which is consistent with our analysis in two-body problem with the first quantization quantum mechanics approach in Section 2.1. One can see that for very negative E, these two f-functions approach each other. Hence, one can find when a_s becomes small and positive, the solution of Eq. 6.8 approaches the solution for Eq. 6.5 for two-body problem in vacuum. In this limit, the size of the two-body bound state becomes much smaller compared with the inter-particle distance, and hence, it does not make much difference if these two fermions are in a many-body environment or by themselves alone. It can also be shown that the bound state behaves as a diatomic bosonic molecule in this limit, therefore, at low temperature these bosonic molecules can Bose–Einstein condense (BEC).

Below we shall first briefly review the BCS theory, and in the next Section 6.2, we will discuss the crossover from a BCS state to a BEC state.

The BCS Mean-Field Hamiltonian. The BCS mean-field theory follows from a standard mean-field approach summarized in the Box 6.1. Here we only focus the center-of-mass $\mathbf{q} = 0$ scatterings sector in Eq. 6.1, and the reason we only include $\mathbf{q} = 0$ sector will be explained later. By taking $\mathbf{q} = 0$, the interaction term in Eq. 6.1 can be written as

$$\frac{g}{V} \sum_{\mathbf{kk'q}} \hat{c}^\dagger_{\mathbf{k}+\frac{q}{2}\uparrow} \hat{c}^\dagger_{-\mathbf{k}+\frac{q}{2}\downarrow} \hat{c}_{-\mathbf{k'}+\frac{q}{2}\downarrow} \hat{c}_{\mathbf{k'}+\frac{q}{2}\uparrow} = \frac{g}{V} \left(\sum_{\mathbf{k}} \hat{c}^\dagger_{\mathbf{k}\uparrow} \hat{c}^\dagger_{-\mathbf{k}\downarrow} \right) \left(\sum_{\mathbf{k'}} \hat{c}_{-\mathbf{k'}\downarrow} \hat{c}_{\mathbf{k'}\uparrow} \right), \quad (6.10)$$

which is the general form of interaction potential discussed in Box 6.1 by choosing $\hat{A} = -\sum_{\mathbf{k}} c_{-\mathbf{k}\downarrow} c_{\mathbf{k}\uparrow}$. Then, following the procedure discussed in Box 6.1, the BCS mean-field Hamiltonian becomes

$$H_{\text{BCS}} = \sum_{\mathbf{k}\sigma} (\epsilon_{\mathbf{k}} - \mu) c^\dagger_{\mathbf{k}\sigma} c_{\mathbf{k}\sigma} - \Delta \sum_{\mathbf{k}} c^\dagger_{\mathbf{k}\uparrow} c^\dagger_{-\mathbf{k}\downarrow} - \Delta \sum_{\mathbf{k}} c_{-\mathbf{k}\downarrow} c_{\mathbf{k}\uparrow} - \frac{\Delta^2 V}{g}, \quad (6.11)$$

where Δ is defined as

$$\Delta = -\frac{g}{V} \sum_{\mathbf{k}} \langle c_{-\mathbf{k}\downarrow} c_{\mathbf{k}\uparrow} \rangle. \quad (6.12)$$

We note that the definition of $\hat{A}$ and Δ can be defined up to a free phase factor, and here we choose the phase factor in such a way that Δ can be taken as real positive number.

Box 6.1   **Self-Consistent Mean-Field Theory**

In this book we will employ the self-consistent mean-field theory in this section and in Section 8.1 in discussing the Bose–Hubbard model. Here we will describe the general scheme. Considering a term $g\hat{A}^\dagger \hat{A}$, we write $g\hat{A}^\dagger \hat{A} = \langle g\hat{A}^\dagger \rangle \hat{A} + \hat{A}^\dagger \langle g\hat{A} \rangle + \left(g\hat{A}^\dagger \hat{A} - \langle g\hat{A}^\dagger \rangle \hat{A} - \hat{A}^\dagger \langle g\hat{A} \rangle \right)$. Furthermore, by taking the expectation value of the last term in the bracket and factorizing $\langle \hat{A}^\dagger \hat{A} \rangle \approx \langle \hat{A}^\dagger \rangle \langle \hat{A} \rangle$, we can approximate the last term in the bracket as $-\frac{\langle g\hat{A}^\dagger \rangle \langle g\hat{A} \rangle}{g}$, which is equivalent to ignoring the fluctuation term $g(\hat{A}^\dagger - \langle \hat{A}^\dagger \rangle)(\hat{A} - \langle \hat{A} \rangle)$. With this approximation, we have $g\hat{A}^\dagger \hat{A} \approx \langle g\hat{A}^\dagger \rangle \hat{A} + \hat{A}^\dagger \langle g\hat{A} \rangle - \frac{\langle g\hat{A}^\dagger \rangle \langle g\hat{A} \rangle}{g}$. We can introduce $\langle g\hat{A} \rangle$ as the order parameter, and finally the order parameter has to be determined self-consistently.

This mean-field Hamiltonian H_{BCS} is a quadratic Hamiltonian which can be diagonalized as

$$H_{\text{BCS}} = \sum_{\mathbf{k}} \left[(\hat{c}_{\mathbf{k}\uparrow}^{\dagger}, \hat{c}_{-\mathbf{k}\downarrow}) \begin{pmatrix} \epsilon_{\mathbf{k}} - \mu & -\Delta \\ -\Delta & -(\epsilon_{\mathbf{k}} - \mu) \end{pmatrix} \begin{pmatrix} \hat{c}_{\mathbf{k}\uparrow} \\ \hat{c}_{-\mathbf{k}\downarrow}^{\dagger} \end{pmatrix} + (\epsilon_{\mathbf{k}} - \mu) \right] - \frac{\Delta^2 V}{g}$$

$$= \sum_{\mathbf{k}} \left[\mathcal{E}_{\mathbf{k}} \left(\hat{\alpha}_{\mathbf{k}}^{\dagger} \hat{\alpha}_{\mathbf{k}} - \hat{\beta}_{\mathbf{k}} \hat{\beta}_{\mathbf{k}}^{\dagger} \right) + (\epsilon_{\mathbf{k}} - \mu) \right] - \frac{\Delta^2 V}{g}$$

$$= \sum_{\mathbf{k}} \left[\mathcal{E}_{\mathbf{k}} \left(\hat{\alpha}_{\mathbf{k}}^{\dagger} \hat{\alpha}_{\mathbf{k}} + \hat{\beta}_{\mathbf{k}}^{\dagger} \hat{\beta}_{\mathbf{k}} \right) + (\epsilon_{\mathbf{k}} - \mu) - \mathcal{E}_{\mathbf{k}} \right] - \frac{\Delta^2 V}{g}, \tag{6.13}$$

where

$$\mathcal{E}_{\mathbf{k}} = \sqrt{(\epsilon_{\mathbf{k}} - \mu)^2 + \Delta^2}, \tag{6.14}$$

and $\hat{\alpha}_{\mathbf{k}}$ and $\hat{\beta}_{\mathbf{k}}^{\dagger}$ are a unitary rotation of $\hat{c}_{\mathbf{k}\uparrow}$ and $\hat{c}_{-\mathbf{k}\downarrow}^{\dagger}$ as

$$\hat{\alpha}_{\mathbf{k}} = u_{\mathbf{k}} \hat{c}_{\mathbf{k}\uparrow} - v_{\mathbf{k}} \hat{c}_{-\mathbf{k}\downarrow}^{\dagger} \tag{6.15}$$

$$\hat{\beta}_{\mathbf{k}}^{\dagger} = v_{\mathbf{k}} \hat{c}_{\mathbf{k}\uparrow} + u_{\mathbf{k}} \hat{c}_{-\mathbf{k}\downarrow}^{\dagger}, \tag{6.16}$$

and therefore

$$\hat{\alpha}_{\mathbf{k}}^{\dagger} = u_{\mathbf{k}} \hat{c}_{\mathbf{k}\uparrow}^{\dagger} - v_{\mathbf{k}} \hat{c}_{-\mathbf{k}\downarrow} \tag{6.17}$$

$$\hat{\beta}_{\mathbf{k}} = v_{\mathbf{k}} \hat{c}_{\mathbf{k}\uparrow}^{\dagger} + u_{\mathbf{k}} \hat{c}_{-\mathbf{k}\downarrow}. \tag{6.18}$$

Here both $u_{\mathbf{k}}$ and $v_{\mathbf{k}}$ are real and positive values, and

$$u_{\mathbf{k}}^2 = \frac{1}{2} \left(1 + \frac{\epsilon_{\mathbf{k}} - \mu}{\mathcal{E}_{\mathbf{k}}} \right); \quad v_{\mathbf{k}}^2 = \frac{1}{2} \left(1 - \frac{\epsilon_{\mathbf{k}} - \mu}{\mathcal{E}_{\mathbf{k}}} \right). \tag{6.19}$$

Similar as the Bogoliubov Hamiltonian for bosons discussed in Section 3.3, the new quasi-particles $\hat{\alpha}$ and $\hat{\beta}^{\dagger}$ operators mix the creation and annihilation operators of original fermions. There we have mentioned a unitary transformation of the bosonic creation and annihilation operators do not obey bosonic commutation relations. However, here $\hat{\alpha}$ and $\hat{\beta}^{\dagger}$ is a unitary rotation of original fermionic creation and annihilation operators, but they obey the fermionic commutation relations.

Diagonalizing the Bogoliubov Hamiltonian can determine following results:

• First, it determines the ground state wave function. Since $\mathcal{E}_{\mathbf{k}}$ is always positive, the ground state wave function $|\Psi_{\text{BCS}}\rangle$ has to be a vacuum of quasi-particles $\hat{\alpha}_{\mathbf{k}}^{\dagger}$ and $\hat{\beta}_{\mathbf{k}}^{\dagger}$, that is to say, one needs to find out a wave function that satisfies

$$\hat{\alpha}_{\mathbf{k}} |\Psi_{\text{BCS}}\rangle = 0, \quad \hat{\beta}_{\mathbf{k}} |\Psi_{\text{BCS}}\rangle = 0. \tag{6.20}$$

And it is easy to show that such $|\Psi_{\text{BCS}}\rangle$ is given by

$$|\Psi_{\text{BCS}}\rangle = \prod_{\mathbf{k}} (u_{\mathbf{k}} + v_{\mathbf{k}} \hat{c}_{\mathbf{k}\uparrow}^{\dagger} \hat{c}_{-\mathbf{k}\downarrow}^{\dagger}) |0\rangle. \tag{6.21}$$

With this wave function, it is easy to see that the momentum distribution $n_{\mathbf{k}\sigma} = |v_{\mathbf{k}}|^2$.

- Second, it determines the ground state energy. For the vacuum of the quasi-particles, the ground state energy is also determined by Eq. 6.11 as

$$E_{\mathrm{BCS}} = \sum_{\mathbf{k}} [((\epsilon_{\mathbf{k}} - \mu) - \mathcal{E}_{\mathbf{k}}] - \frac{\Delta^2 V}{g}. \tag{6.22}$$

To determine the value of Δ, we can compute the r.h.s. of Eq. 6.12 with respect to the BCS wave function Eq. 6.21, and it gives

$$\Delta = -\frac{g}{V} \sum_{\mathbf{k}} u_{\mathbf{k}} v_{\mathbf{k}} = -\frac{g}{V} \sum_{\mathbf{k}} \frac{\Delta}{2\mathcal{E}_{\mathbf{k}}}. \tag{6.23}$$

Here we again use the renormalization condition for g to eliminate the divergency in the summation of r.h.s. of Eq. 6.23, which yields

$$-\frac{m}{4\pi \hbar^2 a_s} = \frac{1}{V} \sum_{\mathbf{k}} \left(\frac{1}{2\mathcal{E}_{\mathbf{k}}} - \frac{1}{2\epsilon_{\mathbf{k}}} \right). \tag{6.24}$$

Alternatively, we can also minimize the energy of the BCS ground state energy Eq. 6.22 with respect to Δ, and $\partial E_{\mathrm{BCS}}/\partial \Delta = 0$ yields the same equation as Eq. 6.23. Eq. 6.24 is called the *gap equation*.

- Third, it determines the excitation spectrum and the wave function of excitations. It is straightforward to show that

$$\hat{\alpha}_{\mathbf{k}_0}^\dagger |\Psi_{\mathrm{BCS}}\rangle = \hat{c}_{\mathbf{k}_0\uparrow}^\dagger \prod_{\mathbf{k}\neq\mathbf{k}_0} (u_{\mathbf{k}} + v_{\mathbf{k}} \hat{c}_{\mathbf{k}\uparrow}^\dagger \hat{c}_{-\mathbf{k}\downarrow}^\dagger)|0\rangle \tag{6.25}$$

$$\hat{\beta}_{\mathbf{k}_0}^\dagger |\Psi_{\mathrm{BCS}}\rangle = \hat{c}_{-\mathbf{k}_0\downarrow}^\dagger \prod_{\mathbf{k}\neq\mathbf{k}_0} (u_{\mathbf{k}} + v_{\mathbf{k}} \hat{c}_{\mathbf{k}\uparrow}^\dagger \hat{c}_{-\mathbf{k}\downarrow}^\dagger)|0\rangle. \tag{6.26}$$

The excited energy of these two excited states are $\mathcal{E}_{\mathbf{k}_0}$. Furthermore, we can consider another excited state

$$\hat{\alpha}_{\mathbf{k}_0}^\dagger \hat{\beta}_{\mathbf{k}_0}^\dagger |\Psi_{\mathrm{BCS}}\rangle = (-v_{\mathbf{k}_0} + u_{\mathbf{k}_0} \hat{c}_{\mathbf{k}_0\uparrow}^\dagger \hat{c}_{-\mathbf{k}_0\downarrow}^\dagger) \prod_{\mathbf{k}\neq\mathbf{k}_0} (u_{\mathbf{k}} + v_{\mathbf{k}} \hat{c}_{\mathbf{k}\uparrow}^\dagger \hat{c}_{-\mathbf{k}\downarrow}^\dagger)|0\rangle. \tag{6.27}$$

The excitation energy of this state is $2\mathcal{E}_{\mathbf{k}_0}$. At the Fermi surface when $\epsilon_{\mathbf{k}_0} = \mu$, the excitation energy $\mathcal{E}_{\mathbf{k}_0} = \Delta$ for exciting a single quasi-particle described by Eq. 6.25 and Eq. 6.26, and $2\mathcal{E}_{\mathbf{k}_0} = 2\Delta$ for exciting two quasi-particles described by Eq. 6.27.

It is interesting to discuss the limit $\Delta \to 0$, and in this limit $\mu \to E_{\mathrm{F}}$. For $|\mathbf{k}| < k_{\mathrm{F}}$, $u_{\mathbf{k}} \to 0$ and $v_{\mathbf{k}} \to 1$, and therefore, $\hat{\alpha}_{\mathbf{k}}^\dagger \to \hat{c}_{-\mathbf{k}\downarrow}$ and $\hat{\beta}_{\mathbf{k}}^\dagger \to \hat{c}_{\mathbf{k}\uparrow}$, both of which create hole excitation below the Fermi sea. For $|\mathbf{k}| > k_{\mathrm{F}}$, $u_{\mathbf{k}} \to 1$ and $v_{\mathbf{k}} \to 0$, and therefore, $\hat{\alpha}_{\mathbf{k}}^\dagger \to \hat{c}_{\mathbf{k}\uparrow}^\dagger$ and $\hat{\beta}_{\mathbf{k}}^\dagger \to \hat{c}_{-\mathbf{k}\downarrow}^\dagger$, both of which create particle excitation above the Fermi sea. In the limit $\Delta \to 0$, $\mathcal{E}_{\mathbf{k}}$ also becomes $|\epsilon_{\mathbf{k}} - \mu|$, which is the excitation energy for a hole with $|\mathbf{k}| < k_{\mathrm{F}}$ or for a particle with $|\mathbf{k}| > k_{\mathrm{F}}$ in a free Fermi sea, and $|\Psi_{\mathrm{BCS}}\rangle$ recovers to the filled Fermi sea. These are all consistent with the noninteracting fermions discussed in Section 5.1. When Δ is finite, $\mathcal{E}_{\mathbf{k}}$ is always nonzero and it acquires a minimum value Δ as the excitation gap. The momentum distribution is a smooth function and does not display any discontinuity. Hence, the BCS state does not obey the condition of a Fermi liquid discussed in Section 5.1, and it belongs to the ordered phase discussed in Box 5.1.

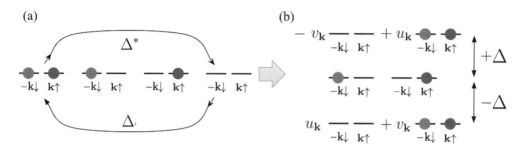

(a)   (b)

Schematic of the BCS mechanism. (a) Consider two modes $\mathbf{k}, \uparrow$ and $-\mathbf{k}, \downarrow$ with $\mathbf{k}$ located at the Fermi surface. The Hilbert space contains four states as a doubly occupied state, two singly occupied states, and a vacuum state. The doubly occupied and the vacuum states are hybridized by Δ. (b) The hybridization lowers the energy of the BCS wave function by Δ. A color version of this figure can be found in the resources tab for this book at cambridge.org/zhai.

This discussion of $\Delta \to 0$ also applies to the momentum regime away from the Fermi surface even for finite Δ case, as long as $|\epsilon_\mathbf{k} - \mu| \gg \Delta$. In other words, for the BCS pairing, the reconstruction of the many-body wave function mainly takes place in an energy window $\sim \Delta$ around the Fermi surface where $\epsilon_\mathbf{k} - \mu = 0$. Hence, below we only focus on the Fermi surface in order to gain an intuitive picture of the BCS pairing.

Physical Interpretation of the BCS Pairing. Here we will offer a physical picture of the BCS pairing, which is illustrated by Figure 6.2. Here we consider an arbitrary momentum $\mathbf{k}_0$ at the Fermi surface, and because of the time-reversal symmetry, $-\mathbf{k}_0$ is also located at the Fermi surface. At this point we should explain why we focus on $\mathbf{q} = 0$ sector of the scattering process. Because when $\mathbf{q} \neq 0$, for a general $\mathbf{k}_0$, if $\mathbf{k}_0 + \mathbf{q}/2$ is located at the Fermi surface, $-\mathbf{k}_0 + \mathbf{q}/2$ is not located at the Fermi surface.[1] As we will see below, for the BCS pairing, it is important that both $\mathbf{k}_0$ and $-\mathbf{k}_0$ are simultaneously located at the Fermi surface. In this case, let us consider a pair of state $\mathbf{k}_0, \uparrow$ and $-\mathbf{k}_0, \downarrow$. Before turning on interactions, there are four zero-energy degenerate states, which are the doubly occupied state $\hat{c}^\dagger_{\mathbf{k}_0\uparrow} \hat{c}^\dagger_{-\mathbf{k}_0\downarrow} |0\rangle$, two singly occupied states $\hat{c}^\dagger_{\mathbf{k}_0\uparrow} |0\rangle$ and $\hat{c}^\dagger_{-\mathbf{k}_0\downarrow} |0\rangle$, and the fully empty state $|0\rangle$, as shown in Figure 6.2(a). The Δ-term in the BCS Hamiltonian Eq. 6.11 couples the doubly occupied state to the fully empty state. Thus, the BCS wave function Eq. 6.21 is a superposition between the doubly occupied state and the fully empty state, whose energy is lowered by Δ due to the level repulsion, as shown in Figure 6.2(b). The BCS Hamiltonian leaves the singly occupied states unchanged, thus, they correspond to excitations Eq. 6.25 and Eq. 6.26, whose energy is higher than the BCS state by the amount of Δ. The energy of another superposition orthogonal to the BCS wave function is increased by Δ due to the level repulsion, thus, its energy is higher than the ground state by 2Δ, which corresponds to the excited state Eq. 6.27 with two quasi-particles.

---

[1] For some specific shape of Fermi surface, it is possible that there exists a nonzero $\mathbf{q}$ such that for most $\mathbf{k}_0$, if $\mathbf{k}_0 + \mathbf{q}/2$ is at the Fermi surface, $-\mathbf{k}_0 + \mathbf{q}/2$ is also at the Fermi surface. In this case, pairing with nonzero $\mathbf{q}$ is also possible. Such a BCS pairing state is called the Fulde-Ferrell-Larkin-Ovchinnikov state, usually short-noted as the FFLO state [57, 103].

This discussion reveals that the key of the BCS wave function is so-called pairing in the momentum space. The precise meaning of pairing in momentum space is that a pair of states with opposite momenta are either both occupied or both empty. In other words, the singly occupied states are excluded. The physical insight is that when the two states are both occupied, two atoms can be scattered into another pair of momenta. And if both states are empty, two atoms in other pair of momenta can be scattered into this state. Thus, these two states can enjoy the interaction energy. However, the singly occupied state can neither be the initial state nor be the final state of the scattering process, therefore, it cannot participate in the scattering processes and cannot benefit from the interaction energy. Here we should particularly stress that the pairing in momentum space is different from our intuitive picture of pairing in real space. In real space, one can image that two particles come close to each other in order to gain attractive interaction energy. However, this is *not* the physical picture of the BCS pairing. As we will show in Section 6.2, the typical size of a Cooper pair is about $\sim 1/k_{\mathrm{F}}$, which is not smaller than the typical interparticle spacing.

The Contact. For the BCS wave function, the momentum distribution $n_{\mathbf{k}}$ is given by $|v_{\mathbf{k}}|^2$, and it can be shown that for sufficiently large k,

$$n_{\mathbf{k}} = |v_{\mathbf{k}}|^2 \rightarrow \frac{m^2 \Delta^2}{\hbar^4 k^4}. \tag{6.28}$$

The momentum distribution $n_{\mathbf{k}}$ behaves $\sim 1/k^4$ at large momentum. It can be shown that this is a general feature of a many-body system with zero-range interaction in three dimensions, and this feature is quite general and holds independent of statistics, temperature, and even independent of whether the system is at equilibrium or not [171, 192]. In fact, this $1/k^4$ tail of momentum distribution is a consequence of the behavior of the short-range wave function discussed in Section 2.1, which says that the wave function should always behaves as $1/|\mathbf{r}_i - \mathbf{r}_j|$ when any $\mathbf{r}_i$ gets close to another $\mathbf{r}_j$. Hence, we can define a quantity $\mathcal{C}$ called *Contact* [171, 192], and it is defined through the large momentum asymptotic behavior of momentum distribution as

$$\mathcal{C} = \lim_{k \to \infty} k^4 n_{\mathbf{k}}. \tag{6.29}$$

In the case of the BCS wave function, with Eq. 6.28, $\mathcal{C}$ is given by

$$\mathcal{C} = \frac{m^2 \Delta^2}{\hbar^4}. \tag{6.30}$$

Now considering the total energy given by Eq. 6.22, we consider $\partial E_{\mathrm{BCS}} / \partial a_{\mathrm{s}}^{-1}$, it leads to

$$\frac{\partial E_{\mathrm{BCS}}}{\partial a_{\mathrm{s}}^{-1}} = \frac{\partial E_{\mathrm{BCS}}}{\partial \Delta} \frac{\partial \Delta}{\partial a_{\mathrm{s}}^{-1}} + \frac{\partial E_{\mathrm{BCS}}}{\partial g^{-1}} \frac{\partial g^{-1}}{\partial a_{\mathrm{s}}^{-1}} = -\frac{m \Delta^2 V}{4 \pi \hbar^2}. \tag{6.31}$$

Here we have used the gap equation $\partial E_{\mathrm{BCS}} / \partial \Delta = 0$ and the renormalization condition Eq. 2.31. By comparing Eq. 6.30 and Eq. 6.31, we find a relation [171, 192]

$$\frac{dE}{da_{\mathrm{s}}^{-1}} = -\frac{\hbar^2 V}{4 \pi m} \mathcal{C}. \tag{6.32}$$

Though we obtain this Eq. 6.32 using the BCS theory as a specific example, this identity holds for any equilibrium state in three dimensions with zero-range interaction. This relation connects the total energy and the large-k asymptotic behavior of momentum distribution.

Since the large momentum part refers to the ultraviolet behavior, and therefore, it should also be related to the short-range behavior. Here we consider the local density-density correlation written as $\langle \hat{n}_\uparrow(\mathbf{r})\hat{n}_\downarrow(\mathbf{r})\rangle$. Using the BCS wave function Eq. 6.21, it is easy to show that

$$g^2 \langle \hat{n}_\uparrow(\mathbf{r})\hat{n}_\downarrow(\mathbf{r})\rangle = \frac{g^2}{V^2}\left(\sum_\mathbf{k} u_\mathbf{k} v_\mathbf{k}\right)^2 = \Delta^2, \tag{6.33}$$

where we have used the gap equation Eq. 6.23. Thus, we have reached another relation that is [171, 192]

$$\frac{\hbar^4}{m^2}\mathcal{C} = g^2 \langle \hat{n}_\uparrow(\mathbf{r})\hat{n}_\downarrow(\mathbf{r})\rangle. \tag{6.34}$$

Again, this relation also holds for general many-body systems in three dimensions with zero-range interaction.

Response to Zeeman Field. As one can easily see from the BCS wave function Eq. 6.21, there are always equal number of spin-up and spin-down fermions in this ground state. Now let us consider adding a Zeeman field term as

$$h\sum_\mathbf{k}(\hat{c}^\dagger_{\mathbf{k}\uparrow}\hat{c}_{\mathbf{k}\uparrow} - \hat{c}^\dagger_{\mathbf{k}\downarrow}\hat{c}_{\mathbf{k}\downarrow}). \tag{6.35}$$

This simply modifies the kinetic energy term in the BCS mean-field Hamiltonian Eq. 6.11 as

$$\sum_{\mathbf{k}\sigma}(\epsilon_\mathbf{k} - \mu)\hat{c}^\dagger_{\mathbf{k}\sigma}\hat{c}_{\mathbf{k}\sigma} \rightarrow \sum_{\mathbf{k}\sigma}(\epsilon_\mathbf{k} - \mu_\sigma)\hat{c}^\dagger_{\mathbf{k}\sigma}\hat{c}_{\mathbf{k}\sigma}, \tag{6.36}$$

where $\mu_\uparrow = \mu - h$ and $\mu_\downarrow = \mu + h$. It is straightforward to show that this modified mean-field Hamiltonian can still be diagonalized, and it yields a very similar diagonalized Hamiltonian as Eq. 6.13, except that

$$\sum_\mathbf{k}\mathcal{E}_\mathbf{k}\left(\hat{\alpha}^\dagger_\mathbf{k}\hat{\alpha}_\mathbf{k} + \hat{\beta}^\dagger_\mathbf{k}\hat{\beta}_\mathbf{k}\right) \rightarrow \sum_\mathbf{k}(\mathcal{E}_\mathbf{k} + h)\hat{\alpha}^\dagger_\mathbf{k}\hat{\alpha}_\mathbf{k} + (\mathcal{E}_\mathbf{k} - h)\hat{\beta}^\dagger_\mathbf{k}\hat{\beta}_\mathbf{k}. \tag{6.37}$$

However, the expression for $u_\mathbf{k}$ and $v_\mathbf{k}$ remain unchanged. The excitation spectrums are changed from $\mathcal{E}_\mathbf{k}$ to $\mathcal{E}_\mathbf{k} \pm h$. This can be understood because the BCS ground state is a spin singlet, and $\hat{\alpha}^\dagger_\mathbf{k}$ and $\hat{\beta}^\dagger_\mathbf{k}$ create excitations of spin-1/2 with $S_z = \pm 1/2$, respectively. Hence, the energy of the BCS ground state cannot be changed by the Zeeman field, and the energies of two quasi-particles are increased or lowered by h, respectively. However, since the minimum of $\mathcal{E}_\mathbf{k}$ is Δ, therefore, as long as $h < \Delta$, both two excitation energies are still positive and the ground state is still the vacuum of two quasi-particles. That is to say, the BCS ground state cannot be affected by the Zeeman field as long as $h < \Delta$. This in fact is not a surprise. Because the spin excitation is gapped for the BCS state, it means that the spin susceptibility is zero and the response of the system to the external Zeeman field vanishes.

As we discussed above, the energy of the BCS wave function does not change with the increasing of h. In contrast to the BCS state, we can also consider the normal state with $\Delta = 0$. For the normal state, fermions nearby the Fermi surface can immediately be polarized by the Zeeman field. In other words, the Fermi surfaces for the spin-up and the spin-down particles are split by the Zeeman field. Thus, the energy of the normal state decreases as the Zeeman field increases. In the absence of the Zeeman field, the energy of the BCS state is lower than the normal state. As the Zeeman field h increases, there exists a critical Zeeman field h_c, above which the normal state has a lower energy than the BCS state. This critical Zeeman field is usually smaller than Δ, and in the weakly interacting limit, $h_c = \Delta/\sqrt{2}$. Then, a first order phase transition from the BCS pairing state to the normal state takes places. This critical field is also known as the *Chandrasekhar–Clogston limit* [29, 37]. Before the transition, the spin-up and spin-down atoms have equal densities and the spin polarization vanishes. There exists a jump of spin polarization at the first order transition, after which the spin polarization increases with the increasing of the Zeeman field.

Here a Zeeman field actually means difference in chemical potential between two spin components,[2] and in ultracold atom experiments, it can be realized by mixing two spin components with different number of atoms in each spin component. By the local density approximation, we have $\mu_\sigma = \mu_\sigma^0 - V(\mathbf{r})$, and $\mu_\uparrow^0 > \mu_\downarrow^0$ supposing that total number of $N_\uparrow$ is greater than $N_\downarrow$. Hence, we have $h = \mu_\uparrow - \mu_\downarrow = \mu_\uparrow^0 - \mu_\downarrow^0$, which is a constant throughout the system, but the averaged density, controlled by $\bar{\mu} = (\mu_\uparrow + \mu_\downarrow)/2 = \frac{1}{2}(\mu_\uparrow^0 + \mu_\downarrow^0) - V(\mathbf{r})$, decreases as r increases. Thus, as r increases, the Fermi energy E_F decreases, and there-fore, Δ decreases. Hence, as shown in Figure 6.3(a1) and (b1), there exists a critical radius R_c at which h/Δ reaches the Chandrasekhar–Clogston limit, and a first order phase transition takes place. In the regime $r < R_c$ denoted by I in Figures 6.3(a1) and (b1), it is the BCS state with equal densities of two spin components and zero spin polarization. At $r = R_c$, spin polarization jumps to a finite value. Then, in the regime denoted by II, the spin polarization continuously increases as r increases, until the density of spin down fermions vanishes. Then the spin polarization remains the constant unity in the regime denoted by III. This phenomenon has been observed in experiments, as shown by Figures 6.3(a2) and (b2). Note that in ultracold atom experiments, what can be measured by the absorption imaging is an integrated density along the light propagation direction, and the integration smears out the jump of polarization. Hence, in the experiment Ref. [159], in order to see the jump associated with the first order transition, they have to use the inverse Abel transformation to reconstruct the three-dimensional density from the integrated column density.

---

[2] Unlike the inevitable spin flip processes in the solid state materials, in ultracold atom systems there is no spin flip term and the total number of atoms in each spin component is conserved individually, an energy offset between these two spin states actually does not play a role of Zeeman field.

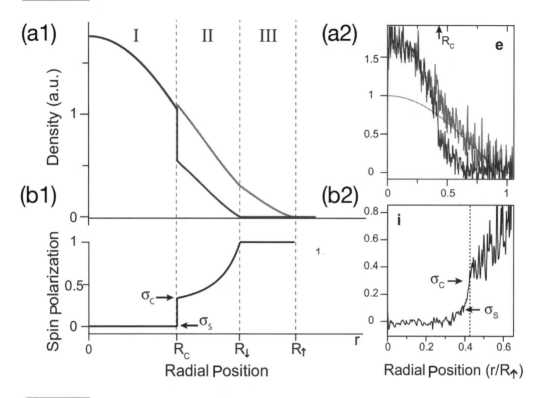

Figure 6.3 Spin-imbalanced Fermi gas. (a) Density for two spin components in a harmonic trap. (b) Spin polarization in a
harmonic trap. (a1, b1) Theoretical expectation. (a2, b2) Experimental observation. Reprinted from Ref. [159]. A color
version of this figure can be found in the resources tab for this book at cambridge.org/zhai.

Pairing Symmetry. Above we have focused on the s-wave pairing because we only consider interaction in the s-wave channel. In general, the BCS state can have different pairing
symmetries. We now consider a general interaction form in the center-of-mass momentum
$\mathbf{q} = 0$ sector and between two spin components, written as

$$-\frac{1}{V} \sum_{\mathbf{kk}'} V_{\mathbf{kk}'} \hat{c}^\dagger_{\mathbf{k}\uparrow} \hat{c}^\dagger_{-\mathbf{k}\downarrow} \hat{c}_{-\mathbf{k}'\downarrow} \hat{c}_{\mathbf{k}'\uparrow}. \tag{6.38}$$

We decompose $V_{\mathbf{kk}'}$ as

$$V_{\mathbf{kk}'} = \sum_i \lambda_i f^i(\mathbf{k}) f^i(\mathbf{k}'), \tag{6.39}$$

where $f^i(\mathbf{k})$ is called the form factor. Let us only focus on the channel with the largest λ^i,
say, denoted by λ, the interaction term can be written as

$$-\frac{\lambda}{V} \sum_{\mathbf{kk}'} f_{\mathbf{k}} f_{\mathbf{k}'} \hat{c}^\dagger_{\mathbf{k}\uparrow} \hat{c}^\dagger_{-\mathbf{k}\downarrow} \hat{c}_{-\mathbf{k}'\downarrow} \hat{c}_{\mathbf{k}'\uparrow} = -\frac{\lambda}{V} \left(\sum_{\mathbf{k}} f_{\mathbf{k}} \hat{c}^\dagger_{\mathbf{k}\uparrow} \hat{c}^\dagger_{-\mathbf{k}\downarrow} \right) \left(\sum_{\mathbf{k}'} f_{\mathbf{k}'} \hat{c}_{-\mathbf{k}'\downarrow} \hat{c}_{\mathbf{k}'\uparrow} \right). \tag{6.40}$$

Introducing

$$\Delta = \frac{\lambda}{V} \sum_{\mathbf{k}} \langle f_{\mathbf{k}} \hat{c}_{-\mathbf{k}\downarrow} \hat{c}_{\mathbf{k}\uparrow} \rangle, \tag{6.41}$$

the BCS mean-field Hamiltonian Eq. 6.11 still takes similar form with a small modification as

$$-\Delta \sum_{\mathbf{k}} c^{\dagger}_{\mathbf{k}\uparrow} c^{\dagger}_{-\mathbf{k}\downarrow} - \Delta^{*} \sum_{\mathbf{k}} c_{-\mathbf{k}\downarrow} c_{\mathbf{k}\uparrow} \rightarrow -\Delta \sum_{\mathbf{k}} f_{\mathbf{k}} c^{\dagger}_{\mathbf{k}\uparrow} c^{\dagger}_{-\mathbf{k}\downarrow} - \Delta^{*} \sum_{\mathbf{k}} f_{\mathbf{k}} c_{-\mathbf{k}\downarrow} c_{\mathbf{k}\uparrow}. \tag{6.42}$$

The following discussion of the BCS state still holds except one needs to replace Δ everywhere with $\Delta f_{\mathbf{k}}$, for instance, the excitation spectrum will be changed to

$$\mathcal{E}_{\mathbf{k}} = \sqrt{(\epsilon_{\mathbf{k}} - \mu)^2 + |\Delta f_{\mathbf{k}}|^2}. \tag{6.43}$$

The symmetry of $f_{\mathbf{k}}$ therefore plays an important role. For instance, if $f(\mathbf{k}) \propto k_x$ or $f(\mathbf{k}) \propto k_y$, then $f(\mathbf{k})$ has the p-wave symmetry in the momentum space, and such BCS state is called the *p-wave pairing*. If $f(\mathbf{k}) \propto k_x k_y$ or $f(\mathbf{k}) \propto k_x^2 - k_y^2$, then $f(\mathbf{k})$ has the d-wave symmetry in the momentum space, and such BCS state is called the *d-wave pairing*. We will mention the d-wave pairing in the Fermi–Hubbard model in Section 8.2. For the p-wave pairing, $f_{\mathbf{k}}$ changes sign once in the entire momentum space and hence there is a nodal line where $f_{\mathbf{k}} = 0$. When this nodal line intersects with the Fermi surface defined as $\epsilon_{\mathbf{k}} = \mu$, the excitation spectrum displays a gapless excitation at this point, around which the excitation spectrum shows a Dirac cone. Similarly, for the d-wave pairing, $f_{\mathbf{k}}$ changes sign twice and there are two nodal lines. Their intersections with the Fermi surface also gives rise to gapless excitations there.

6.2 BCS-BEC Crossover

Although the BCS state is not a Fermi liquid anymore, it retains lots of features as a fermionic system. For negative a_s and when $|a_s| \ll 1/k_F$, the pairing gap is much weaker compared with the Fermi energy, and the reconstruction of many-body state is limited to the momentum space around the Fermi surface. This is pairing in the momentum space discussed in the previous section, and no real space two-body bound state can exist in low-energy. This regime is considered as the *BCS limit*.

On the other hand, for positive a_s, there exists stable two-body bound state. Such a bound state, in contrast to the Cooper pairs, is pairing in real space. Two fermions in each bound state do stay closer in the real space, and the size of the pair is $\sim a_s$. This bound state is considered as a diatomic bosonic molecule. When $a_s \ll 1/k_F$, the size of bound state is much smaller than the interparticle distance, hence, the system can be viewed a gas of bosonic molecules. $a_s \ll 1/k_F$ is also equivalent to $\hbar^2/(m a_s^2) \gg \hbar^2 k_F^2/(2m)$, that is to say, the molecular binding energy is much larger than the Fermi energy. Hence, the system loses its fermionic characters. At low temperature, these bosonic molecules can Bose condense. This is considered as the *BEC limit*.

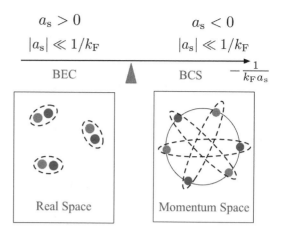

Figure 6.4   Schematic of the BEC and BCS states. $-1/(k_F a_s)$ is used as the tuning parameter; $a_s < 0$ and $|a_s| \ll 1/k_F$ is the BCS limit, where pairing occurs in the momentum space; and $a_s > 0$ and $|a_s| \ll 1/k_F$ is the BEC limit, where pairing occurs in the real space. A color version of this figure can be found in the resources tab for this book at cambridge.org/zhai.

The physics of these two limits look quite differently and these two limits are schematically shown in Figure 6.4. It is an important observation that one can tune the parameter $-1/(k_F a_s)$ to connect these two limits. Hence, a natural question is that, as one tunes $-1/(k_F a_s)$ from the BCS limit to the BEC limit, whether there exists a phase transition in between, or it is a smooth crossover. The answer to this question crucially depends on whether there exists a wave function that can smoothly interpolate the wave function of BCS pairing in momentum space and the wave function for Bose condensation of molecules. In fact, it has been pointed out that this BCS wave function can also capture the physics of molecule BEC [52, 106], because

$$|\Psi_{\text{BCS}}\rangle \propto \prod_{\mathbf{k}} \left(1 + \frac{v_{\mathbf{k}}}{u_{\mathbf{k}}} \hat{c}_{\mathbf{k}\uparrow}^\dagger \hat{c}_{-\mathbf{k}\downarrow}^\dagger\right) |0\rangle = \exp\left\{\sum_{\mathbf{k}} \frac{v_{\mathbf{k}}}{u_{\mathbf{k}}} \hat{c}_{\mathbf{k}\uparrow}^\dagger \hat{c}_{-\mathbf{k}\downarrow}^\dagger\right\} |0\rangle. \qquad (6.44)$$

Here we have used the fact that $(\hat{c}_{\mathbf{k}\uparrow}^\dagger)^n |0\rangle = (\hat{c}_{-\mathbf{k}\downarrow}^\dagger)^n |0\rangle = 0$ for $n \geq 2$. Introducing

$$\tilde{g}_{\mathbf{k}} = \frac{v_{\mathbf{k}}}{u_{\mathbf{k}}} = \frac{\Delta}{\mathcal{E}_{\mathbf{k}} + (\epsilon_{\mathbf{k}} - \mu)} \qquad (6.45)$$

and a normalization factor $\mathcal{A} = \sum_{\mathbf{k}} |\tilde{g}_{\mathbf{k}}|^2$, we define $g_{\mathbf{k}} = \tilde{g}_{\mathbf{k}}/\sqrt{\mathcal{A}}$ and introduce operator $\hat{b}^\dagger$ as a Cooper pair operator

$$\hat{b}^\dagger = \sum_{\mathbf{k}} g_{\mathbf{k}} \hat{c}_{\mathbf{k}\uparrow}^\dagger \hat{c}_{-\mathbf{k}\downarrow}^\dagger. \qquad (6.46)$$

It is clear that $\hat{b}$ describes a composite object of two fermions and $\mathbf{k}$ is the relative momentum. Therefore, we can define the Cooper pair wave function as

$$\psi(\mathbf{r}) = \frac{1}{\sqrt{(2\pi)^3}} \int d^3\mathbf{k} \, g_{\mathbf{k}} e^{i\mathbf{k}\cdot\mathbf{r}}. \qquad (6.47)$$

In order that the wave function Eq. 6.44 can describe a Bose condensation, it is important for $\hat{b}$ to obey the bosonic commutation relation. It is straightforward to show that

$$[\hat{b}, \hat{b}^\dagger] = \sum_{\mathbf{k}} |g_{\mathbf{k}}|^2 (1 - \hat{n}_{\mathbf{k}\uparrow} - \hat{n}_{-\mathbf{k}\downarrow}). \tag{6.48}$$

One can see that in the r.h.s. of Eq. 6.48, if $\hat{n}_{\mathbf{k}\uparrow}$ and $\hat{n}_{-\mathbf{k}\downarrow}$ can be ignored, then we reach $[\hat{b}, \hat{b}^\dagger] = 1$ and the boson commutation relation is satisfied. Later we will discuss when $\hat{n}_{\mathbf{k}\uparrow}$ and $\hat{n}_{-\mathbf{k}\downarrow}$ can be ignored, and we shall also show that, under the same condition, $\psi(\mathbf{r})$ becomes a tightly bounded wave function with size $\sim a_s$. The many-body wave function Eq. 6.44 can then be written as $\exp\{\sqrt{\mathcal{A}}\hat{b}^\dagger\}|0\rangle$. When $\hat{b}$ is a bosonic operator, this is a coherent state representation of the Bose condensate wave function as we have discussed in Section 3.1. Therefore, it is possible that the BCS wave function Eq. 6.21 can be continuously evolved into a wave function for a BEC state. This is called the *BCS-BEC crossover*.

Here we should note a key insight for the BCS-BEC crossover is the change of chemical potential. Chemical potential corresponds to the energy cost of adding a particle into the system. In a Fermi gas, particle can only be added above the Fermi surface, and therefore, the chemical potential in the BCS limit is close to the Fermi energy. However, in the BEC limit, if one adds a pair of fermions with opposite spins, they form a bound state with binding energy $-\hbar^2/(ma_s^2)$. Hence, the chemical potential should be half of this binding energy. In order to take this change of the chemical potential into account, one needs to include another equation for conservation of the total number of fermions

$$n = \frac{1}{2V} \sum_{\mathbf{k}} \left(1 - \frac{\epsilon_{\mathbf{k}} - \mu}{\mathcal{E}_{\mathbf{k}}}\right). \tag{6.49}$$

We should solve gap equation Eq. 6.24 together with the number equation Eq. 6.49. Note that $n = k_F^3/(6\pi^2)$, using k_F as the momentum unit, and correspondingly, $1/k_F$ as the length unit and E_F as the energy unit, the two equations Eq. 6.24 and Eq. 6.49 can be written into a dimensionless form. The only tunable dimensionless parameter is $-1/(k_F a_s)$. By solving these two equations, we can determine Δ/E_F and μ/E_F as functions of $-1/(k_F a_s)$. The results are shown in Figure 6.5. One can see that from the BCS side to the BEC side, the

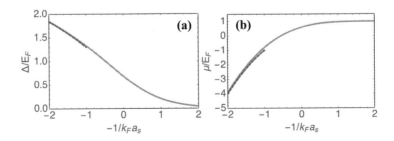

Figure 6.5 BEC-BCS crossover. (a) The pairing gap Δ/E_F and (b) the chemical potential μ/E_F as a function of $-1/(k_F a_s)$. The dashed line in (a) shows $\Delta/E_F = 4/\sqrt{3\pi k_F a_s}$, and the dashed line in (b) shows $\mu/E_F = -1/(k_F a_s)^2$. A color version of this figure can be found in the resources tab for this book at cambridge.org/zhai.

pairing gap Δ/E_F monotonically increases and the chemical potential μ/E_F monotonically decreases.

We divide the system into three different regime. The BCS limit with $-1/(k_F a_s) \gg 1$, the BEC limit with $-1/(k_F a_s) \ll -1$ and the unitary regime where the scattering length is very large and $-1/(k_F a_s) \sim 0$. Below we will first show that the mean-field theory can recover the right physics in the BCS limit and the BEC limit, respectively, and finally we will address the unitary regime.

The BCS Limit. In this limit we expect $\Delta \ll \mu$ so that the number equation can be approximated by that of a free Fermi gas, and thus $\mu \simeq E_F$. Substituting $\mu = E_F$ into the gap equation Eq. 6.24, one can obtain an approximate solution [139]

$$\Delta = \frac{8E_F}{e^2} \exp\left\{ \frac{\pi}{2k_F a_s} \right\}, \tag{6.50}$$

where $e \approx 2.71828$ is the Euler's number. Here the gap is exponentially small in terms of $1/(k_F a_s)$. $\pi/(2k_F a_s)$ can be rewritten as $1/(\mathcal{D}(E_F)U)$, where $\mathcal{D}(E_F) = mk_F/(2\hbar^2\pi^2)$ is the density-of-state at the Fermi energy, and $U = 4\pi\hbar^2 a_s/m$. This is consistent with a weakly interacting picture that shows only fermions nearby the Fermi surface are significantly affected by pairing, and therefore, only the density-of-state at the Fermi energy matters. As we show in Figure 6.6(a) the momentum distribution deviates from the momentum distribution of free Fermi gas in an energy window $\sim \Delta$ around the Fermi energy. In Figure 6.6(b), we show that the size of Cooper pair wave function is of the order of $\sim 1/k_F$.

The BEC Limit. In this limit, each spin-up fermion form a bound pair with a spin-down fermion. When μ is negative and assuming $(-\mu) \gg \Delta$, we have $|\epsilon_\mathbf{k} - \mu| \geqslant -\mu \gg \Delta$. Below we will discuss how to recover the BEC limit properly in orders of Δ.

• The zeroth order of Δ recovers the two-body bound state.

To the zeroth order of Δ, $\mathcal{E}_\mathbf{k}$ becomes $\epsilon_\mathbf{k} - \mu$ and the gap equation Eq. 6.24 becomes

$$-\frac{m}{4\pi\hbar^2 a_s} = \sum_\mathbf{k} \left(\frac{1}{2\epsilon_\mathbf{k} - 2\mu} - \frac{1}{2\epsilon_\mathbf{k}} \right). \tag{6.51}$$

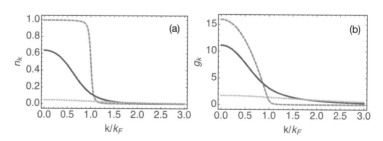

Figure 6.6 The BCS-BEC crossover. (a) The momentum distribution n_k and (b) the pair wave function g_k. The blue dashed line is for the BCS side $-1/(k_F a_s) = 2$, the red solid line is for the unitary regime $-1/(k_F a_s) = 0$, and the green dotted line is for the BEC side $-1/(k_F a_s) = -2$. A color version of this figure can be found in the resources tab for this book at cambridge.org/zhai.

By equaling 2μ in Eq. 6.51 as E in the two-body equation Eq. 6.5, we can see that Eq. 6.51 is equivalent to the two-body equation Eq. 6.5, and therefore, we obtain

$$\mu/E_F = -(1/(k_F a_s))^2. \tag{6.52}$$

This is consistent with our expectation of chemical potential being half of two-body binding energy. We have compared this approximated result with the solution of the mean-field equation in Figure 6.5(b). By taking $\mathcal{E}_{\mathbf{k}} = \epsilon_{\mathbf{k}} - \mu$, we also obtain $v_{\mathbf{k}}^2 \approx 0$ and $n_{\mathbf{k}\sigma} \approx 0$, as we can also see in Figure 6.6(a). This satisfies the requirement that we can ignore $n_{\mathbf{k}\uparrow}$ and $n_{-\mathbf{k}\downarrow}$ in Eq. 6.48, and hence $\hat{b}$ and $\hat{b}^\dagger$ satisfy the boson commutation relation. In this regime, one can see that the momentum distribution is a broad smooth function, as shown in Figure 6.6(a), with no feature reminiscent of the Fermi surface. With the same approximation, we can see that the pair wave function Eq. 6.45 behaves as

$$\tilde{g}_{\mathbf{k}} = \frac{v_{\mathbf{k}}}{u_{\mathbf{k}}} = \frac{\Delta}{\mathcal{E}_{\mathbf{k}} + (\epsilon_{\mathbf{k}} - \mu)} \approx \frac{\Delta}{-2\mu + 2\epsilon_{\mathbf{k}}} \approx \frac{\Delta}{-2\mu}\left(1 - k^2 a_s^2\right). \tag{6.53}$$

This gives rise to a pair wave function whose size is $\sim a_s$ in real space, consistent with the two-body bound state.

• The first order of Δ reveals the interaction effects between molecules.

To the leading order of Δ, the number equation Eq. 6.49 gives

$$\frac{16}{3\pi} = \left(\frac{\Delta}{E_F}\right)^2 \sqrt{\frac{E_F}{-\mu}}. \tag{6.54}$$

Substituting $\mu/E_F = -(1/(k_F a_s))^2$ into Eq. 6.54, we find $\Delta = 4E_F/\sqrt{3\pi k_F a_s}$. We have also compared this approximated formula with the solution to mean-field equations in Figure 6.5(a). Hence, $\Delta/(-\mu) \sim (k_F a_s)^{3/2} \ll 1$ in the BEC limit. This justifies the approximation of small Δ expansion that we have taken. To the first order of Δ, the gap equation Eq. 6.24 becomes

$$-\frac{\pi}{4k_F a_s} = -\frac{\pi}{4}\sqrt{\frac{-\mu}{E_F}} - \frac{\sqrt{-\mu/E_F}\pi}{64}\left(\frac{\Delta}{-\mu}\right)^2, \tag{6.55}$$

and this gives the first-order correction to the chemical potential as [59]

$$\frac{\mu}{E_F} = -\frac{1}{(k_F a_s)^2}\left[1 - \frac{1}{8}\left(\frac{\Delta}{-\mu}\right)^2\right] = -\frac{1}{(k_F^2 a_s^2)} + \frac{2k_F a_s}{3\pi}. \tag{6.56}$$

Here we introduce the chemical potential of a diatomic molecule μ_m as 2μ, because adding a diatomic molecule corresponds to adding two atoms, and

$$\mu_m = -\frac{\hbar^2}{m a_s^2} + \frac{4\pi\hbar^2(2a_s)}{(2m)}n_m, \tag{6.57}$$

where we have used n_m to present the density of molecules, and $n_m = n_\uparrow = n_\downarrow = k_F^3/(6\pi^2)$. The first term in Eq. 6.57 is the binding energy of a single molecule, and the second term is proportional to the density of molecules, the physical meaning of which is the interaction between molecules. Compared with the mean-field equation-of-state of interacting bosons discussed in Section 3.2, it shows that the interaction between molecules is repulsive, and the scattering length between molecules can be defined as $a_m = 2a_s$.

A more rigorous value of the molecule-molecule scattering length can be obtained by solving a four-body problem with two spin-up and two spin-down fermions, which gives $a_m = 0.6a_s$ [59]. Despite of difference in the coefficient, this mean-field theory captures the key feature that a_m is proportional to a_s. This is in fact originated from the Pauli exclusion principle and the composite nature of the diatomic molecules. Note that here each molecule is a composite object of two fermions with opposite spins, and the size of molecule is about a_s. Therefore, when the distance between two molecules is close to the size of each molecule, the Pauli exclusion principle starts to play a role, and molecules start to feel repulsion. That is to say, roughly speaking, we can view each molecule as a hard core boson with the hard core radius about a_s. As discussed in Section 2.1, the scattering length between such hard core particles is about the radius of the hard core, and in this case, it is proportional to a_s.

Excitation and Superfluidity. Above we have discussed that the ground state of the BCS wave function can qualitatively recover the physics of the BEC limit. Here we shall focus on excitation modes. We have discussed the excitation of breaking of the Cooper pairs, and the excitation energy is given by the mean-field theory discussed above. This excitation creates singly occupied fermion mode, therefore, it is fermionic excitation. When $\mu > 0$, the minimum of the excitation gap occurs at $\epsilon_{\mathbf{k}} = \mu$ and is given by Δ. When $\mu < 0$, the minimum of the excitation gap occurs at $k = 0$ and is given by $\sqrt{\mu^2 + \Delta^2}$. From Figure 6.5(a), we can conclude that the excitation gap of this fermionic mode monotonically increases from the BCS limit to the BEC limit. We schematically show in Figure 6.7 the different behavior of this fermionic mode in the BCS and the BEC sides, respectively.

Recalling the discussion of Bose condensate in Section 3.2, we always emphasize the phonon mode of the BEC with a linear and gapless dispersion. The physical meaning of

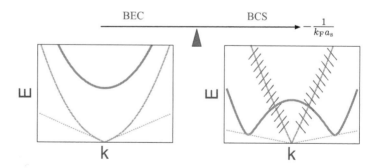

Figure 6.7 Schematic of the excitation across the BCS-BEC crossover. Schematic of the excitation spectrum on the BCS side and the BEC side. The solid line stands for the fermionic excitation of pair breaking, and the dashed line stands for bosonic excitation of center-of-mass motion of pairs. Note that the bosonic excitation becomes heavily damped once it enters the particle-hole continuum of the fermionic modes, as indicated by the slash lines. The dotted line stands for the slope that determines the critical velocity. A color version of this figure can be found in the resources tab for this book at cambridge.org/zhai.

this mode is the motion of bosons, and in this case, it should be the finite momentum motion of the diatomic molecules. The corresponding bosonic mode should also exist in the BCS regime, which describes the finite center-of-mass momentum motion of the Cooper pairs. However, such a mode is not included in the mean-field theory described above, because the mean-field theory starts with the Hamiltonian that only includes the scattering processes in the zero center-of-mass momentum sector. Here we take a phenomenological hydrodynamic approach, which is given by [59]

$$\frac{\partial^2 \delta n}{\partial t^2} = \frac{1}{m} \nabla \left[n \nabla \left(\frac{\partial \mu}{\partial n} \delta n \right) \right]. \tag{6.58}$$

This leads to a low-energy phonon dispersion $\omega = ck$ with $mc^2 = n(\partial \mu / \partial n)$. In the BCS regime, we take $\mu \approx E_F$ and c will saturate to the value of noninteracting Fermi gas as $v_F/\sqrt{3}$. In the BEC regime, $\mu = \mu_m/2$ with μ_m given by Eq. 6.57, hence, $c = \sqrt{4\pi \hbar^2 a_m n_m / 2m}$, which increases as a_s increases at the positive a_s side. To rigorously deduce this mode for the entire BCS-BEC crossover regime, one has to go beyond the mean-field theory and perform the random phase approximation (RPA) calculation [6]. This calculation will show that the sound velocity monotonically decreases from the BCS limit to the BEC limit, as we show schematically in Figure 6.7.

To show that the Fermi gas in the BCS-BEC crossover is indeed a superfluid, we recall the Landau criterion Eq. 3.31 for the critical velocity. In Figure 6.7 we draw the dotted line to indicate the slope for determining the critical velocity. In the BCS regime, the bosonic excitation has a relatively large velocity, and the fermionic excitation has a small gap. Thus, the low-energy response is dominated by the fermionic quasi-particles, and the critical velocity is determined by the fermionic excitation. Hence, $v_c = \Delta / \sqrt{2m\mu/\hbar^2}$, which increases from the BCS limit toward unitary regime. In the BEC regime, the velocity for bosonic mode becomes small, and fermionic quasi-particles acquire a large gap. Thus the low-energy response is dominated by this bosonic mode. The critical velocity is determined by the phonon velocity of the molecular BEC, which also increases from the BEC side toward unitary regime. Therefore, critical velocity reaches a maximum in the intermediate regime where the superfluidity is the most robust. A finite critical velocity is another hallmark evidence of superfluidity. Similar as discussed in Section 3.2, critical velocity can be measured by looking at the thermal excitations excited by a moving object, which can be a weak lattice optical potential in ultracold atom experiments. The critical velocity has been measured for different interaction strengths, as shown in Figure 6.8. Indeed, it shows a peak at the intermediate regime between the BCS and the BEC regimes.

Transition Temperature. Above we have discussed the zero-temperature case and here we will discuss the finite temperature situation, and we focus on the transition temperature from the normal state to the Fermi superfluid phase. Let us first start with the BCS mean-field Hamiltonian. Above T_c, when $\Delta = 0$, this mean-field Hamiltonian returns to the Hamiltonian for noninteracting Fermi gas, whose free energy is denoted by $\mathcal{F}_0$. Considering the normal to the superfluid transition as a second-order one, the order parameter

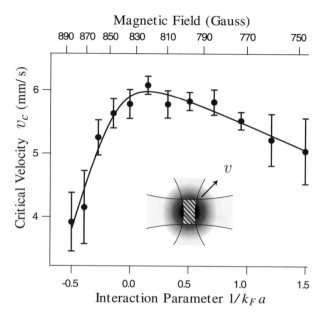

Figure 6.8 The critical velocity across the BEC-BCS crossover. Experimental measured critical velocity at different values of $-1/(k_F a_s)$. Reprinted from Ref. [120].

Δ can be arbitrarily small on the vicinity of the phase transition, and therefore, we can consider the Δ-term as perturbations. Using the finite temperature perturbation theory [101], it is straightforward to obtain the free energy expansion as

$$\mathcal{F} = \mathcal{F}_0 + \left(\sum_{\mathbf{k}} -\frac{1 - 2f(\epsilon_{\mathbf{k}} - \mu)}{2(\epsilon_{\mathbf{k}} - \mu)} - \frac{V}{g} \right) \Delta^2 + \dots, \tag{6.59}$$

where f is the Fermi distribution function. This gives a standard example of the Landau phase transition, where Δ plays a role as the order parameter. In the Landau theory, the free energy can be expanded in terms of order parameter as $\mathcal{F} = r|\Delta|^2 + b|\Delta|^4$. Here one can show b is always positive. When $r > 0$, the minimum of the free-energy occurs at $\Delta = 0$, and when $r < 0$, the minimum of the free-energy occurs at finite Δ. Hence the transition takes place at $r = 0$. A more systematic study can also include the spatial and temporal fluctuations of order parameter Δ and can give rise to the full Ginzburg–Landau function. We will come back to revisit this topic in the discussion of the Bose–Hubbard model in Section 8.1.

In this case, the phase transition is determined by $r = 0$, which is explicitly given by

$$r = V \left[\frac{1}{V} \sum_{\mathbf{k}} \left(-\frac{1 - 2f(\epsilon_{\mathbf{k}} - \mu)}{2(\epsilon_{\mathbf{k}} - \mu)} + \frac{1}{2\epsilon_{\mathbf{k}}} \right) - \frac{m}{4\pi \hbar^2 a_s} \right] = 0. \tag{6.60}$$

This is known as the *Thouless criterion* for the BCS transition temperature. In the BCS limit, if we fix μ at E_F, it gives a T_c as

$$k_B T_c = \frac{8\gamma}{\pi e^2} E_F \exp\left\{\frac{\pi}{2k_F a_s}\right\}, \tag{6.61}$$

where $\gamma = 0.5772$ denotes the Euler–Mascheroni constant. This shows that the ratio of $k_B T_c$ to the zero-temperature gap Δ given by Eq. 6.50 is a universal constant γ/π.

One of the important lessons we learn from above discussion of the zero temperature situation is that we need to seriously consider the change of chemical potential from the BCS regime to the BEC regime. Thus, we need the number equation to determine the chemical potential. Note that the BCS mean-field Hamiltonian recovers free fermion Hamiltonian above T_c, it gives the number equation

$$n_\sigma = n_{\text{free}} = \sum_{\mathbf{k}} \frac{1}{e^{(\epsilon_{\mathbf{k}} - \mu)/(k_B T)} + 1}. \tag{6.62}$$

This number equation, together with Eq. 6.60, determines a T_c as shown by the dashed line in Figure 6.9. However, this result does not look physical, as it keeps increasing toward the BEC side. As we have discussed in Section 3.1, for a given density of bosons, the Bose–Einstein condensation temperature is fixed as

$$k_B T_{\text{BEC}} \approx 3.31 \frac{\hbar^2 n_m^{2/3}}{2m} \approx 0.218 E_F. \tag{6.63}$$

This inconsistency is because the mean-field Hamiltonian only captures the pair formation, and therefore, the dashed line can be interpreted as the temperature scale that pairs start to form. However, if this temperature is too high, these bosonic pairs cannot Bose condense, and therefore, the system is still not a superfluid. Thus, to obtain the superfluid

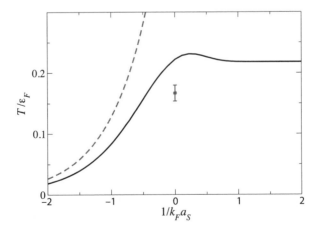

Figure 6.9 The superfluid transition temperature across the BCS-BEC crossover. T_c/T_F as a function of $1/(k_F a_s)$. The dashed line is the result without including the pair fluctuations, and the solid line includes the pair fluctuations. The point with error bar is the experimental results measured by Ref. [97]. Reprinted from Ref. [135]. A color version of this figure can be found in the resources tab for this book at cambridge.org/zhai.

transition temperature properly, we need to capture the possibility of noncondensed pairs. To this end, one needs to incorporate the thermal fluctuation of pairs, or in other words, one needs to include the noncondensed pairs with finite center-of-mass momentum. This can be done by including an extra term in the number equation to count for the contribution from these noncondensed pairs, which modifies the number equation as

$$n_\sigma = n_{\text{fluc}} + n_{\text{free}}. \tag{6.64}$$

The simplest way to compute n_{fluc} approximately is to include the Gaussian fluctuations, or equivalently to say, to include the summation of all ladder diagrams [128, 151]. We have performed the summation of all ladder diagrams in the discussion of the two-body problem in Section 2.2. Summation over the ladder diagram is exact for the two-body problem, but it is an approximation for a many-body problem. This approximation is known as the *Noziéres–Schmitt-Rink (NSR) approach* [128, 151]. We will not discuss this in details here, and the result of T_c from the NSR calculation is shown by the solid line in Figure 6.9. In the BCS limit $n_{\text{free}} \gg n_{\text{fluc}}$ at T_c, and the contribution from pairs fluctuation is insignificant. The NSR approach recovers the result of Eq. 6.61. In the BEC limit, $n_{\text{fluc}} \gg n_{\text{free}}$ at T_c, and approximately, one can show that

$$n_{\text{fluc}} = \frac{1}{V} \sum_{\mathbf{k}} \frac{1}{e^{\left(\frac{\hbar^2 \mathbf{k}^2}{4m} - \frac{\hbar^2}{ma_s^2} - 2\mu\right)/(k_B T)} - 1}. \tag{6.65}$$

In the BEC limit, $2\mu \to -\hbar^2/(ma_s^2)$, $n \approx n_{\text{fluc}}$ determines the Bose–Einstein condensation temperature of these diatomic molecules, as given by Eq. 6.63. Indeed the solid line in Figure 6.9 approaches a constant consistent with Eq. 6.63.

Experimental Observations. In ultracold atom systems, one can use the Feshbach resonance discussed in Section 2.4 as a major tool to tune the scattering length and to study the BCS-BEC crossover. There have been extensive experimental studies of the BEC-BCS crossover. Aside from measuring the critical velocity discussed above, here we list a few important ones:

- Pairwise Projection and the Time-of-Flight Imaging: As discussed in Section 3.1, the hallmark of Bose condensation is the macroscopic occupation of the zero momentum state. The way to measure this macroscopic occupation is to measure the momentum distribution through the time-of-flight imaging. In the BCS state, the Cooper pairs also have zero center-of-mass momentum. However, if one directly performs the time-of-flight imaging at the BCS side, the Cooper pairs immediately break when the density drops significantly during the time-of-flight. Hence, the experimental method is to sweep the magnetic field from the BCS side to the BEC side. On one hand, the sweep speed should be slow enough to ensure efficiently converting each Cooper pair into a tightly bound molecule. On the other hand, the sweep speed should be fast enough to avoid collisions between pairs, such that the momentum of molecule after projection equals to the center-of-mass momentum of Cooper pair before projection. Hence, the condensate fraction measured after magnetic field sweeping can reveal the center-of-mass momentum distribution of pairs before magnetic field sweeping. In this way, we can measure the

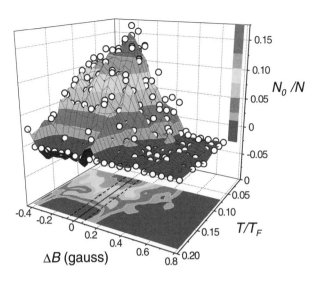

Figure 6.10 The condensate fraction across the BCS-BEC crossover. The molecular condensate fraction N_0/N measured after sweeping the magnetic field reveals the fraction of zero-momentum Cooper pairs before sweeping. Here N_0/N is measured as a function of initial magnetic field and temperature before sweeping. Negative ΔB is the BEC side, and positive ΔB is the BCS side. $\Delta B = 0$ corresponds to the Feshbach resonance point. Reprinted from Ref. [148]. A color version of this figure can be found in the resources tab for this book at cambridge.org/zhai.

condensate fraction for different interaction $-1/(k_F a_s)$ and different temperature T/T_F before sweeping, and the results are shown in Figure 6.10. It shows that for a given temperature, the condensed fraction is higher at the BEC side. One can also see a critical temperature for a given scattering length, which increases from the BCS limit to the BEC limit, and saturates in the BEC regime.

- Vortex Lattices: As we discussed in Section 4.2, another hallmark of the superfluid is the existence of stable quantized vortices, and these vortices form a triangular Abrikosov lattice. Such vortex lattices have also been observed in the entire BEC-BCS crossover regime, as shown in Figure 6.11.

- Radio-Frequency Spectroscopy Measurement of the Pairing Gap: As discussed in Section 5.2, the radio-frequency spectroscopy brings an atom in $|\downarrow\rangle$ to the third internal state without changing its momentum. Here, the momentum-resolved radio-frequency spectroscopy can be calculated via the Fermi Golden rule as

$$\mathcal{S}(\mathbf{k}, \omega) = \frac{2\pi}{\hbar} \sum_f |\langle f| \hat{c}^\dagger_{\mathbf{k},3} \hat{c}_{\mathbf{k},\downarrow} |\Psi_{\text{BCS}}\rangle|^2 \delta(\hbar\omega - (\epsilon_\mathbf{k} + \mathcal{E}_\mathbf{k}))$$

$$= \frac{2\pi}{\hbar} |v_\mathbf{k}|^2 \delta(\hbar\omega - (\epsilon_\mathbf{k} + \mathcal{E}_\mathbf{k})), \qquad (6.66)$$

and the integrated radio-frequency spectroscopy is given by

$$\mathcal{S}(\omega) = \int d^3\mathbf{k}\, \mathcal{S}(\mathbf{k}, \omega). \qquad (6.67)$$

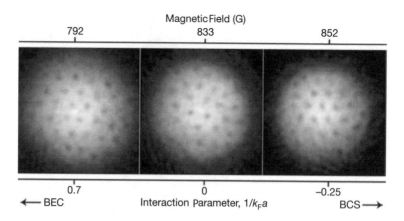

Figure 6.11 Vortex lattice across the BEC-BCS crossover. Triangular vortex lattices have been observed in three different scattering lengths with $1/(k_F a_s) = 0.7, 0$, and -0.25. Reprinted from Ref. [196].

The results of both integrated and momentum-resolved radio-frequency spectroscopy are shown in Figure 6.12. Compared with the excitation frequency of the normal gas, denoted by zero in Figure 6.12, the threshold frequency of the spectroscopy is shifted by the presence of pairing. One can see that the threshold frequency increases from the BCS limit to the BEC limit.

- Equation-of-State: The equation-of-state can be determined through measuring the pressure $\mathcal{P}$, with the help of the local density approximation. The experimental results are shown in Figure 6.13. Here we define $\mathcal{P} = \mathcal{P}_0 h$, where $\mathcal{P}_0$ refers to the pressure of noninteracting gas. h should be a function of the scattering length. In Figure 6.13 we have introduced $\tilde{\mu} = \mu$ for negative a_s and $\tilde{\mu} = \mu - E_b/2$ for positive a_s, where $E_b = -\hbar^2/(m a_s^2)$. In Figure 6.13, h is plotted as a function of dimensionless parameter δ, defined as $\hbar/(\sqrt{2m\tilde{\mu}} a_s)$. In the BCS side, $\mu \approx E_F$ and therefore $\sqrt{2m\tilde{\mu}}/\hbar \approx k_F$, and δ is essentially the same as $1/(k_F a_s)$. In the BEC side, since we have removed the contribution of the two-body binding energy in $\tilde{\mu}$, $h(\delta)$ only counts for the interaction effects between molecules. Because at the BEC side the scattering length between molecules a_m increases when a_s increases, the beyond mean-field LHY effect discussed in Section 3.3 becomes visible when $a_m^3 n$ becomes larger. Indeed, the experimental results shown in Figure 6.13 are closer to the equation-of-state of an interacting Bose gas including the LHY corrections.

The Unitary Regime. So far our discussions are mainly focused on the BCS regime with $-1/(k_F a_s) \gg 1$ or the BEC regime with $-1/(k_F a_s) \ll -1$. Here we move to discuss the unitary regime with $-1/(k_F a_s) \sim 0$, where the two-body interaction potential is tuned to the vicinity of a Feshbach resonance, and the scattering length diverges. For the ground state, since we have shown that the mean-field theory can correctly recover physics in the two limits, and since the entire process is a crossover, we can believe that the physics in the intermediate regime is also fairly captured, at least qualitatively. For instance, when a_s diverges, the mean-field theory predicts that both μ/E_F and Δ/E_F are constant, which

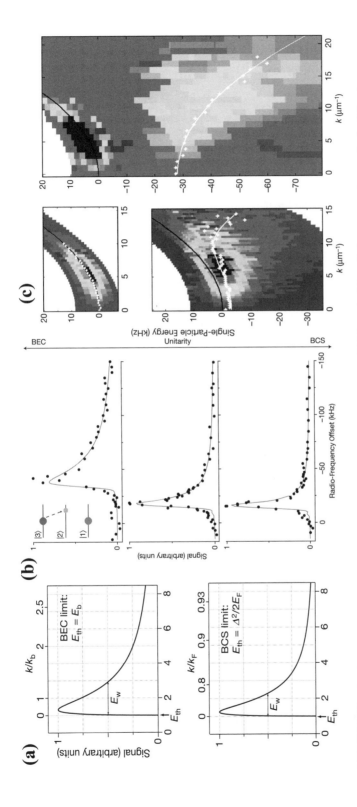

Figure 6.12

The radio-frequency spectroscopy. (a–b) Integrated radio-frequency spectroscopy. (a) Schematic of the radio frequency spectroscopy for the BCS regime (lower one) and the BEC regime (upper one). (b) Measured spectroscopy for $-1/(k_F a_s) = -0.4$ (upper one), 0 (middle one), and 0.3 (lower one). (c) Momentum-resolved radio frequency spectroscopy for weakly interacting case (left and upper one), unitary with $-1/(k_F a_s) = 0$ (left and lower one), and the BEC side with $-1/(k_F a_s) = -1$ (right one). (a) and (b) are reprinted from Ref. [155], and (c) is reprinted from Ref. [166]. A color version of this figure can be found in the resources tab for this book at cambridge.org/zhai.

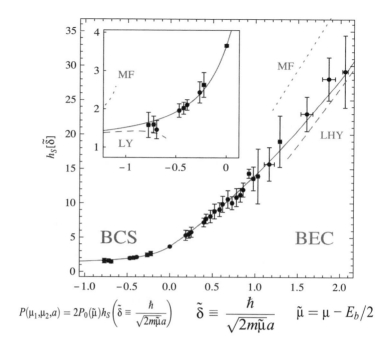

$$P(\mu_1,\mu_2,a) = 2P_0(\tilde{\mu})h_S\left(\tilde{\delta} \equiv \frac{\hbar}{\sqrt{2m\tilde{\mu}a}}\right) \qquad \tilde{\delta} \equiv \frac{\hbar}{\sqrt{2m\tilde{\mu}a}} \qquad \tilde{\mu} = \mu - E_b/2$$

Figure 6.13   The equation-of-state across the BEC-BCS crossover. The pressure $P = P_0h_S$, where P_0 is the pressure of the noninteracting single-component Fermi gas; h_S is the vertical axis. See text for the definition of horizontal axes δ. The solid lines are fitting functions. The two dashed lines stand for the equation-of-state for interacting bosons at the mean-field level, denoted by MF, and the equation-of-state including the LHY correction, denoted by LHY. Reprinted from Ref. [125]. A color version of this figure can be found in the resources tab for this book at cambridge.org/zhai.

means that the pairing energy is proportional to the kinetic energy. However, beyond the ground state physics, this regime is difficult to understand theoretically, at least for the following two reasons:

- In the superfluid state, as far as the low-energy physics is concerned, in the BCS side only the fermionic excitations are important, which can be well studied by the BCS mean-field approach discussed above, and in the BEC side only the bosonic excitations are important, which can be well studied by the Bogoliubov theory as discussed in Section 3.3. However, in the unitary regime, both the fermionic and the bosonic excitations are important, and they need to be treated on equal footing.

- In the normal state, when $-1/(k_Fa_s) \gg 1$, the system is a weakly interacting Fermi gas and k_Fa_s can be treated as perturbation parameter. When $-1/(k_Fa_s) \ll -1$, although attraction in two-body interaction potential is quite strong, the main contribution of the interaction effect is counted by forming the two-body bound state. Once we subtract the binding energy in the total energy, the system can be considered as a weakly interacting Bose gas. As we discussed in Section 3.3, $a_m^3 n_m$ can be treated as perturbation parameter, where a_m is the scattering length between molecules and n_m is the density of molecules. In the unitary regime, the system is neither a weakly interacting Fermi gas nor a weakly

| Box 6.2 | Holographic Duality and AdS/CFT |

Holographic duality states that some quantum field theory in spacetime D dimension can have a gravity theory dual in spacetime $D+1$ dimension [69]. Here "dual" means that the correlation functions in the field theory can always be equivalent to some correlation functions in the gravity theory. Usually, the field theory possesses the conformal symmetry, known as the conformal field theory (CFT), and the gravity theory is usually defined in the anti–de Sitter (AdS) geometry, since the symmetry of AdS geometry is compatible with the conformal symmetry. Therefore, the holographic duality is also known as the AdS/CFT correspondence. This duality is also a "strong-weak" duality. Usually the field theory is a strongly interacting theory and the interaction is so strong that there is an absence of well-defined quasi-particles. But hopefully, the dual gravity theory is a weak coupling classical gravity. Thus, we can compute the correlation functions in the gravity side and use the mapping to determine the correlation functions in the field theory side. That is how the holographic duality can help us study quantum many-body systems. So far, the holographic duality has been firmly established only in a very few examples. One example in string theory is the duality between supersymmetric Yang–Mills and an AdS supergravity theory. Another example in condensed matter physics is the duality between the Sachdev–Ye–Kitaev model and AdS_2 gravity. Nevertheless, we have many examples of quantum many-body systems that have emergent conformal symmetry and are absent of quasi-particle description, such as the unitary Fermi gas discussed here and the quantum critical regime discussed in Section 8.1. It is an interesting open research subject whether some properties of these systems are close to that predicted by the holographic duality.

interacting Bose gas. In the unitary regime, the total interaction energy is proportional to the total kinetic energy, which makes the system intrinsically nonperturbative.

The challenges in solving strongly interacting quantum systems such as the unitary Fermi gas call for new ideas, and the holographic duality discussed in Box 6.2 is one of such new ideas. When a_s diverges, the system lacks of any relevant length scales and the Fermi energy is the only energy scale of the problem. Thus, the unitary Fermi gas is one of the examples of scale invariant quantum many-body systems, as discussed in Box 3.2. Moreover, it can be shown that the unitary Fermi gas possesses the Schrödinger symmetry, which is a nonrelativistic version of the conformal symmetry and is a symmetry group larger than the usual Galilean symmetry [162, 161]. In addition, because of the strong interaction effects in the unitary Fermi gas, it is still an open issue whether the system is a Fermi liquid or non-Fermi liquid right above the superfluid transition temperature. If there is no well-defined quasi-particles in the low-temperature normal state of a unitary Fermi gas, together with the presence of the conformal symmetry, it is interesting to ask whether the system shares similar properties as predicted by the holographic duality.

One prediction from the holographic theory is a universal lower bound for η/s. Here η is the shear viscosity, which is defined as the ratio of the shear force to the velocity gradient in a fluid, and s is the entropy. In a quantum theory, it is not obvious that these two quantities are related. However, if a quantum theory has a gravity dual, these two quantities are related in the dual gravity theory. With the holographic dual to an Einstein gravity, it can be shown that

$$\frac{\eta}{s} = \frac{\hbar}{4\pi k_{\mathrm{B}}}. \tag{6.68}$$

It is very intriguing to note that this equality holds as long as the quantum theory has a gravity dual, and this ratio only depends on $\hbar$ and k_{B}, and is independent of the details of physical models. In general situations, it is speculated that for all fluids, η/s is always bounded by

$$\frac{\eta}{s} \geqslant \frac{\hbar}{4\pi k_{\mathrm{B}}}. \tag{6.69}$$

It is also speculated that, the more a quantum system is close to a theory with gravity dual, the more its η/s is close to the lower bound. In Figure 6.14 (a) we have shown the experimental measurements of η/s on unitary Fermi gas over different temperatures. It is found that the measured η/s indeed approaches the bound when the system enters quantum degenerate regime, but it does not violate the bound. In Figure 6.14 (b), we also compare η/s for different fluids, and all fluids we know so far obey the bound. It is interesting to note that the unitary Fermi gas and the quark-gluon plasma are two systems whose η/s is the closest to the bound. These two systems are the coldest and the hottest systems in human's laboratory. However, both of them share the same feature of being strongly interacting Fermi gas, whose interaction energy is comparable to the kinetic energy.

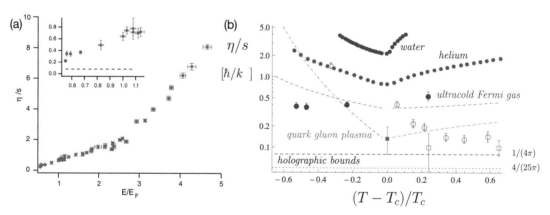

Figure 6.14 η/s and the holographic prediction. (a) η/s measured for the unitary Fermi gas as a function of temperature. Here the horizontal axis E determines temperature. The inset makes a comparison with the dashed line indicating the bound predicted by holographic duality. (b) η/s for different kinds of fluids. It is plotted in units of $\hbar/k_{\mathrm{B}}$ ($\hbar/k$ in the figure vertical axes) so $1/(4\pi)$ indicated by the horizontal dashed line shows the bound predicted by holography. (a) is reprinted from Ref. [26], and (b) is reprinted from Ref. [3]. A color version of this figure can be found in the resources tab for this book at cambridge.org/zhai.

Exercises

6.1 Solve the Cooper problem Eq. 6.8 and compare the solution with the solution of the two-body problem in a vacuum of Eq. 6.5 for positive a_s.

6.2 Calculate the ground state energy of the BCS mean-field Hamiltonian as a function of Δ, and minimize this ground state energy with respect to Δ to obtain the gap equation.

6.3 Discuss the self-consistent solution of the BCS problem for the Hamiltonian

$$H_{BCS} = \sum_{\mathbf{k},\sigma}(\epsilon_{\mathbf{k}} - \mu_\sigma)c^\dagger_{\mathbf{k}\sigma}c_{\mathbf{k}\sigma} + \sum_{\mathbf{k}}\Delta c^\dagger_{\mathbf{k}\uparrow}c^\dagger_{-\mathbf{k}\downarrow} + \sum_{\mathbf{k}}\Delta^* c_{-\mathbf{k}\downarrow}c_{\mathbf{k}\uparrow} \qquad (6.70)$$

when $\mu_\uparrow \neq \mu_\downarrow$, and compare the energy of the BCS state with the free Fermi gas for different $h = \mu_\uparrow - \mu_\downarrow$.

6.4 Derive Eq. 6.59 using the second-order perturbation theory.

6.5 Compute the radio-frequency spectroscopy using the mean-field theory of the BCS-BEC crossover, and discuss the threshold energy for the BCS limit and the BEC limit, respectively.

Part IV

Optical Lattices

7

Noninteracting Bands

Learning Objectives

- Introduce the basic concepts of single-particle dispersion in a lattice.
- Introduce maximally localized Wannier wave functions and their importance in constructing interacting models.
- Introduce the basic idea of quantum simulation.
- Introduce the Dirac point in the honeycomb lattice and the concept of a semimetal.
- Illustrate the Dirac and the Weyl points as topological defects in the momentum space.
- Introduce the concept of the symmetry protection for the topological stability of the Dirac point.
- Introduce the Su–Schrieffer–Heeger model and the concept of symmetry-protected topology.
- Introduce the Haldane model and the topological phase transition in the Haldane model.
- Summarize the common mathematical structures between topological characters of a band theory and topological excitation in a Bose condensate.
- Discuss the physical consequences of topological invariants for a band insulator in both near-equilibrium transport and far-from-equilibrium quench dynamics.
- Discuss the equivalence between neural atoms in a moving lattice and charged particles in an electric field.
- Introduce the Floquet theory and the Floquet effective Hamiltonian from the high-frequency expansion.
- Discuss various applications of periodical-driven optical lattices.
- Introduce the Hamiltonian for an atom-cavity system and highlight the role of the Langevin force term.
- Introduce the superradiant quantum phase transition of a Bose condensate coupled to cavity light.

7.1 Basic Band Theory

As we have discussed in Section 1.3, when atoms are placed in a pair of counter-propagating laser beams along $\hat{x}$ and with wave vector k_0, and these two lasers have the same polarization, atoms can experience a scalar light shift given by $V_x \cos^2(k_0 x)$, where V_x is proportional to the strength of the laser fields. Similarly, one can apply another two pairs of counterpropagating laser beams along $\hat{y}$ and along $\hat{z}$ directions, respectively. Inside these laser fields, the single-particle Hamiltonian is then given by

$$\hat{H} = -\frac{\hbar^2 \nabla^2}{2m} + V_{\text{lat}}(\mathbf{r}) \tag{7.1}$$

$$V_{\text{lat}}(\mathbf{r}) = V_x \cos^2(k_0 x) + V_y \cos^2(k_0 y) + V_z \cos^2(k_0 z), \tag{7.2}$$

where V_x, V_y, and V_z are controlled by the laser intensities of these three pairs of laser beams. This Hamiltonian describes a particle moving in a three-dimensional cubic lattice. To understand the physics of optical lattices, we first need to review the basics of the single-particle dispersion in such a periodical potential, namely, the band theory.

One-Dimensional Lattices. Since the Hamiltonian of Eq. 7.2 is separable along these three different directions, for simplicity, let us first consider a one-dimensional Hamiltonian

$$\hat{H} = -\frac{\hbar^2 \partial_x^2}{2m} + V_x \cos^2(k_0 x). \tag{7.3}$$

In the absence of the lattice, the system possesses the translational symmetry, that is, an operation $x \rightarrow x + x_0$ can leave the Hamiltonian invariant for any x_0, and the eigenstate can be chosen as the plane wave e^{ikx}. Here the momentum k is a good quantum number and can take any real value. In the presence of the lattice, the system is no longer translational invariant but it still maintains a discrete translational symmetry, that is to say, x_0 can only choose a set of discrete values of $\pi n/k_0$ with n being any integer. Due to the lattice potential, a plane wave state e^{ikx} can be mixed with all other plane waves $e^{i(k+2nk_0)x}$, where n can be any integer. Thus, we restrict k between $-k_0$ and k_0, which is called the *first Brillouin zone* of this case, and k is now called the *quasi-momentum*. For each quasi-momentum k, the mixing of all states $\{e^{i(k+nk_0)x}\}$ leads to infinite many eigenstates, which are labeled by another quantum number m called the *band index*. After the restriction, the quasi-momentum is still a good quantum number, which is in fact ensured by the discrete translational symmetry.

More explicitly, the eigenstate of this Schrödinger equation can always be written as a Bloch wave function labeled by the quasi-momentum k and the band-index m,

$$\psi_k^m(x) = \sum_n u_n^m(k) e^{i(k+2nk_0)x}. \tag{7.4}$$

$u_n^m(k)$ can be obtained by diagonalizing the Hamiltonian, that is,

$$\frac{\hbar^2}{2m}(k + 2nk_0)^2 u_n^m(k) + \frac{V_x}{4} u_{n-1}^m(k) + \frac{V_x}{4} u_{n+1}^m(k) = \mathcal{E}_k^m u_n^m(k), \tag{7.5}$$

where the upper index m denotes the mth eigenvalue of this matrix with the eigen-values sorted from the lower to the higher. The eigenenergy $\mathcal{E}_k^m$ is the single-particle dispersion for the mth band in the lattice.

The weak lattice regime is called the *free particle limit* and the lattice effect can be treated perturbatively. The effect is most significant at the edge of the Brillouin zone where the kinetic energies of the two plane wave states, $e^{-ik_0 x}$ and $e^{ik_0 x}$, are degenerate in the absence of lattices. Since their momenta are differed by $2k_0$, the lattice potential can couple them. The lattice effect manifests itself as a degenerate perturbation, which lifts the

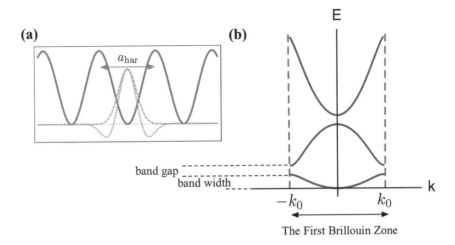

Figure 7.1 (a) Schematic of the band theory. The solid line is the schematic of the real space lattice structure. The dotted line shows the lowest Wannier wave function, and the dashed line shows a Gaussian function as the approximate envelope of the Wannier wave function. (b) Schematic of the band structure. The band dispersion is plotted in the first Brillouin zone. Both the bandwidth of the lowest band and the band gap between the lowest and the first excited band are marked. A color version of this figure can be found in the resources tab for this book at cambridge.org/zhai.

Box 7.1   **Different Kinds of Insulators**

Insulator refers to a state whose charge excitation is gapped. Therefore, it cannot conduct mass current, or charge current if particles are charged, when the applied potential gradient is smaller than the charge gap. There are different kinds of insulators based on the origin of the charge gap. If the charge gap is originated from the band structure, as discussed in this chapter, it is called a band insulator. If the charge gap is originated from the interaction effect, as we will discuss in Chapter 8 for the Hubbard model, it is called a Mott insulator. A band insulator can be further classified as a topological trivial band insulator and a topological nontrivial band insulator, depending on the topological invariants of the occupied band, as we will discuss in Section 7.3. Note that this discussion does not include the disorder effect.

degenerate and opens up a band gap proportional to V_x, as one can see from a typical band structure shown in Figure 7.1(b).

In this case, the band gap refers to an energy window where the density-of-state vanishes. For fermions in such a lattice, when the chemical potential lies inside the band gap, the particle and the hole excitation always cost finite energy. Therefore, changing density of the system costs finite energy, which is called the *charge gap*. A system with finite charge gap is called an *insulator*, and we have summarized different kinds of insulators in Box 7.3. A system with vanishing charge gap is a *metallic state*. For instance, the noninteracting fermions and the Fermi liquid discussed in Section 5.1 are metallic states. When an infinitesimal small potential gradient is applied to the system, a metallic state can always conduct charge, but an insulator cannot.

From the Bloch wave function one can construct the Wannier wave function as

$$w_m(x - x_i) = \frac{1}{\sqrt{N_s}} \sum_k e^{-ikx_i} \psi_k^m(x), \qquad (7.6)$$

where N_s is the total number of sites, and $x_i = (2i + 1)\pi/(2k_0)$ labels the site index, as shown in Figure 7.1(a). Equivalently, we can write

$$\psi_k^m(x) = \frac{1}{\sqrt{N_s}} \sum_i e^{ikx_i} w_m(x - x_i). \qquad (7.7)$$

It is important to note that all the wannier functions form an orthonormal and complete bases, that is,

$$\int w_m^*(x - x_i) w_{m'}(x - x_j) dx = \delta_{mm'} \delta_{ij}. \qquad (7.8)$$

Maximally Localized Wannier Wave Function. Here we should mention that, given a set of the Bloch wave functions, the construction of the Wannier wave functions are not unique. It is because we can add a momentum dependent phase into each Bloch wave function, that is,

$$\tilde{\psi}_k^m(x) = e^{i\theta_m(k)} \psi_k^m(x), \qquad (7.9)$$

where $\theta_m(k)$ can be any smooth function of k satisfying the periodic boundary condition in the first Brillouin zone, that is, $\theta_m(-k_0) = \theta_m(k_0)$. By replacing $\psi_k^m(x)$ by $\tilde{\psi}_k^m(x)$ in the Eq. 7.6, it yields a different set Wannier wave functions with different shape. In the later discussion of interacting atoms in optical lattices, we expand the field operators in the Wannier wave function bases. For this purpose, we want to make the Wannier wave function as localized as possible. This is because when the Wannier function is sufficiently localized, the kinetic energy can be captured mainly by the short-range hopping and the interaction can be captured mainly by the on-site interaction, such that the Hamiltonian takes the simplest form in this bases. To this end, we need to choose a proper function of $\theta_m(k)$ such that the resulting Wannier wave functions are maximally localized. The maximally localized Wannier wave function is an important research topic in the band theory, and we will not go into details here [91]. In the simple case discussed here, in order for the Wannier wave function to be maximally localized, $\theta_m(k)$ should be chosen in such a way that the phases of the Bloch wave functions are uniform in k.

When the lattice potential becomes deep enough, one can expand the lattice potential around the bottom of each minimum x_i, and up to the quadratic order we obtain a harmonic potential

$$\hat{H} = -\frac{\hbar^2 \partial_x^2}{2m} + V_x k_0^2 (x - x_i)^2. \qquad (7.10)$$

It is convenient to introduce the recoil energy $E_R = \hbar^2 k_0^2/(2m)$ as the energy unit, and we denote $V_x = \alpha_x E_R$, then it is easy to show that the effective harmonic frequency is

$$\hbar \omega = 2 \sqrt{\alpha_x E_R}, \qquad (7.11)$$

and the effective harmonic length is given by

$$a_{\text{har}} = \frac{1}{k_0 \alpha_x^{1/4}}. \tag{7.12}$$

The deep lattice regime is called the *tight binding limit*. It can be shown that the envelope of the maximally localized Wannier wave function for the lowest band $w_0(x - x_i)$ is very close to the Gaussian wave function with the harmonic length a_{har}, as shown in Figure 7.1(a).

Nevertheless, one should pay attention to the difference between a Gaussian wave function and the actual Wannier wave function. Because the Gaussian wave function is always positive definite, two Gaussian wave functions with different x_i cannot be orthogonal with each other. Nevertheless, two Wannier wave functions with different x_i are orthogonal with each other. Thus, to be more precise, only the envelope of $w_0(x - x_i)$ is close to a Gaussian. Away from the center x_i the actual Wannier wave function displays oscillatory behavior between positive and negative values, as also shown in Figure 7.1(a).

The harmonic potential approximation also shows that the band gap can be roughly estimated as $\hbar\omega \propto \sqrt{V_x}$, which is different from the weak lattice regime discussed above. One should also caution that this harmonic potential approximation only works for a few lowest energy bands. It is because when $n > \sqrt{\alpha_x}/2$, $n\hbar\omega > V_x$, which means that the eigenenergies of the harmonic levels exceed the lattice potential height, and therefore, the harmonic expansion Eq. 7.10 is no longer valid. In other words, for a given lattice potential, the tight-binding approximation works for low-lying bands, and the free particle approximation works for high-energy bands.

High-Dimensional Lattices. The discussion below can be generalized straightforwardly to two-dimensional square or three-dimensional cubic lattices. For a two-dimensional square lattice, the band index m is characterized by two integer numbers as (m_x, m_y). For the ground band $m = (0, 0)$, and two of the first excited bands are $m = (0, 1)$ and $m = (1, 0)$, which are degenerate if $V_x = V_y$. They are called the p_x and the p_y bands because their corresponding maximally localized Wannier wave functions have the same symmetry as the p_x and p_y orbital in the centrifugal potential problem. Similarly, for a three-dimensional cubic lattice, the band index m is labeled by (m_x, m_y, m_z).

In Figure 7.2 we plot the density-of-state for a three-dimensional cubic lattice. There are two notable new features in this plot that are worth emphasizing.

- Direct versus Indirect Band Gap: Considering two neighboring bands, if the minimum energy state of the upper band and the maximum energy state of the lower band occur at the same quasi-momentum, it is called a *direct band gap*. If they occur in different quasi-momentum, it is called an *indirect band gap*. Taking the band gap between the ground and the first excited band as an example, in one dimension, the bottom of the first excited band and the top of the ground band are both at the edge of the Brillouin zone. This is a direct band gap. In this case, since the gap opens for any finite value of the lattice potential, there is always a real gap defined by vanishing density-of-state in certain energy window. When the dimension is higher than two, indirect band can occur, and the bottom of the first excited band does not necessarily occur at the same quasi-momentum as the top of the ground band. Under this situation, the lowest energy

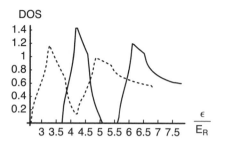

DOS

3 3.5 4 4.5 5 5.5 6 6.5 7 7.5 $\frac{\epsilon}{E_R}$

Figure 7.2   The density-of-state (DOS) for an optical lattice. The density-of-state (DOS) is plotted as a function of energy ϵ/E_R for a three-dimensional cubic lattice. Here we have taken $V_x = V_y = V_z = V_0$, and $V_0/E_R = 2$ for the dashed line, and $V_0/E_R = 3$ for the solid line.

state of the first excited band can be lower than the highest energy state of the ground band, as shown by the dashed line in Figure 7.2 when the lattice is sufficiently shallow. Therefore, there is no real gap in terms of vanishing density-of-states. So a band gap with vanishing density-of-state exists only when the strength of the lattice is larger than certain critical value, as shown by the solid line in Figure 7.2.

- Van Hove Singularity: As one can also see from Figure 7.2, another notable feature is that there exist kinks in the density-of-state where the derivative is not continuous. This is known as the *Van Hove singularity*. It happens when $|\nabla \epsilon^m(k)| = 0$ for certain k-points of the equal energy contour, that is to say, when the band dispersion has a local maximum, or a local minimum or a saddle point. In the simple one-dimensional lattice discussed above, such a singularity can only happen at the band edge. For a two-dimensional or three-dimensional lattice, such singularities can also happen inside a band.

Momentum Mapping versus Band Mapping. There are two different ways to perform the time-of-flight measurement of the momentum distribution for atoms in optical lattices, which are known as the *momentum mapping* and the *band mapping*, respectively. For the momentum mapping, the optical lattices and the external harmonic trap are turned off abruptly and simultaneously. In this case, the measurement projects the quantum state with lattice into the eigenstate in free space, that is, the plane wave state. Thus, the occupation n_k^m of a Bloch state ψ_k^m transfers into occupation in a series plane wave states with momentum $k + 2nk_0$, which are given by $n_k^m |u_n^m(k)|^2$. That is to say, for a given plane wave momentum p, one can always find out an integer n such that $p - 2nk_0$ lies between $-k_0$ and k_0, the occupation n_p measured by this time-of-flight is given by

$$n_p = \sum_m |u_n^m(p - 2nk_0)|^2 n_{p-2nk_0}^m. \qquad (7.13)$$

The momentum distribution shown in Figure 8.4 of Section 8.1 is a momentum mapping experiment. As one can see from (b-e) of Figure 8.4, when bosons are condensed in the lowest energy state in the optical lattices, that is, when the state with the band index $m = 0$ and the quasi-momentum $\mathbf{k} = 0$ is macroscopically occupied, the momentum distribution measured in this way not only shows peak at $\mathbf{k} = 0$ but also shows peaks at all the reciprocal momentum vectors.

For the band mapping, first the optical lattice potential is slowly turned off, after which the external harmonic potential is turned off abruptly to measure the momentum distribution. During the first stage, $u_n^m(k)$ gradually approaches either zero or unity because the coupling between different plane wave states is provided by the lattice potential. As the lattice potential is gradually turned off, the Bloch state ψ_k^m adiabatically approaches one of the plane wave state $e^{i(k+2nk_0)x}$. For instance, the Bloch state ψ_k^m of the lowest band with $m = 0$ is adiabatically connected to the plane wave state e^{ikx} with k lying in the first Brillouin zone between $-k_0$ and k_0, and ψ_k^m of the second band with $m = 1$ is adiabatically connected to the plane wave state $e^{i(k+k_0)x}$ for $k > 0$ and $e^{i(k-k_0)x}$ for $k < 0$, that is to say, the plane wave momentum lies in the second Brillouin zone. In this way, the Bloch states in the higher bands are mapped to the plane wave states in the higher Brillouin zone. Now the question is whether the occupation n_k^m of the Bloch state ψ_k^m before turning off the lattice can also be transferred into the occupation of the corresponding plane wave state after turning off the lattice. This crucially depends on the speed of how fast the lattice potential is turned off. First of all, the speed has to be slow enough such that the lattice decreases adiabatically with respect to the energy difference between different Bloch bands. This is always difficult for certain momentum points where the energy separation between Bloch bands becomes vanishing small as the lattice potential vanishes, for instance, the edge of the first Brillouin zone. Thus, the band mapping is always not so accurate around these points. Another condition is that the speed has to be faster than all the other time scales that relax the momentum distribution, for instance, the period of the harmonic trap or the interatomic collision time. Under these two conditions, the population of different Brillouin zone in the time-of-flight reflects the initial population of different bands. Thus, this scheme is called the *band mapping*. Figure 7.3 shows a band mapping measurement of noninteracting fermions in optical lattices [99]. One can see that as the characteristic density increases from the left to the right, the Fermi surface changes from a circular shape to a square shape reflecting the geometry of the first Brillouin zone [99].

Bloch Oscillation. In free space, the single-particle dispersion is $\epsilon_k = \hbar^2\mathbf{k}^2/(2m)$, which leads to the velocity $\mathbf{v} = (1/\hbar)\partial\epsilon_{\mathbf{k}}/\partial\mathbf{k} = \hbar\mathbf{k}/m$. The velocity always monotonically

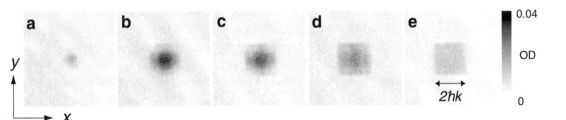

Figure 7.3   Time-of-flight measurement with band mapping. Band mapping measurement of the quasi-momentum distribution for noninteracting fermions in an optical lattice. As the characteristic density increases from (a) to (e), the Fermi surface geometry also changes from a small circle to a square. Reprinted from Ref. [99].

increases with the increasing of k. However the situation is very different with lattice. As the velocity is still defined as

$$\mathbf{v} = \frac{1}{\hbar} \frac{\partial \mathcal{E}_{\mathbf{k}}^m}{\partial \mathbf{k}}, \tag{7.14}$$

it vanishes at the band edge. Taking the one-dimensional band structure as an example, as k varies from the band bottom toward the band edge, the velocity first increases from zero and then decreases back to zero at the band edge. As we will discuss later in this section, in this tight binding limit the band dispersion of the lowest band can be well approximated by $\mathcal{E}_k = -2J \cos(\pi k/k_0)$, where $4J$ is the bandwidth. Taking this tight binding dispersion as an example, we will have

$$\frac{v}{v_R} = \frac{J\pi}{E_R} \sin\left(\frac{\pi k}{k_0}\right), \tag{7.15}$$

where $v_R = \hbar k_0/m$. Eq. 7.15 shows that the velocity v is a periodic function in k. which has a dramatical consequence known as the *Bloch oscillation*.

We consider a constant force $F\hat{x}$ applied to the particle. In solid state materials, this can be realized by applying an electric field to electrons. In ultracold atom systems, this can be realized by applying a gravity field. Or alternatively, when the optical lattice is moving with a constant acceleration, in the comoving frame the lattice becomes stationary but the atoms experience an inertial force proportional to the acceleration, which simulates charged particle placed in the electric field. This will be discussed in detail in Section 7.4.

Let us first consider the case with nonzero band gap, and the force F is weak enough that the interband transition induced by this force is negligible. In this case, by a semiclassical analysis, we have

$$\hbar \frac{dk}{dt} = F. \tag{7.16}$$

Thus

$$k(t) = k_0 + \frac{F}{\hbar} t, \tag{7.17}$$

where k_0 is the initial quasi-momentum at $t = 0$. Substituting Eq. 7.17 into Eq. 7.15, we find that the velocity shows an oscillatory behavior as

$$\frac{v(t)}{v_R} = \frac{J\pi}{E_R} \sin\left[\frac{\pi}{k_0}\left(k_0 + \frac{F}{\hbar} t\right)\right]. \tag{7.18}$$

This oscillatory behavior in v can also lead to a spatial oscillation, and this is known as the Bloch oscillation. In the solid state materials, it is very difficult to observe the Bloch oscillation of electrons because the scattering between electrons or between electrons and impurities can change the momentum of electron in a time scale much faster than the Bloch oscillation period. The ultracold atom systems can be impurity free and the interaction between atoms can also be suppressed by keeping the system dilute enough and by keeping the scattering length between atoms small enough. Thus, it offers an ideal platform to observe the Bloch oscillation. Indeed, as shown in Figure 7.4, the Bloch oscillation has been observed by loading atoms in an accelerated moving optical lattice, and the mean velocity is directly measured by the time-of-flight measurements after certain duration of

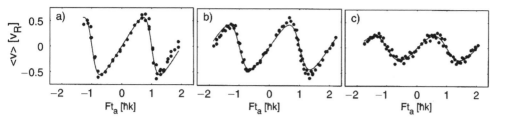

Figure 7.4 Experimental measurement of the Bloch oscillation. Mean value of atom velocity $\langle v \rangle$, in units of v_R, is measured as a function of the acceleration time t. Three different lattice potentials are considered, with (a) $V = 1.4E_R$, (b) $V = 2.3E_R$, and (c) $V = 4.4E_R$. The horizontal axis Ft is in units of $\hbar k_0$, and the negative Ft is measured by changing the sign of F. Reprinted from Ref. [16].

lattice acceleration [16]. One can see that the oscillation amplitude decreases with the increasing of lattice potential, and this is because the bandwidth J/E_R decreases as the lattice potential increases, which is consistent with Eq. 7.18.

The Hubbard Model. We consider interacting bosons and fermions in optical lattice, and derive the Hamiltonian that is well known as the *Hubbard model*. We will show below that, by expanding the field operators in terms of the maximally localized Wannier wave function bases, both the kinetic energy and the interacting model can take the simplest form. First considering interacting bosons in a three-dimensional cubic lattice, $\hat{\psi}(\mathbf{r})$ is the field operator for bosons and we expand

$$\hat{\psi}(\mathbf{r}) = \sum_{m,\mathbf{R}_i} \hat{b}_{m,i} w_m(\mathbf{r} - \mathbf{R}_i), \tag{7.19}$$

where $\hat{b}_{m,i}$ is the annihilation operator of bosons with site index i and band index m. The most general form of the lattice model reads

$$\hat{H} = -\sum_{ijm} J_{ij}^m \hat{b}_{m,i}^\dagger \hat{b}_{m,j} + \frac{1}{2} U_{ijkl}^{mnm'n'} \sum_{iji'j'}^{mnm'n'} \hat{b}_{m,i}^\dagger \hat{b}_{m',i'}^\dagger \hat{b}_{n,j} \hat{b}_{n',j'} - \mu \sum_{i,m} \hat{b}_{m,i}^\dagger \hat{b}_{m,i}, \tag{7.20}$$

where

$$J_{ij}^m = -\int d^3\mathbf{r}\, w_m(\mathbf{r} - \mathbf{R}_i) \left(-\frac{\hbar^2 \nabla^2}{2m} + V_{\text{lat}}(\mathbf{r}) \right) w_m(\mathbf{r} - \mathbf{R}_j) \tag{7.21}$$

and

$$U_{ijkl}^{mnm'n'} = \frac{4\pi \hbar^2 a_s}{m} \int d^3\mathbf{r}\, w_m(\mathbf{r} - \mathbf{R}_i) w_n(\mathbf{r} - \mathbf{R}_j) w_{m'}(\mathbf{r} - \mathbf{R}_k) w_{n'}(\mathbf{r} - \mathbf{R}_l). \tag{7.22}$$

Now we shall discuss how this general Hamiltonian can be simplified.

- The interaction energy is estimated as $\sim \hbar^2 a_s/(ma_{\text{har}}^3)$, where a_{har} denotes the typical width of the Wannier wave function, and $a_{\text{har}} = 1/(k_0 \alpha^{1/4})$ for a potential depth $V = \alpha E_R$. The band gap is of the order $\hbar^2/(ma_{\text{har}}^2)$. Here we focus on the situation that the interaction potential is away from a scattering resonance, that is, $k_F a_s \ll 1$ with k_F defined as $(6\pi^2 n)^{1/3}$. And because for a typical density of ultracold atomic gas, $k_F \sim k_0$.

Hence we have $k_0 a_\mathrm{s} \ll 1$. Thus, the ratio of the interaction energy to the band gap is of the order $\sim a_\mathrm{s}/a_\mathrm{har} \sim k_0 a_\mathrm{s} \ll 1$.

- In the ultracold regime, temperature can also be much smaller than the band gap, and the thermal population in the higher bands can also be safely ignored. Because of these two reasons, we can only keep the lowest band with $m = 0$. Hereafter we shall ignore the band index.
- Since the wannier wave function for the lowest band is as localized as a Gaussian wave function, the hopping matrix element J_{ij} decays as $\sim e^{-|\mathbf{R}_i - \mathbf{R}_j|^2/a_\mathrm{har}^2}$. So we can only keep J_{ij} for the nearest neighboring sites, short-noted as J.
- For similar reasons as discussed above, compared with the on-site interaction term, all the other interactions between two different sites will be suppressed by an exponential factor. And compared with the nearest hopping term, the interaction between two nearest neighboring sites is suppressed by a factor of $\sim a_\mathrm{s}/a_\mathrm{har} \sim k_0 a_\mathrm{s}$. Therefore, except for the on-site interaction, all the other interaction terms can be safely ignored. Hence, we only keep $U_{ijkl}^{mnm'n'}$ with $m = n = m' = n' = 0$ and $i = j = k = l$, short-noted by U.

With these justifications, we arrive at a simple single-band Bose–Hubbard model (BHM)

$$\hat{H}_\mathrm{BHM} = -J \sum_{\langle ij \rangle} \hat{b}_i^\dagger \hat{b}_j + \frac{U}{2} \sum_i \hat{n}_i (\hat{n}_i - 1) - \mu \sum_i \hat{n}_i. \tag{7.23}$$

With similar analysis, for spin-$1/2$ fermions, we can also reach a single-band Fermi–Hubbard model (FHM), with one more condition that the filling of fermions is always smaller than unity, otherwise the Pauli exclusion principle can also push fermions to populate the higher bands. The Hamiltonian for the FHM is given by

$$\hat{H}_\mathrm{FHM} = -J \sum_{\langle ij \rangle, \sigma} \hat{c}_{i\sigma}^\dagger c_{j\sigma} + U \sum_i \hat{n}_{i\uparrow} \hat{n}_{i\downarrow} - \mu \sum_{i\sigma} \hat{n}_{i\sigma}. \tag{7.24}$$

Here we can introduce a lattice version of the Fourier transformation, for instance, for the BHM in a three-dimensional cubic lattice,

$$\hat{b}_i = \frac{1}{\sqrt{N_\mathrm{s}}} \sum_{\mathbf{k}} \hat{b}_{\mathbf{k}} e^{i\mathbf{k} \cdot \mathbf{r}_i}, \tag{7.25}$$

where $\mathbf{r}_i = n_{i,x} \hat{\mathbf{a}}_x + n_{i,y} \hat{\mathbf{a}}_y + n_{i,z} \hat{\mathbf{a}}_z$, $\hat{\mathbf{a}}_x = (1, 0, 0)a$, $\hat{\mathbf{a}}_x = (0, 1, 0)a$ and $\hat{\mathbf{a}}_z = (0, 0, 1)a$, and $\{n_{i,x}, n_{i,y}, n_{i,z}\}$ are three integers, with $a = \pi/k_0$. Here $\mathbf{k}$ are limited in the quasi-momentum inside the first Brillouin zone. Then the kinetic energy becomes

$$-J \sum_{\langle ij \rangle} \hat{b}_i^\dagger \hat{b}_j = -2J \sum_{\mathbf{k}} (\cos(\mathbf{k} \cdot \mathbf{a}_x) + \cos(\mathbf{k} \cdot \mathbf{a}_y) + \cos(\mathbf{k} \cdot \mathbf{a}_z)) \hat{b}_{\mathbf{k}}^\dagger \hat{b}_{\mathbf{k}}. \tag{7.26}$$

In this case, the bandwidth is $12J$. J decreases as the lattice depth increases, and when $U \gg J$, the interaction energy overwhelms the kinetic energy and the system enters the strongly interacting regimes. We summarize different ways to entering strongly interacting regimes in Box 7.2.

The above discussion is a good example for discussing the central idea of *Quantum Simulation*. The final goal of the quantum simulation is to understand real material. Using the high-Tc cuprate superconductivity as an example, it is widely believed that the FHM in the

Box 7.2	Ways to Entering the Strongly Interacting Regimes

Here we should remark that there are two different ways of entering the strongly interacting regime. In Section 6.2, we consider the situation that the kinetic energy term is not changed, but the interaction energy is increased significantly, by bringing the system to the vicinity of two-body resonances. In the optical lattices, in contrast, the interaction potential is always away from a resonance, and the scattering length remains small compared with the inter-particle distance. However, we suppress the kinetic energy using the lattice effect. In both cases, the interaction energy eventually becomes comparable to or larger than the kinetic energy. In the conventional perturbative treatment of the interacting many-body system, we use the interaction strength as the small parameter to perform the perturbation expansion. Such an approach fails in the strongly interacting regime. Therefore, studying strongly interacting quantum many-body systems is a major research direction nowadays.

strongly interacting regime can describe many properties of the high-Tc superconductor. However, on one hand, because of the complicated material structure, there is no ab-initio derivation of the FHM from the high-Tc materials, and therefore the connections between experimental observations in materials and the properties of the FHM are usually uncertain. On the other hand, there is neither an exact solution nor reliable theoretical treatment of the strongly interacting FHM away from the half-filling, except for in one dimension. The applications of numerical methods are also limited. The quantum Monte Carlo simulation fails for the FHM in most regimes because of the fermion sign problem, and the exact diagonalization is limited for very small size with the capability of classical computers. Therefore, one can neither obtain reliable theory for the FHM nor reliably relate the results of the FHM to experimental observations in high-Tc materials, which makes the studies of high-Tc materials so difficult. From the discussion above, we see that by using ultracold atoms in optical lattices, one can reliably build up a physical system described by the FHM with high accuracy, because here the FHM can be derived step by step from the microscopic model with controlled approximations. Therefore, the experimental results of this system can be unambiguously related to the properties of the FHM. For instance, one can try to determine experimentally whether the ground state of the FHM in certain regime is superconducting or not. In summary, we can build a quantum system to simulate a model that cannot be solved by classical computers, which can be viewed as a special purpose quantum computer called *quantum simulator*.

7.2 Dirac Semimetal

As we have discussed in Section 4.2, topology refers to global property of a system that is invariant under continuous deformation. In Sections 4.2 and 4.4, we have discussed various examples of topological excitations in a Bose condensate. These topological properties are characterized by the homotopy group of mapping from the real space to the space of condensate wave function, as summarized in Table 4.1. From the discussion of the band theory in the previous section, we know that for each given band, there is a Bloch wave function at each quasi-momentum in the first Brillouin zone. For each band it defines a mapping from the first Brillouin zone to the space of the Bloch wave function, therefore, we can also discuss the topological properties of these mappings, which will be the focus of this and the next session.

Honeycomb Lattices. As we will explain below, the honeycomb lattice is one of the simplest lattice geometry that can exhibit nontrivial topological band structure. By straightforward interference of counterpropagating pairs of lasers, we can realize a two-dimensional square lattice. Honeycomb lattice can be realized by adding an extra pair of laser beams, for instance, by using the three pairs of counter-propagating laser beams X, Y and $\bar{X}$, as shown in the Figure 7.5(a). The X and Y beams are phase locked and they can also interfere with each other, and the $\bar{X}$ beam is slightly detuned, which creates an extra one-dimensional lattice. Thus, the total lattice potential is given by

$$V(x,y) = - V_{\bar{X}} \cos^2(k_0 x + \theta/2) - V_Y \cos^2(k_0 y)$$
$$- V_X \cos^2(k_0 x) - 2\alpha\sqrt{V_x V_y} \cos(k_0 x) \cos(k_0 y) \cos\phi, \qquad (7.27)$$

where all three lattice depths V_X, V_Y and $V_{\bar{X}}$, as well as parameters θ, ϕ and α, can be independently adjusted by controlling the relative phases and the polarization of lasers. This allows one to realize different lattice configurations including chequerboard, dimerized and honeycomb lattices. In the situation when $V_{\bar{X}}, V_Y > V_X, \theta = \pi$, the potential minima for the first two terms occur at a square lattice at $(x,y) = ((2n-1)\pi/(2k_0), m\pi/k_0)$, where n and m are integers. The last two terms shift the minima and the shift is mostly dominated by the last term. When $\phi = 0$, for even or odd m, the last term shifts minima along $\hat{x}$ to opposite directions. It is easy to show that such a distortion leads to an energy landscape as shown in Figure 7.5(b), where the energy minima form a honeycomb lattice.

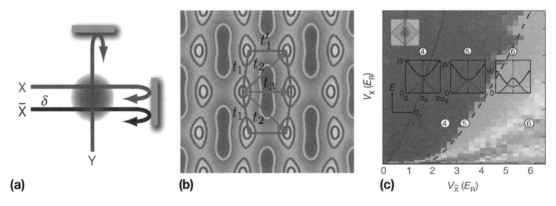

(a)   (b)   (c)

Figure 7.5 Honeycomb lattice realized in the experiment. (a) The laser configuration for creating honeycomb lattice. (b) The potential energy landscape whose minima form a honeycomb lattice. Here t_1, t_2, and t_3 label J_1, J_2, and J_3 discussed in the main text. (c) The merging transition of the Dirac points, by tuning V_X and $V_{\bar{X}}$. Reprinted from Ref. [173] with modifications. A color version of this figure can be found in the resources tab for this book at cambridge.org/zhai.

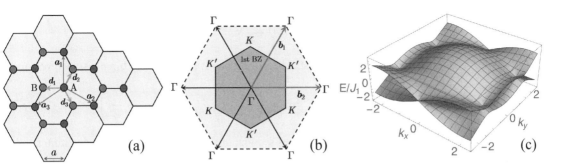

Figure 7.6 Band structure of the honeycomb lattice. (a) The Bravais lattice and the primitive vectors of a honeycomb lattice in the real space. (b) the Brillouin zone of the honeycomb lattice used in this section. (c) The dispersion of the tight-binding model that shows two Dirac points. A color version of this figure can be found in the resources tab for this book at cambridge.org/zhai.

Dirac Point. A major difference between the honeycomb lattice and the simple square or cubic lattice is the number of sites in each the unit cell. A unit cell is the smallest unit of a lattice, and the whole lattice can be constructed by repeating discrete translations of the unit cell. It is easy to see that for a square or cubic lattice, there is only one site in each unit cell, but for a honeycomb lattice, a unit cell must contain two sites, denoted by A and B, as shown in Figure 7.6(a). In other words, the honeycomb lattice is not invariant under the translation of a single site. Let us start with a honeycomb lattice where all nearest neighboring bonds are of equal spacing, denoted by a. We choose the primitive vectors of the Bravais lattice as

$$\mathbf{a}_1 = \left(0, \sqrt{3}\right) a, \quad \mathbf{a}_2 = \left(\frac{3}{2}, -\frac{\sqrt{3}}{2}\right) a, \quad \mathbf{a}_3 = \left(-\frac{3}{2}, -\frac{\sqrt{3}}{2}\right) a. \qquad (7.28)$$

Note that here only $\mathbf{a}_1$ and $\mathbf{a}_2$ are independent, and $\mathbf{a}_3 = -\mathbf{a}_1 - \mathbf{a}_2$. The reciprocal lattice vectors are given by the relation $\mathbf{a}_i \cdot \mathbf{b}_j = 2\pi \delta_{ij}$ as

$$\mathbf{b}_1 = \frac{2\pi}{a} \left(\frac{1}{3}, \frac{1}{\sqrt{3}}\right), \quad \mathbf{b}_2 = \frac{2\pi}{a} \left(\frac{2}{3}, 0\right). \qquad (7.29)$$

With these reciprocal lattice vectors, the first Brillouin zone can be constructed as shown in Figure 7.6(b), where

$$\mathbf{K} = \frac{2\pi}{a} \left(0, \frac{2}{3\sqrt{3}}\right), \quad \mathbf{K}' = \frac{2\pi}{a} \left(0, -\frac{2}{3\sqrt{3}}\right) \qquad (7.30)$$

are two unequivalent corners of the Brillouin zone. Note that the choice of the first Brillouin zone is not unique. Figure 7.6(b) is just one choice that is commonly used by most literatures.

Now we consider a tight-binding model of the honeycomb lattice. First let us only include the nearest neighboring hopping, which only occurs between A and B sublattices. The tight-binding Hamiltonian is given by

$$\hat{H} = -J_1 \sum_{\langle ij \rangle} \left(\hat{c}_{B,j}^\dagger \hat{c}_{A,i} + \text{h.c.} \right), \tag{7.31}$$

where $\langle ij \rangle$ denotes all the nearest neighboring bonds connected by three displace vectors as

$$\mathbf{d}_1 = (-1, 0)\, a, \quad \mathbf{d}_2 = \left(\frac{1}{2}, \frac{\sqrt{3}}{2} \right) a, \quad \mathbf{d}_3 = \left(\frac{1}{2}, -\frac{\sqrt{3}}{2} \right) a. \tag{7.32}$$

As we have discussed in the previous section, quasi-momentum is a good quantum number because of the discrete translational symmetry. Hence, when we write the Hamiltonian in the quasi-momentum space, the bases have to preserve the discrete translational symmetry such that the quasi-momentum is a good quantum number. Thus, for the honeycomb lattice, this tight-binding Hamiltonian has to be written in a two-component spinor bases representing the two sites in each unit cell. Here we take two sites connected by $\mathbf{d}_1$ as a unit cell, and therefore the tight-binding model can be written as

$$\hat{H} = \sum_{\mathbf{k}} \left(\hat{c}_A^\dagger (\mathbf{k}), \hat{c}_B^\dagger (\mathbf{k}) \right) H(\mathbf{k}) \left(\begin{array}{c} \hat{c}_A (\mathbf{k}) \\ \hat{c}_B (\mathbf{k}) \end{array} \right), \tag{7.33}$$

where the matrix is given by

$$H(\mathbf{k}) = \left(\begin{array}{cc} 0 & -J_1 \left(1 + e^{i\mathbf{k}\cdot\mathbf{a}_3} + e^{-i\mathbf{k}\cdot\mathbf{a}_2} \right) \\ -J_1 \left(1 + e^{-i\mathbf{k}\cdot\mathbf{a}_3} + e^{i\mathbf{k}\cdot\mathbf{a}_2} \right) & 0 \end{array} \right). \tag{7.34}$$

It is straightforward to check that this $H(\mathbf{k})$ is a periodic function in the Brillouin zone, that is, $H(\mathbf{k} + \mathbf{b}_1) = H(\mathbf{k})$ and $H(\mathbf{k} + \mathbf{b}_2) = H(\mathbf{k})$.

The facts that the honeycomb lattice has two sites in each unit cell, and the resulting two-component representation of the wave function, are of crucial importance for this discussion, as well as the later discussion of topological band insulator in Section 7.3. Because $H(\mathbf{k})$ is a 2×2 matrix, it can be expanded in terms of the Pauli matrix as $H(\mathbf{k}) = \mathbf{B}(\mathbf{k}) \cdot \boldsymbol{\sigma}$, and it represents a pseudo-spin in a Zeeman field. In other words, only when each unit cell contains more than one site, we can introduce a pseudo-spin degree of freedom to describe the internal structure within each unit cell, and the Hamiltonian can be viewed as a Zeeman field acting on this pseudo-spin degree of freedom. Thus, the energy of the upper and the lower bands are given by the strength of the Zeeman field as $\pm|\mathbf{B}(\mathbf{k})|$, and the spin wave function is determined by the direction of the Zeeman field $\hat{\mathbf{B}}(\mathbf{k}) = \mathbf{B}(\mathbf{k})/|\mathbf{B}(\mathbf{k})|$. Here the important feature is that the Zeeman field depends on the quasi-momentum $\mathbf{k}$. The momentum dependence of $\mathbf{B}(\mathbf{k})$ is also reminiscent of the spin-orbit coupling effect discussed in Sections 1.3 and 4.5.

In this case, it is easy to see that

$$B_x(\mathbf{k}) = -J_1 \left(1 + \cos(\mathbf{k} \cdot \mathbf{a}_3) + \cos(\mathbf{k} \cdot \mathbf{a}_2) \right) \tag{7.35}$$

$$B_y(\mathbf{k}) = J_1 \left(\sin(\mathbf{k} \cdot \mathbf{a}_3) - \sin(\mathbf{k} \cdot \mathbf{a}_2) \right), \tag{7.36}$$

and $B_z(\mathbf{k}) = 0$. So the band structure can be obtained as

$$E_\pm (\mathbf{k}) = \pm|\mathbf{B}(\mathbf{k})| = \pm\sqrt{B_x^2(\mathbf{k}) + B_y^2(\mathbf{k})}. \tag{7.37}$$

It is straightforward to check that for $\mathbf{K}$ or $\mathbf{K}'$ points, both $B_x(\mathbf{K})$ and $B_y(\mathbf{K})$ vanish, and since $B_z(\mathbf{k})$ is always zero in this model, the energies of upper and lower bands are degenerate at $\mathbf{K}$ and $\mathbf{K}'$ points. Hence, the band gap is closed, as shown in Figure 7.6(c). Expanding the dispersion nearby $\mathbf{K}$ or $\mathbf{K}'$ point, one can find a linear dispersion. Nearby the $\mathbf{K}$ and $\mathbf{K}'$ point, the linear dispersion is $E(\mathbf{k}) = \pm 3J_1|\mathbf{q}|a/2$, where $\mathbf{q} = \mathbf{k} - \mathbf{K}$ or $\mathbf{k} - \mathbf{K}'$. $\mathbf{K}$ and $\mathbf{K}'$ points are therefore called the *Dirac points*.

When fermions are loaded in such a honeycomb lattice and when the number of fermions in each spin component equals to the number of unit cell, the lower band is completely filled and the Fermi surface is right at the Dirac point. This state is not an insulator because there is no band gap in the system, and the charge excitation is not gapped. However, this state is also somewhat different from the normal metallic phase because the density-of-state at the Fermi surface is zero. This kind of state is called a *semimetal*. Since here the semimetal behavior is caused by the Dirac point, this state is therefore called a *Dirac semimetal*.

Stability of the Dirac Point. We have shown that a pair of Dirac points appear in the honeycomb lattice model with only nearest neighboring hoppings. Now the question is whether the Dirac points are stable against small perturbations. For instance, in the experimentally realized honeycomb lattice discussed above, the lattice is stretched along one of the spatial direction and this honeycomb lattice does not possess the C_3 symmetry, as shown in Figure 7.5(b). In this experimental geometry, the tight-binding model is different from Eq. 7.31 for at least two aspects. (i) The hopping amplitude along the horizontal nearest neighboring site, say, denoted by J_1', is larger than the other two nearest hopping amplitudes denoted by J_1 [173]; and (ii) the next nearest tunneling processes such as J_2 and J_3 denoted in Figure 7.5(b) are not negligible [173]. Nevertheless, these effects only shift the positions of the Dirac points but cannot open up the gap, as long as these extra terms are not strong enough. This is because introducing the difference between J_1 and J_1', as well as adding the J_3 term, only modifies the function form of $B_x(\mathbf{k})$ and $B_y(\mathbf{k})$, and introducing the J_2 term adds a term proportional to identity matrix $\mathbf{I}$ in the Hamiltonian. With all these distortions, $B_z(\mathbf{k})$ is still zero. Considering the equations $\mathbf{B}_x(\mathbf{k}) = 0$ and $\mathbf{B}_y(\mathbf{k}) = 0$, since one is looking for solutions from two equations with two variables (k_x and k_y), there always exist two real solutions in the neighborhood of original model Eq. 7.36. Therefore, the momentum $\mathbf{k}$ that satisfies $B_x(\mathbf{k}) = 0$ and $B_y(\mathbf{k}) = 0$ is still a Dirac point. Adding these perturbations, only the locations of the Dirac points are shifted away from $\mathbf{K}$ and $\mathbf{K}'$ points and the velocity of the Dirac dispersions are modified. The system is still a semimetal phase when the lower-band is completely filled. This demonstrates the stability of the Dirac semimetal phase against these perturbations. Only when the honeycomb lattice is strongly distorted such that these two Dirac points are moved to the same point in the first Brillouin zone, they will merge and annihilate each other. Mathematically, it corresponds to the situation that the solutions to these two equations are no longer real. After the annihilation of the Dirac points, the band dispersion becomes fully gapped, and the free Fermi system at half-filling undergoes a transition from a semimetal to an insulator. The moving and merging of the Dirac points have been observed in Ref. [173], as shown in Figure 7.5(c).

Now we ask the question that whether there exists such terms that can open up the band gap at the Dirac point with infinitesimal small strength. Based on the analysis above, we can see that if there exists a $B_z(\mathbf{k})\sigma_z$ term in $\hat{H}(\mathbf{k})$, it can immediately open up the band gap, because generally a momentum (k_x, k_y) cannot make all three functions of $B_x(\mathbf{k})$, $B_y(\mathbf{k})$ and $B_z(\mathbf{k})$ simultaneously vanish. In other word, the stability of the Dirac point is guaranteed by the fact that $B_z(\mathbf{k})$ has to vanish.

Dirac Point as Topological Defect. When $B_z(\mathbf{k})$ is always zero, we consider the Hamiltonian written as $B_x(\mathbf{k})\sigma_x + B_y(\mathbf{k})\sigma_y$. At different momentum, the pseudo-spin is polarized to different direction given by $\hat{\mathbf{B}}$. In this case, it is very important to notice that the pseudo-spins always lie along the equator and has no $\hat{z}$ component. If we draw a closed loop in the momentum space, we can define a mapping from this closed loop to the Bloch wave function space. In this case, such mappings can be classified by $\Pi_1(S^1)$ and the topological invariant is the winding number as discussed in Section 4.2. Mathematically, here the winding number can be computed as

$$w = \frac{1}{\pi} \int_C dk \hat{\mathbf{B}}_x(k) \partial_k \hat{\mathbf{B}}_y(k), \tag{7.38}$$

where $\int_C$ denotes the integration along the loop.

Around $\mathbf{K}$ and $\mathbf{K}'$ points, we can expand the Hamiltonian nearby the $\mathbf{K}$ and $\mathbf{K}'$ point in terms of $\mathbf{q} = \mathbf{k} - \mathbf{K}$ or $\mathbf{q} = \mathbf{k} - \mathbf{K}'$. To the leading order of $\mathbf{q}$, the Hamiltonian can be simplified as

$$\mathbf{H} = \frac{3J_1 a}{2} \left(\pm q_y \sigma_x + q_x \sigma_y \right). \tag{7.39}$$

The pseudo-spin texture around $\mathbf{K}$ point for the lower band is schematically shown in the inset of Figure 7.7. Thus, if one draws a closed loop around either $\mathbf{K}$ or $\mathbf{K}'$ point, the pseudo-spins wind around the equator to complete a closed circle either clock-wise or counterclockwise. For loops enclosed either $\mathbf{K}$ or $\mathbf{K}'$ points, the winding number is either $+1$ or -1. If one tries to shrink the loop, it will eventually encounter a singularity. Note that as long as everywhere $|\mathbf{B}(\mathbf{k})| \neq 0$, the pseudo-spin direction is always well defined and the mapping is always deformed continuously when the loop is deformed continuously. Therefore, the topological invariant cannot be changed. Hence, the singularity is defined by $|\mathbf{B}(\mathbf{k})| = 0$ at a particular momentum point, such that the pseudo-spin direction is no longer well defined at that point. On the other hand, $|\mathbf{B}(\mathbf{k})| = 0$ means nothing but the upper and lower bands are degenerate. Therefore, we have established the connection between the Dirac points and the topological defect in momentum space.

This topological interpretation can help us to further understand the stability of the Dirac points as follows:

- First, the effects such as different nearest hopping strengths and next nearest hopping can be viewed as continuous deformation of the pseudo-spin configuration. It cannot change topological invariant unless the two topological defects with opposite winding numbers meet each other in momentum space and annihilate each other. This is similar as the annihilation of two vortices with opposite charge in a Bose condensate, as discussed in Section 4.2. When two Dirac points annihilate each other, there is no topological defect

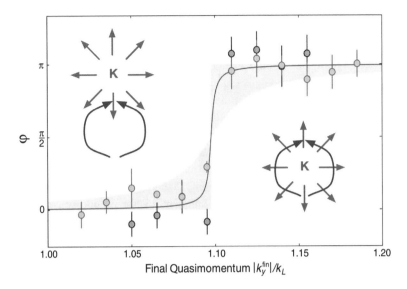

Figure 7.7 Interference in momentum space. The relative phase revealed by the interference when a BEC is dragged through two different paths in the momentum space, as indicated by the insets. The relative phase differs by π depending on whether the loop encloses the **K** point. Reprinted from Ref. [51] with modification. A color version of this figure can be found in the resources tab for this book at cambridge.org/zhai.

in the momentum space. That is to say, everywhere $|\mathbf{B}(\mathbf{k})|$ is finite and the system is an insulator when the lower band is completely filled. Thus, the semimetal to insulator transition is described by the annihilation of two topological defects.

- Second, it is important to keep pseudo-spin always lying along the equator. If there exists a nonzero B_z component such that the Bloch wave function manifold is the entire Bloch sphere S^2 instead of S^1, because $\Pi_1(S^2)$ is always trivial, as we have discussed in Section 4.2, the Dirac point loses its topological protection and it can immediately become unstable. This is consistent with above discussion that infinitesimal small $B_z(\mathbf{k})$ term can gap out the Dirac point.

- Third, when B_z component is nonzero, if we consider three dimensions instead of two dimensions, there still exists stable semimetal phase. One can see that with three momentum variables $\{k_x, k_y, k_z\}$, generally there exists a solution for all three components of $\mathbf{B}(\mathbf{k})$ vanishing simultaneously. Hence, at these specific momentum points, $|\mathbf{B}(\mathbf{k})|$ vanishes, and the upper and lower bands are degenerate. These points are called the *Weyl points*. When the Fermi surface lies at the Weyl points, it gives rise to the *Weyl Semimetal* phase. Topologically, we can consider a two-dimensional surface, and consider topological classification of the mappings from this surface to the Bloch wave function space. Such mappings are described by $\Pi_2(S^2)$ in this case and are characterized by the Chern number as its topological invariant. When this two-dimensional surface encloses the Weyl point, the mapping is nontrivial and the Chern number is nonzero. Hence, the surface in the momentum space must enclose a topological defect where $|\mathbf{B}(\mathbf{k})| = 0$ and the spin direction is no longer well defined, and the topological defect in this case is the Weyl point. This mathematical structure is similar as the monopole discussed in Section 4.4.

Finally, we should mention that this nontrivial winding of pseudo-spins around $\mathbf{K}$ or $\mathbf{K}'$ points can be detected experimentally. Imaging that one drags an atom adiabatically around a closed loop in the momentum space, the change of pseudo-spin gives rise to a Berry phase, which depends on the solid angle expanded by the pseudo-spin around the loop. Therefore, the phase is π for a loop enclosing the $\mathbf{K}$ point and the phase is zero for a loop not enclosing the $\mathbf{K}$ point. This has been observed experimentally and the results are shown in Figure 7.7. We should also note that this is in analogy with the emergent synthetic gauge field discussed in Section 1.2. There it is the real spin of atom that is polarized by a spatial varying Zeeman field, and here it is the pseudo-spin of atoms that is polarized by a momentum space varying "Zeeman field."

Symmetry Protection. Above, we have discussed that the stability of the Dirac semimetal phase is guaranteed by that $B_z(\mathbf{k})$ is zero everywhere in the momentum space. Here we should emphasize that a vanishing $B_z(\mathbf{k})$ is protected by *symmetry*. First, we consider the spatial inversion symmetry with the inversion center located at the center of a honeycomb plaquette, and this symmetry operation interchanges A and B sites and simultaneously changes $\mathbf{k} \to -\mathbf{k}$. If the Hamiltonian is invariant under the inversion symmetry, that is to say,

$$\hat{H}(\mathbf{k}) \to \sigma_x \hat{H}(-\mathbf{k}) \sigma_x = \hat{H}(\mathbf{k}). \tag{7.40}$$

This symmetry requires

$$B_x(\mathbf{k}) = B_x(-\mathbf{k}), \quad B_y(\mathbf{k}) = -B_y(-\mathbf{k}), \quad B_z(\mathbf{k}) = -B_z(-\mathbf{k}). \tag{7.41}$$

Second, we consider the time-reversal symmetry $\mathcal{T}$, which also changes $\mathbf{k} \to -\mathbf{k}$. And because time-reversal symmetry is accompanied by taking the complex conjugation, the three Pauli matrices here change as[1]

$$\mathcal{T}\sigma_x\mathcal{T}^{-1} = \sigma_x; \quad \mathcal{T}\sigma_y\mathcal{T}^{-1} = -\sigma_y; \quad \mathcal{T}\sigma_z\mathcal{T}^{-1} = \sigma_z. \tag{7.42}$$

Thus, if we require the Hamiltonian is also invariant under the time-reversal symmetry, that is to say,

$$\mathcal{T}\hat{H}(-\mathbf{k})\mathcal{T}^{-1} = \hat{H}(\mathbf{k}), \tag{7.43}$$

it leads to

$$B_x(\mathbf{k}) = B_x(-\mathbf{k}), \quad B_y(\mathbf{k}) = -B_y(-\mathbf{k}), \quad B_z(\mathbf{k}) = B_z(-\mathbf{k}). \tag{7.44}$$

Therefore, when we require that the Hamiltonian obeys both the spatial inversion symmetry and the time-reversal symmetry, by combining Eq. 7.41 and Eq. 7.44, we obtain that $B_z(\mathbf{k})$ has to vanish everywhere. Meanwhile, $B_x(\mathbf{k})$ is an even function of $\mathbf{k}$ and $B_y(\mathbf{k})$ is an odd function of $\mathbf{k}$. From these discussion, we conclude that the stability of the Dirac point is

---

[1] Here we should note that the time-reversal symmetry acting on the pseudo-spin is different from the time-reversal symmetry acting on the angular momentum or the real spin. Here the pseudo-spin represents two Wannier orbitals, therefore, the time-reversal symmetry does not invert the spin direction. However, the time-reversal symmetry does revert the spin direction when acting on the angular momentum or the real spin.

guaranteed by the requirement of the simultaneous presence of both the spatial inversion symmetry and the time-reversal symmetry. If a term breaks one of these two symmetries, it can immediately open up the band gap, and can lead to an insulator phase at half-filling, as we will discuss in the Haldane model in Section 7.3. Unlike the Dirac semimetal, the Weyl semimetal does not require symmetry protection. That is the main fundamental difference between these two semimetal phases. In Section 7.3, we will also come back to revisit this symmetry protection when we discuss the Su–Schrieffer–Heeger model.

7.3 Topological Band Insulator

The Su–Schrieffer–Heeger Model. The Su–Schrieffer–Heeger (SSH) model is a one-dimensional model with two sites in each unit cell [170]. Similar as the honeycomb lattice case, we denote these two sites as A and B sublattices. The hopping always occurs between two adjacent A and B sites. The strength of hopping within each unit cell is denoted by J_1 and the strength of hopping between two neighboring unit cells is denoted by J_2, as shown in Figure 7.8. The model is therefore written as

$$\hat{H} = -J_1 \left(\sum_i \hat{c}_{A,i}^\dagger \hat{c}_{B,i} + \text{h.c.} \right) - J_2 \left(\sum_i \hat{c}_{A,i+1}^\dagger \hat{c}_{B,i} + \text{h.c.} \right). \tag{7.45}$$

This Hamiltonian can be written in the momentum space as

$$\hat{H} = \sum_k \left(\hat{c}_A^\dagger(k), \hat{c}_B^\dagger(k) \right) H(k) \begin{pmatrix} \hat{c}_A(k) \\ \hat{c}_B(k) \end{pmatrix}, \tag{7.46}$$

where the matrix is given by

$$H(k) = \begin{pmatrix} 0 & J_1 + J_2 e^{-ika} \\ J_1 + J_2 e^{ika} & 0 \end{pmatrix}. \tag{7.47}$$

Here we have set the distance between two unit cells as a, and thus the lattice spacing as $a/2$. The first Brillouin zone ranges from $k = -\pi/a$ to $k = \pi/a$, and since $k = -\pi/a$ and

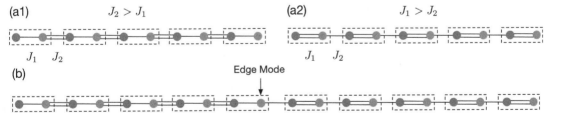

(a1) $J_2 > J_1$ **(a2)** $J_1 > J_2$

J_1 J_2 J_1 J_2

(b)   Edge Mode

Figure 7.8   Schematic of the two phases of the SSH model. (a1) Topological nontrivial case with $J_2 > J_1$. (a2) Topological trivial case with $J_2 < J_1$. (b) Edge state emerges when connecting A1 and A2. Here the double line denotes bonds with larger hopping amplitude, and a single line denotes bonds with smaller hopping amplitude. A color version of this figure can be found in the resources tab for this book at cambridge.org/zhai.

$k = \pi/a$ are equivalent, the geometry of the first Brillouin zone is in fact a one-dimensional circle. This Hamiltonian Eq. 7.47 satisfies $H(k + 2\pi/a) = H(k)$.

Similar as the discussion in the previous section, $H(k)$ can be expanded in terms of the Pauli matrix as $H(k) = B_x(k)\sigma_x + B_y(k)\sigma_y$, where

$$B_x(k) = J_1 + J_2 \cos(ka) \tag{7.48}$$

$$B_y(k) = J_2 \sin(ka). \tag{7.49}$$

The energy of two bands is given by $E_\pm(k) = \pm|\mathbf{B}(k)|$, and the Bloch wave function is determined by a vector $\hat{\mathbf{B}}(k) = \mathbf{B}(k)/|\mathbf{B}(k)|$. Thus, $\hat{\mathbf{B}}(k)$ lives on a one-dimensional circle, and this Hamiltonian can be viewed as a mapping from a one-dimensional circle of the Brillouin zone to the one-dimensional circle of $\hat{\mathbf{B}}(k)$ space. As discussed in Section 4.2 and Section 7.2, the topology of this mapping is characterized by the homotopy group $\Pi_1(S_1) = Z$, whose topological invariant is the winding number and can take all integer values. However, we should note the difference between the discussion here and the discussion in Section 7.2. Here it is the mapping from the entire Brillouin zone to the Bloch wave function space, but in Section 7.2, the mapping is defined from a loop in the Brillouin zone to the Bloch wave function space. The difference between these two situations will be discussed in detail later.

When $J_1 > J_2$, it is easy to see that B_x is always positive for all k, and therefore, the image of this mapping only covers a portion of the circle, which can be continuously deformed to a single point. Hence, the winding number for this case is zero and it is a topological trivial case. When $J_2 > J_1$, the image of this mapping can fully cover the circle of $\hat{B}(k)$ space, and the winding number equals to unity. This is a topological nontrivial case. Similar as Eq. 7.38, the winding number is defined as

$$w = \frac{1}{\pi} \int_{-\pi}^{\pi} dk\, \hat{\mathbf{B}}_x(k)\partial_k \hat{\mathbf{B}}_y(k). \tag{7.50}$$

It is straightforward to show that, with this formula and the definition of $\hat{B}_x(k)$ and $\hat{B}_y(k)$ given above, $w = 1$ for $J_2 > J_1$ and $w = 0$ for $J_2 < J_1$.

Thus, $J_2 > J_1$ and $J_2 < J_1$ represent two distinct topological phases. The transition between them takes place at $J_1 = J_2$, and this transition is characterized by a change of the topological invariant. This is now known as the *topological phase transition*. We should note that this phase transition is different from the Landau phase transition that we have discussed before, because here there is no change of order parameter before and after the phase transition. And we should also distinguish this transition from the Kosterlitz–Thouless transition discussed in Section 4.2, which is a transition driven by the deconfinement of topological defects, and there is also no topological distinction between two phases at different sides of the phase transition, as we summarized in Box 4.1. Note that, if $J_1 \neq J_2$, $|\mathbf{B}(k)|$ is always nonzero everywhere in the Brillouin zone, and there is always a finite gap separating the upper and the lower bands. Only when $J_1 = J_2$, both $B_x(k)$ and $B_y(k)$ vanish at $k = \pm\pi/a$, and the gap between the upper and the lower bands vanishes at the band edge. In fact, gap closing is the generic feature of the topological transition. Similar as what we have discussed in Section 7.2, only when $|\mathbf{B}(k)| = 0$, $\hat{\mathbf{B}}(k)$ is no longer well defined, which breaks the continuity in deforming the mappings and allows the change of topology number.

Here, another important point is that because $B_z(k)$ is always zero, we can present the Hamiltonian at each momentum as a two-dimensional vector $\hat{\mathbf{B}}(k) = \{\hat{B}_x(k), \hat{B}_y(k)\}$. Otherwise, the wave function has to be represented by a three-dimensional vector living on a S^2 surface, and $\Pi_1(S^2)$ is always trivial. In other words, once a nonzero $B_z(k)$ term is allowed, $J_2 > J_1$ regime and $J_2 < J_1$ regime can be smoothly connected by continuously deforming the mapping without closing the gap, and therefore these two regimes are no longer topologically distinct. Hence, in order to allow nontrivial topological classification in this case, the σ_z term has to vanish. In Section 7.2, we have discussed that in the case of honeycomb lattice, the presence of both the time-reversal symmetry and the refection symmetry can lead to vanishing σ_z term, and there, for the inversion symmetry, the inversion center is the center of a honeycomb plaquette. With similar discussion, here we can also show that the presence of the time-reversal symmetry and the inversion symmetry also leads to vanishing σ_z term. Here the inversion center should be taken as the center between A and B sites. In summary, it is the presence of both the time-reversal and the inversion symmetry that guarantees the nontrivial topological classification here. This phenomenon is called *symmetry-protected topological phases*, usually short-noted as SPT.

The Haldane Model. Let us now continue the discussion in Section 7.2 of the tight-binding model in a honeycomb lattice. As discussed in Section 7.2, when we impose the requirement that the Hamiltonian has both the inversion and the time-reversal symmetry, the σ_z-term has to vanish. The Dirac points are stable against weak perturbations as long as the perturbation does not break any of these two symmetries, and the system remains as a semimetal at half-filling. In other words, in order to open up a band gap at half-filling, one can add σ_z term into the Hamiltonian, which inevitably breaks either the inversion or time-reversal symmetry, or both.

A simple way to add σ_z term is to introduce a potential energy difference M between A and B sublattices, which corresponds to a $M\sigma_z$ term in the Hamiltonian. The M-term is an even function in $\mathbf{k}$, and from the discussion in the Section 7.2, it breaks the inversion symmetry but respects the time-reversal symmetry.

In a seminal paper, Haldane proposed an alternative way to open up the gap [67]. Haldane introduces the next nearest neighbor hopping, whose strength is denoted by J_2. The next nearest neighboring hoppings take place either among A sites themselves or among B sites themselves. Without loss of generality, we introduce a phase ϕ_A for hopping between A sites and a phase ϕ_B for hopping between B sites, as shown in Figure 7.9. Explicitly, this term is written as

$$J_2 \sum_{\langle\langle ij \rangle\rangle} \left(e^{i\phi_A} \hat{c}_{A,j}^\dagger \hat{c}_{A,i} + e^{i\phi_B} \hat{c}_{B,j}^\dagger \hat{c}_{B,i} + \text{h.c.} \right), \tag{7.51}$$

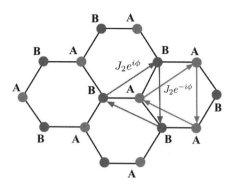

Figure 7.9   Schematic of the Haldane model: The arrows denote the next nearest hopping among A sites and among B sites, respectively, and these two hopping matrix elements have opposite phases. A color version of this figure can be found in the resources tab for this book at cambridge.org/zhai.

where $\langle\langle..\rangle\rangle$ denotes the next-nearest-neighbor hopping. It is straightforward to show that this introduces a new term in $H(\mathbf{k})$ as

$$\begin{pmatrix} 2J_2 \sum_\alpha \cos\left(\mathbf{k}\cdot\mathbf{a}_\alpha + \phi_A\right) & 0 \\ 0 & 2J_2 \sum_\alpha \cos\left(\mathbf{k}\cdot\mathbf{a}_\alpha + \phi_B\right) \end{pmatrix}. \tag{7.52}$$

If $\phi_A = \phi_B$, this term does not introduce a σ_z component. In general, a σ_z component exists as long as $\phi_A \neq \phi_B$. Especially, if we take $\phi_A = -\phi_B = -\phi$, Eq. 7.52 can be written as

$$2J_2 \cos\phi \sum_\alpha \cos\left(\mathbf{k}\cdot\mathbf{a}_\alpha\right)\mathbf{I} + 2J_2 \sin\phi \sum_\alpha \sin\left(\mathbf{k}\cdot\mathbf{a}_\alpha\right)\sigma_z. \tag{7.53}$$

In this case, the σ_z term is an odd function in $\mathbf{k}$, which breaks the time-reversal symmetry and respects the inversion symmetry.

Now we add both the M-term and the next-nearest-neighbor hopping term into the tight-binding Hamiltonian discussed in Section 7.2. Here the strength of the next-nearest-neighbor hopping is denoted by J_2, and we take opposite phases for such hoppings among A sites and among B sites, as shown in Figure 7.9. This model is referred to as the *Haldane model*. The total Hamiltonian of the Haldane model is written as

$$H = -J_1 \sum_{\langle ij\rangle} \left(\hat{c}_{B,j}^\dagger \hat{c}_{A,i} + \text{h.c.}\right) + J_2 \sum_{\langle\langle ij\rangle\rangle} \left(e^{-i\phi}\hat{c}_{A,j}^\dagger \hat{c}_{A,i} + e^{i\phi}\hat{c}_{B,j}^\dagger \hat{c}_{B,i} + \text{h.c.}\right)$$

$$+ M \sum_i \left(\hat{c}_{A,i}^\dagger \hat{c}_{A,i} - \hat{c}_{B,i}^\dagger \hat{c}_{B,i}\right). \tag{7.54}$$

Now in momentum space, $H(\mathbf{k})$ becomes

$$H(\mathbf{k}) = E_0(\mathbf{k})\mathbf{I} + \mathbf{B}(\mathbf{k})\cdot\sigma, \tag{7.55}$$

where

$$E_0(\mathbf{k}) = 2J_2 \cos\phi \sum_\alpha \cos\left(\mathbf{k}\cdot\mathbf{a}_\alpha\right). \tag{7.56}$$

$B_x(\mathbf{k})$ and $B_y(\mathbf{k})$ still behave the same as Eq. 7.36, and $B_z(\mathbf{k})$ term becomes non-zero, which is now given by

$$B_z(\mathbf{k}) = M + 2J_2 \sin\phi \sum_\alpha \sin(\mathbf{k} \cdot \mathbf{a}_\alpha) . \tag{7.57}$$

Now at each momentum $\mathbf{k}$ we have introduced a three-dimensional $\mathbf{B}(\mathbf{k})$ vector, and the eigenstates of the upper and the lower bands can be described by pseudo-spins that are either parallel or antiparallel to the direction of $\hat{\mathbf{B}}(\mathbf{k})$ field, where $\hat{\mathbf{B}}(\mathbf{k}) = \mathbf{B}(\mathbf{k}) / |\mathbf{B}(\mathbf{k})|$ lies in a Bloch sphere. Thus, the Hamiltonian defines a mapping from the first Brillouin zone of the momentum space to the S^2 Bloch sphere. Strictly speaking, the two-dimensional Brillouin zone is a torus. But in many cases, one can ignore the difference between sphere and torus, and such a mapping can also be classified by the second homotopy group $\Pi_2(S^2)$. As discussed couple times before, it is known that $\Pi_2(S^2) = Z$ and is characterized by the Chern number. The Chern number in this case is defined as

$$C = \frac{1}{2\pi} \int_{BZ} d^2k \, \Omega(\mathbf{k}) \tag{7.58}$$

$$\Omega(\mathbf{k}) = \frac{1}{2} \left(\frac{\partial \hat{\mathbf{B}}}{\partial k_x} \times \frac{\partial \hat{\mathbf{B}}}{\partial k_y} \right) \cdot \hat{\mathbf{B}}. \tag{7.59}$$

Here $\Omega(\mathbf{k})$ is the local Berry curvature, and the Chern number is the integration of the local Berry curvature over the entire Brillouin zone. For these two bands, the one with spins paralleled to $\hat{\mathbf{B}}(\mathbf{k})$ has the Chern number as defined by Eq. 7.59, and the one with spins antiparallel to $\hat{\mathbf{B}}(\mathbf{k})$ has an opposite Chern number.

The Chern number describes how many times that the spin vector $\hat{\mathbf{B}}$ covers the Bloch sphere when the momentum $\mathbf{k}$ scans through the entire Brillouin zone. If the spin vector only covers part of the Bloch sphere, it can be shrink to a point and the Chern number is zero. This corresponds to a topologically trivial state. If C is a nonzero integer, this state is a topological nontrivial state. In general situations, it is straightforward to calculate the Chern number using Eq. 7.59, which determines whether a model is topologically trivial or not. In case of the Haldane model, there is a short-cut to determine the Chern number. That is, in order for Chern number to be nonzero, the Bloch vector must at least cover both the north pole and the south pole once. Generally, this is a necessary but not sufficient condition. However, it turns out to be a sufficient condition in the case of the Haldane model. Note that $\hat{\mathbf{B}}$ can point to either the north or the south pole only when $B_x = B_y = 0$, and in this model, $B_x = B_y = 0$ only at $\mathbf{K}$ and $\mathbf{K}'$ points. That is to say, in order to cover both the north and the south poles, $B_z(\mathbf{K})$ and $B_z(\mathbf{K}')$ must take opposite sign, that is, $B_z(\mathbf{K})B_z(\mathbf{K}') < 0$. This gives the condition $-3\sqrt{3}|J_2 \sin\phi| < M < 3\sqrt{3}|J_2 \sin\phi|$, under which C equals $+1$ or -1.

Therefore, the phase diagram of the Haldane model is presented in Figure 7.10. Different phases are distinguished by different topological Chern number. Note that in the topological trivial phase $B_z(\mathbf{K})B_z(\mathbf{K}') > 0$, and in topological nontrivial phase, $B_z(\mathbf{K})B_z(\mathbf{K}') < 0$. Hence, at the transition point $B_z(\mathbf{K})B_z(\mathbf{K}') = 0$, which means either $B_z(\mathbf{K})$ or $B_z(\mathbf{K}')$ has to vanish. In fact, as shown in Figure 7.10, along the phase boundary $M = 3\sqrt{3}J_2 \sin\phi$,

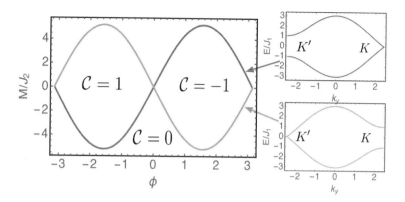

Figure 7.10 The phase diagram for the Haldane model. The phase diagram is plotted in terms of ϕ and M/J_2. The Chern number C of the lower band is marked in different regimes. The solid lines are the phase boundary between topological trivial and nontrivial phases, at which at least one of the **K** or **K'** points becomes gapless, as shown in the left column. A color version of this figure can be found in the resources tab for this book at cambridge.org/zhai.

$B_z(\mathbf{K}) = 0$ and the band gap vanishes at **K** point. Along the other half of the phase boundary $M = -3\sqrt{3}J_2 \sin\phi$, $B_z(\mathbf{K'}) = 0$ and the band gap vanishes at **K'** points. Therefore, the same as in the SSH model, the band gap closes at the phase boundary of the topological phase transition. Moreover, we note that across one of these two lines, the Chern number changes by one and there is also only one gapless mode when the system is located at the phase boundary. However, at the point with $M = 0$ and $\phi = 0$, the model returns to the case discussed in Section 7.2, where both **K** and **K'** are gapless and there are two gapless modes. Actually, across this particular point, the Chern number changes from $+1$ to -1 and the change of Chern number is two. In fact, this is a general feature of the topological transition that the number of gapless modes at the phase boundary equals to the change of topological invariants across this phase boundary.

There is an important difference between the Haldane model and the SSH model, that is, the topological classification in the SSH model requires symmetry protection and the topological classification of the Haldane model does not. This is due to their difference in dimensionality. Thus, the lesson is that the dimensionality and the symmetry play a crucial role in the topology classification of an band insulator. We will not discuss this in detail here and the readers can find more detailed information from Ref. [35, 145], where the discussions are also not restricted to two band models.

The Haldane model has been realized experimentally with a method of periodically driving the lattice, as we will discuss in the following section [85, 194]. They have performed two measurements to determine the phase diagram. The first measurement is based on the Landau–Zener tunneling during the Bloch oscillation. The Bloch oscillation has been introduced in Section 7.1. If the band is everywhere gapped, fermions should stay in the lowest band after a full cycle of the Bloch oscillation as long as the velocity is small enough during the oscillation. However, if there is a point where the gap is closed, fermions can always tunnel to the higher band when they pass through the gapless point during the Bloch oscillation, even though the velocity is sufficiently small. When the interband transition happens, fermions in the higher band can be detected in the second Brillouin zone by the band mapping method discussed in Section 7.1. Given that one of the $\mathbf{K}$ and $\mathbf{K'}$ points becomes gapless at the topological transition, the Landau–Zener tunneling rate to the higher band should display a maximum at the topological phase boundary, as shown in Figure 7.11(a). In this way, a phase diagram has been constructed experimentally, as shown in Figure 7.11(b).

The second measurement is based on the transverse drift during the Bloch oscillation. In the presence of local Berry curvature in momentum space, the semiclassical equation-of-motion Eq. 7.14 should be modified as [182]

$$\mathbf{v} = \frac{1}{\hbar} \frac{\partial \mathcal{E}(\mathbf{k})}{\partial \mathbf{k}} + \frac{d\mathbf{k}}{dt} \times \mathbf{\Omega}(\mathbf{k}). \tag{7.60}$$

The local Berry curvature in the momentum space plays the similar role as a magnetic field in the real space. Suppose that the oscillation is along the $\hat{x}$ direction, it gives rise to a Hall effect manifested as a drift along the $\hat{y}$ direction. If the total Berry curvature is zero, the drift after an entire Bloch cycle is canceled and the net drift vanishes. If the total Berry curvature is not zero for a topological nontrivial case, the net drift remains finite. Moreover, the drift will be opposite if one applies an opposite force to invert the direction of the Bloch oscillation. Thus, one can extract a differential drift from two measurements with opposite forces, as shown in Figure 7.11(d), and a phase diagram based on the transverse drift measurement has also been constructed in Figure 7.11(c). The phase diagrams constructed from these two measurements are qualitatively consistent with each other, and they are also consistent with the theory.

Summary of Topological Band Theory. In the previous section and this section, we have discussed topological nature of the semimetal and the band insulator, respectively. Here we summarize these two types of topological states. First of all, there are two common features:

1. Both require at least two sites in each unit cell. The topology of the band theory is defined by the mapping from the Brillouin zone of quasi-momentum space to the Bloch wave function. Note that there is no correction between the global $U(1)$ phases of the Bloch wave function at different momenta, in other words, we can choose an arbitrary phase for the Bloch wave function at each momentum. That is to say, we cannot use the global $U(1)$ phase degree of freedom to define the topology. Therefore, the Bloch wave function has to possess certain internal structure, and the minimum requirement is to have two sites at each unit cell. In the cases of two sites unit cell, the Hamiltonian can be written as a pseudo-spin in a quasi-momentum dependent Zeeman field, and the Hamiltonian can be generally written as $\hat{H} = \sum_{\mathbf{k}} \mathbf{B}(\mathbf{k}) \cdot \boldsymbol{\sigma}$.

2. The energy of the upper and the lower band is given by $\pm|\mathbf{B}(\mathbf{k})|$. As long as $|\mathbf{B}(\mathbf{k})|$ is nonzero for all $\mathbf{k}$, the upper and the lower bands are separated. And also in this case, one can define a spin direction $\hat{\mathbf{B}}(\mathbf{k}) = \mathbf{B}(\mathbf{k})/|\mathbf{B}(\mathbf{k})|$ everywhere in the Brillouin zone. The pseudo-spin directions of these two bands are parallel and antiparallel to $\hat{\mathbf{B}}(\mathbf{k})$, respectively. Thus, the change of Hamiltonian corresponds to a continuous deformation of the mapping, which cannot change the topological invariant. The change of topology

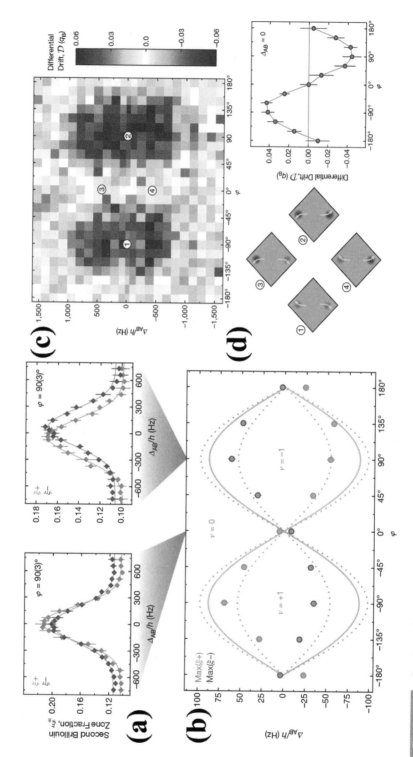

Figure 7.11

Experimental measurements of the phase diagram of the Haldane model. (a) The Landau–Zener tunneling rate after a cycle of Bloch oscillation as a function of Δ, for two different ϕ. (b) The phase diagram in terms of Δ and ϕ extracted by the Landau–Zener tunneling measurement. (c) The phase diagram extracted from the drift experiment. (d) The drift after a full Bloch cycle as a function of ϕ for the $M = 0$ case. φ in the figure is ϕ in the text. Reprinted from Ref. [85]. A color version of this figure can be found in the resources tab for this book at cambridge.org/zhai.

Table 7.1 Different types of topological features in bands		
	Semimetal	Topological insulator
Mapping	From a surface S_{d-1} in BZ to BWF space	From entire BZ T_d to BWF space
Homotopy group	$\Pi_{d-1}(\mathcal{M})$	$\Pi_d(\mathcal{M})$
Examples in 1D	—	$\Pi_1(S^1)$: SSH model (winding number)
	—	
Examples in 2D	$\Pi_1(S^1)$: Dirac semimetal (winding number)	$\Pi_2(S^2)$: Haldane model (Chern number)
Examples in 3D	$\Pi_2(S^2)$: Wely semimetal (Chern number)	— —

Note: "BZ" stands for "Brillouin Zone"; "BWF" stands for " Bloch Wave Function"; d is the spatial dimension.

is possible only when $|\mathbf{B}(\mathbf{k})| = 0$ at a certain quasi-momentum, where $\hat{\mathbf{B}}(\mathbf{k})$ is no longer well defined, and the continuity of the deformation breaks at this particular point. This means that the change of topological invariant must be accompanied by closing the band gap between the upper and the lower bands.

The semimetal and the topological band insulator belong to two different ways of characterizing the band topology, as we summarized in Table 7.1. The semimetal in d dimension is characterized by mapping from a $d-1$-dimensional surface in the momentum space to the Bloch wave function space. When this topological number is nonzero, it means that at least one singularity must be enclosed by the surface, where the band gap closes. The topological insulator is characterized by mapping from the entire d-dimensional Brillouin zone to the Bloch wave function space. If two different insulator states have different topological number, the transition between them must cross a band gap closing point.

Table 7.1 should be compared with Table 4.1 in Section 4.4. In Section 4.4 we have discussed topological excitations in a Bose condensate, characterized by mappings from the real space to the wave function space. In Table 4.1 of Section 4.4, we have discussed two types of topological excitations. One is topological defects such as vortex and monopole, and the other is nonsingular topological objects such as Skyrmion. Comparing the topological bands of fermions discussed here and the topological excitation in a BEC discussed in Sections 4.2 and 4.4, although their physics contents have no relations, they share common mathematical structures. The semimetals should be compared with topological defects. More explicitly, the Dirac point should be compared with vortex in two dimensions and the Weyl point should be compared with monopole in three dimensions. Topological band insulator should be compared with nonsingular topological object, since the wave function is well defined everywhere. In particular, the topological Haldane model should be compared with Skyrmion in two dimensions.

Edge State and Quantized Hall Conductance. Here we further elaborate the physical consequences of nontrivial topological invariants of a band insulator. For the SSH model discussed above, an intuitive picture is shown in Figure 7.8(b) where we connect two phases of the SSH model together. Here double line denotes bonds with strong hopping and single line denotes bonds with weak hopping. Imaging that we adiabatically tune all the weak hopping to zero and only keep strong hopping, all the sites connected by double lines acquires a finite gap, and the site marked by *edge mode* in Figure 7.8(b) becomes isolated and remains as a zero-energy state. This is known as the gapless edge modes, or *zero-energy edge mode*. This is precisely the zero-energy edge mode from topology. Numerical calculation can show that when the weak hopping is finite, although the wave function will be broaden, its wave function remains localized and its energy remains as zero.

Now imaging we smoothly connect two insulator states of the Haldane model with different topological invariants together, we can view the spatial coordinate as a tuning parameter, and the physics at the spatial boundary between two insulators corresponds to the physics at the phase boundary between two bulk phases. Since the topological nature forces the presence of gapless modes at the phase boundary, a natural consequence is the existence of gapless mode localized at the spatial boundary. Furthermore, since one can always view vacuum as a trivial insulator, a further corollary is that an insulator with nonzero Chern number, say C, always has $|C|$ number of gapless modes at its edge. Since the existence of the gapless modes follows directly from the topological requirement, it is stable against perturbations, as long as the perturbation does not breaks the symmetry of the bulk Hamiltonian in case that the topology needs symmetry protection. Here we should emphasize that the significance is not the presence of the edge state, but the stability of the gapless edge state against perturbations. This connection between the physics at the edge and the topology of the bulk Hamiltonian is now well known as the *bulk-edge correspondence*.

In Section 5.1 we have discussed three conditions that can lead to quantized conductance, and the three conditions are one-dimensional geometry, discrete mode and the absence of the backward scattering. Considering a two-dimensional model with nonzero Chern number, there are $|C|$ number of gapless edge states residing inside the gap. When one performs transport measurement on this system, since the bulk is gapped, only these edge modes can conduct particles. These edge states are one-dimensional conducting channels, and these modes are discrete. Moreover, because of their chiral property, the backward scatterings are absent. Hence, these three conditions can be satisfied, and therefore, each edge state contributes a quantized conductance $1/h$, and the total system displays a Hall conductance of $|C|/h$. Moreover, since the existence of these edge modes are stable against perturbations, the quantized Hall conductance is also stable against perturbations. In fact, it is a quite nontrivial effect that an insulator can conduct charge and the conductance is quantized. It is named as the *quantum anomalous Hall effect* that a topological nontrivial insulator can display quantum Hall effect without applying external magnetic field. The quantized Hall conductance is the hallmark experimental evidence of a material with filled topological bands in condensed matter systems.

Quench Dynamics. The transport measurement is quite common in condensed matter systems, and it is a near-equilibrium probe that essentially determines the physics at equilibrium. In ultracold atom systems, many Hamiltonians are realized by utilizing atom-light interactions, which can be easily changed in a time scale of microsecond. Moreover, the ultracold atom systems are quite dilute and the typical relaxation time scale can be millisecond. Hence, the Hamiltonian can be changed in a time scale much faster than the relaxation time, and therefore, it can be viewed as an instantaneous change of the Hamiltonian from an initial one to a final one. After this sudden change, the state remains as the equilibrium state of the initial Hamiltonian and is not the equilibrium state of the final Hamiltonian. The subsequent dynamics is the quantum evolution governed by the final Hamiltonian. This is known as *quench dynamics*. If the final Hamiltonian is quite different from the initial one, the quench dynamics is a far-from-equilibrium dynamics. Quench dynamics has now become a quite common experimental tools in ultracold atom physics, which provides a unique opportunity for ultracold atom systems to go beyond traditional condensed matter paradigm.

Here we describe a manifestation of topological bands in the quench dynamics [179], using the Haldane model as an example. Let us consider an initial Hamiltonian $\hat{H}^i = \sum_{\mathbf{k}} \mathbf{B}^i(\mathbf{k}) \cdot \boldsymbol{\sigma}$ as a topological trivial one, and the initial wave function $|\xi^i(\mathbf{k})\rangle$ at quasi-momentum $\mathbf{k}$ is an eigenstate of $\hat{H}^i$. We consider a sudden quench that changes the Hamiltonian to $\hat{H}^f = \sum_{\mathbf{k}} \mathbf{B}^f(\mathbf{k}) \cdot \boldsymbol{\sigma}$, and the subsequent quantum dynamics at each $\mathbf{k}$ is therefore determined by

$$|\xi(\mathbf{k}, t)\rangle = e^{-i\mathbf{B}^f(\mathbf{k}) \cdot \boldsymbol{\sigma} t} |\xi^i(\mathbf{k})\rangle. \tag{7.61}$$

With this wave function, we can calculate the pseudo-spin dynamics as

$$\mathbf{s}(\mathbf{k}, t) = \langle \xi(\mathbf{k}, t)| \boldsymbol{\sigma} |\xi(\mathbf{k}, t)\rangle. \tag{7.62}$$

Because a vector in a Bloch sphere is a faithful representation of the pseudo-spin-$1/2$ wave function of a two-band model, Eq. 7.62 completely defines the quench dynamics. Initially the pseudo-spin at quasi-momentum $\mathbf{k}$ is antiparallel to $\mathbf{B}^i(\mathbf{k})$ but it is not parallel or antiparallel to $\mathbf{B}^f(\mathbf{k})$, thus, this quench dynamics is in fact described by precession of the pseudo-spin around $\mathbf{B}^f(\mathbf{k})$. Therefore, it is also periodical in time. Hence, Eq. 7.62 defines a mapping from the three-dimensional periodical $\{\mathbf{k}, t\}$ space to the two-dimensional Bloch space. This mapping is known as the *Hopf map*. It turns out that the topology of such a mapping can be described by the third homotopy group $\Pi_3(S^2)$ and a profound mathematical theorem states that

$$\Pi_3(S^2) = \Pi_2(S^2) = Z. \tag{7.63}$$

The topological invariant for the Hopf map is called the *Hopf invariant*, or the *linking number*. The linking number is schematically shown in Figure 7.12. Let us consider a given vector in the Bloch sphere, one can image that its inverse image is a loop in the three-dimensional $\{\mathbf{k}, t\}$ space. Considering two different vectors in the Bloch sphere, their inverse imagings are two loops in the $\{\mathbf{k}, t\}$ space. Then, one can ask whether two loops have nonzero linking number. Figure 7.12(a) shows an example with linking number zero and Figure 7.12(b) shows an example with linking number one.

It is easy to see that the linking number defined in this way is a topological invariant. If one wants to continuously deform the case of Figure 7.12(a) to the case of Figure 7.12(b),

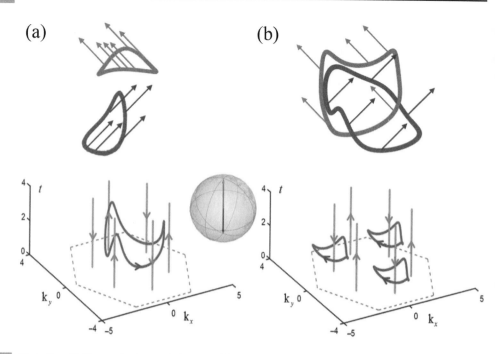

Figure 7.12 Illustration of linking number in quench dynamics. (a) Case with linking number being zero. (b) Case with nonzero linking number. Reprinted from Ref. [179]. A color version of this figure can be found in the resources tab for this book at cambridge.org/zhai.

there must be an instant that these two lines cross. However, such a cross is forbidden because these two lines are defined as the inverse images of two different vectors, which means that the pseudo-spin points to different directions at different lines. This argument shows that the linking number cannot be changed under continuous deformation and is a topological invariant. It can be further proved that for the case that $\hat{H}^i$ is topologically trivial, the linking number defined in this way in the quench dynamics is always equal to the Chern number of the final Hamiltonian $\hat{H}^i$ [179]. This relation establishes an unambiguous relation between the far-from-equilibrium dynamics and the topology of the Hamiltonian that governs the quench dynamics [179].

This relation has been confirmed experimentally [172], and the results are shown in Figure 7.13. In this experiment, they measure the real time dynamics of the Bloch vector by the quantum state tomography. They focus on the inverse images of the north and the south poles in the Bloch sphere, which show up as vortices of the azimuthal angle of the Bloch vector and display opposite vorticity. In Figure 7.13(a), the solid lines are the trajectory of one type of vortex, say, the inverse image of the north pole. The dots are actually straight lines along the time direction, which represent the inverse images of the south pole. One can see that for cases (ii) and (iii) in Figure 7.13(a), the solid line encloses the dot, which displays linking number one as in case (b) of Figure 7.12. The linking number is compared with the calculated Chern number and spectrum of the final Hamiltonian shown in Figures 7.13(b) and (c), and it is found that the change of the linking number measured in this way is consistent with the topological transition of the final Hamiltonian [172].

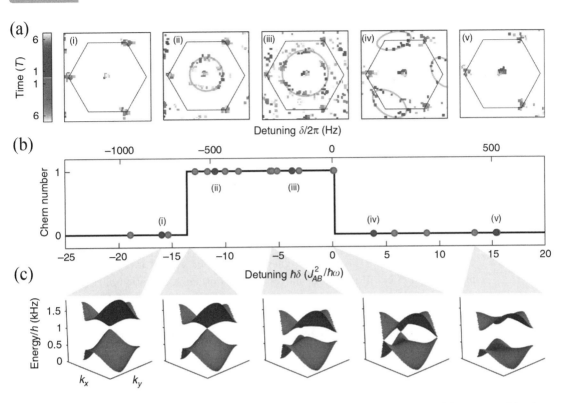

Figure 7.13   Experimental observation of the linking number in quench dynamics. (a) Different cases of the trajectory of the phase vortices, which corresponds to the inverse images of the north and the south poles. (b) Different linking number as a function of parameter in the final Hamiltonian. (c) Illustration of the corresponding band structure of the final Hamiltonian. Reprinted from Ref. [172]. A color version of this figure can be found in the resources tab for this book at cambridge.org/zhai.

7.4 Periodical-Driven Lattice

Moving Lattice. To realize topological phases with ultracold atoms in optical lattices, a common tool is to drive the optical lattices periodically. As we know, optical lattices are formed by interferences of counter propagating laser beams. By time-periodically modulating the relative phase between these two counter-propagating lasers, the interference pattern oscillates in space periodically, which realizes a shaking optical lattice. And periodically modulating the relative phase can be realized by either using an acousto-optic modulator (AOM) or by mechanically shaking the mirror that reflects the laser.

Before we consider periodically shaking, let us first consider a general situation that the relative phase is a time dependent function as $kf(t)$, where k is the laser wave length. Considering one-dimensional lattice as an example, the Hamiltonian is written as

$$H = -\frac{\hbar^2}{2m}\frac{\partial^2}{\partial x^2} + V\cos^2(k(x+f(t))). \tag{7.64}$$

234

Noninteracting Bands

Now if we make a Galilean transformation to the comoving frame with $x' = x + f(t)$ and $t' = t$, then the lattice becomes stationary. However, because

$$i\hbar\partial_t = i\hbar\partial_{t'} + i\hbar\frac{\partial x'}{\partial t}\partial_{x'}, \tag{7.65}$$

the Hamiltonian in the comoving frame can be written as[2]

$$\hat{H} = \frac{1}{2m}(-i\hbar\partial_x + mf'(t))^2 + V\cos^2(kx). \tag{7.66}$$

This is equivalent to the situation that a charged particle is placed in an electric field with $A_x = -mf'(t)$, and the electric field

$$E_x = -\frac{\partial A_x(t)}{\partial t} = mf''(t). \tag{7.67}$$

Hence, if the lattice moves with a constant velocity v, $f(t) = vt$ and therefore, A_x is a constant and $E_x = 0$. A constant gauge field can be gauged away and does not cause any physical effect. If the lattice moves with a constant acceleration a, $f(t) = at^2/2$, and therefore, $A_x = -mat$ and $E_x = ma$. This realizes a constant force, which can be used to excite the Bloch oscillation as discussed in Section 7.1. In a general situation, E_x always equals the instantaneous acceleration. In fact, this is nothing but an inertia force when we move to the comoving frame.

Let us now consider a two-dimensional lattice, two laser beams along $\hat{x}$ has a relative phase modulating as $kb\sin(\omega t)$ and two laser beams along $\hat{y}$ has a relative phase modulating as $kb\sin(\omega t + \phi)$. Then, the single-particle Hamiltonian can therefore be written as

$$H = -\frac{\hbar^2\nabla^2}{2m} + V\left[x + b\sin(\omega t), y + b\sin(\omega t + \phi)\right], \tag{7.68}$$

where b is the shaking amplitude, and ω is the shaking frequency. By making a coordinate transformation $x' = x + b\sin(\omega t)$ and $y' = y + b\sin(\omega t + \phi)$, the new Hamiltonian in the comoving frame reads

$$H(x, y, t) = \frac{1}{2m}\left[-i\hbar\partial_x - A_x(t)\right]^2 + \frac{1}{2m}\left[-i\hbar\partial_y - A_y(t)\right]^2 + V(x, y), \tag{7.69}$$

where

$$A_x(t) = -m\omega b\cos(\omega t), \quad A_y(t) = -m\omega b\cos(\omega t + \phi). \tag{7.70}$$

This vector potential is equivalent to a time-periodical electric field of a laser field propagating along $\hat{z}$. In other words, in this way we can simulate a laser field applied to electrons in solid. If $\phi = 0$, it corresponds to a linear polarized light, and if $\phi = \pi/2$, it corresponds to a circular polarized light. However, despite the mathematical equivalence, there is an important difference between the solid state setting and the ultracold atom setting. As we will discuss below, it is important to reach the parameter regime where the driving frequency is much larger than the bandwidth. In the solid state setting, it usually requires a very high frequency laser because the bandwidth of a material is usually of the order of electron volt, which makes the experimental realization quite challenging. Nevertheless, in

---

[2] Here we ignored $\prime$ since we will work on the comoving frame afterward.

the ultracold atom systems, the typical bandwidth of an optical lattice is of the order of a few thousand Hertz, and it is easy to mechanically shake the lattice with a frequency much larger than a few thousand Hertz.

Peierls Substitution. For the tight-binding model discussed in Section 7.1, we have shown that the hopping integral is given by

$$J_{ij} = -\int d^3\mathbf{r}\, w^*(\mathbf{r} - \mathbf{R}_i) \left(\frac{\mathbf{p}^2}{2m} + V_{\text{lat}}(\mathbf{r})\right) w(\mathbf{r} - \mathbf{R}_j). \tag{7.71}$$

Now let us perform a gauge transformation on the original Wannier wave function in absence of gauge field, and define a new set of Wannier wave functions as

$$\tilde{w}(\mathbf{r} - \mathbf{R}_j) = e^{-\frac{i}{\hbar}\int_{\mathbf{R}_j}^{\mathbf{r}} \mathbf{A}(\mathbf{r}')\cdot d\mathbf{r}'} w(\mathbf{r} - \mathbf{R}_j). \tag{7.72}$$

We expand the field operator in terms of the new Wannier wave function bases, and the hopping integral becomes

$$\tilde{J}_{ij} = -\int d^3\mathbf{r}\, \tilde{w}^*(\mathbf{r} - \mathbf{R}_i) \left(\frac{(\mathbf{p} - \mathbf{A})^2}{2m} + V_{\text{lat}}(\mathbf{r})\right) \tilde{w}(\mathbf{r} - \mathbf{R}_j). \tag{7.73}$$

It is easy to show that

$$\left(\frac{(\mathbf{p} - \mathbf{A})^2}{2m} + V_{\text{lat}}(\mathbf{r})\right) \tilde{w}(\mathbf{r} - \mathbf{R}_j) = e^{\frac{i}{\hbar}\int_{\mathbf{R}_j}^{\mathbf{r}} \mathbf{A}(\mathbf{r}')\cdot d\mathbf{r}'} \left(\frac{\mathbf{p}^2}{2m} + V_{\text{lat}}(\mathbf{r})\right) w(\mathbf{r} - \mathbf{R}_j), \tag{7.74}$$

and by exploring the approximation that

$$e^{-\frac{i}{\hbar}\int_{\mathbf{R}_i}^{\mathbf{r}} \mathbf{A}(\mathbf{r}')\cdot d\mathbf{r}'} e^{\frac{i}{\hbar}\int_{\mathbf{R}_j}^{\mathbf{r}} \mathbf{A}(\mathbf{r}')\cdot d\mathbf{r}'} \approx e^{\frac{i}{\hbar}\int_{\mathbf{R}_j}^{\mathbf{R}_i} \mathbf{A}(\mathbf{r}')\cdot d\mathbf{r}'}, \tag{7.75}$$

we can obtain that

$$\tilde{J}_{ij} \approx e^{\frac{i}{\hbar}\int_{\mathbf{R}_j}^{\mathbf{R}_i} \mathbf{A}(\mathbf{r}')\cdot d\mathbf{r}'} J_{ij}. \tag{7.76}$$

That is to say, the hopping integral is only modified by a phase factor, and this is known as the *Peierls substitution*.

We need to be careful about the validity of the approximation Eq. 7.75.

- If **A** has spatial dependence, Eq. 7.75 means that one ignores the magnetic flux through the triangle formed by $\mathbf{r}$, $\mathbf{R}_i$ and $\mathbf{R}_j$. Note that the Wannier wave function is taken as maximally localized, such that $\mathbf{r}$ has to be nearly both $\mathbf{R}_i$ and $\mathbf{R}_j$, otherwise it does not contribute significantly to the integral. Hence, for hopping between neighboring sites, the effective area of the triangle is usually smaller than the area of a unit cell. This approximation is good when magnetic field strength through each unit cell is small.

- If **A** has a temporal dependence, there will also be an extra term originating from the $i\hbar\partial/\partial t$ term, which provides coupling between different bands, similar as the discussion of $\mathbf{d} \cdot \mathbf{E}$ coupling made in Section 1.3. This contribution can be ignored under following two conditions: (i) The strength of this term is much smaller compared with the band separation; and (ii) if **E** is time periodical with a frequency ω, ω does not match the interband transition energy.

In the case of Eq. 7.69, **A** only depends on time and does not depend on the spatial coordinate. Therefore, when the Peierls substitution is applicable, the kinetic term can be written as

$$\hat{H}_0 = -\sum_{ij} J_{ij} e^{i\mathbf{A}(t)\cdot\mathbf{l}_{ij}/\hbar} \hat{b}_i^\dagger \hat{b}_j + \text{h.c.}, \tag{7.77}$$

where $\mathbf{l}_{ij} = \mathbf{R}_i - \mathbf{R}_j$ is the relative position between i and j sites, and $\hat{b}_i$ is the annihilation operator expanded under the new Wannier orbitals $\tilde{w}(\mathbf{r} - \mathbf{R}_i)$.

Floquet Hamiltonian. A general method to treat the time-periodical problem is called the *Floquet theory*. The central idea of the Floquet theory is to derive a time-independent Hamiltonian that can effectively describe this time-periodical process. Considering an observer who is only allowed to observe the system at integer times of the driven period, one can find out a time-independent Hamiltonian H_{eff}, such that this observer cannot distinguish whether the quantum evolution of this system is governed by this H_{eff} or by the original time-periodical Hamiltonian. Since the time interval between two observations are $T = 2\pi/\omega$, thus, T has to be much shorter than all the time scales of the problem, such that this discrete sets of observations can capture all the essential physics of the system and H_{eff} is a faithful representation of the system. Hence, $\hbar\omega$ has to be much larger than all the other energy scales of the problem.

To be more precise, considering a Schrödinger equation $i\hbar\partial_t\psi = \hat{H}(t)\psi(t)$ of a time-periodical system with $\hat{H}(t + T) = \hat{H}(t)$, the time evolution operator is given by

$$\hat{U}(t) = \hat{T}\exp\left[-\frac{i}{\hbar}\int_0^t dt'\hat{H}(t')\right], \tag{7.78}$$

where $\hat{T}$ is the time-ordering operator. Considering the evolution for one driven period from αT $(0 \leqslant \alpha < 1)$ to $\alpha T + T$, the evolution operator is given by

$$\hat{U}(T,\alpha) = \hat{T}\exp\left[-\frac{i}{\hbar}\int_{\alpha T}^{T+\alpha T} dt\hat{H}(t)\right], \tag{7.79}$$

and $\hat{U}(T)$ only depends on T and α. Now we introduce the effective Hamiltonian $\hat{H}_{\text{eff}}$ such that

$$\hat{U}(T,\alpha) = \exp\left(-\frac{i}{\hbar}\hat{H}_{\text{eff}}T\right). \tag{7.80}$$

In this way, the time-independent Hamiltonian $\hat{H}_{\text{eff}}$ can reproduce exactly the same time evolution of the real periodical system at integer periods of time. We can diagonalize $\hat{U}(T)$ as

$$\hat{U}(T)|\varphi_n\rangle = e^{-\frac{i}{\hbar}\epsilon_n T}|\varphi_n\rangle. \tag{7.81}$$

where ϵ_n is defined in the range of $-\pi/T$ and π/T and is called the *quasi-energy*, and $|\varphi_n\rangle$ is the corresponding eigen-wave function. Usually we can numerically evaluate the Floquet operator $\hat{U}(T)$ and determine its eigenvalues via Eq. 7.81.

Keeping in mind that $\hbar\omega$ is the largest energy scale of the problem, we can also determine H_{eff} by the $1/\omega$ expansion. We first expand $\hat{H}(t)$ as

$$\hat{H}(t) = \sum_n e^{in\omega t}\hat{H}_n, \qquad (7.82)$$

and by $1/\omega$ expansion it is straightforward to deduce $\hat{H}_{\text{eff}}$ as

$$\hat{H}_{\text{eff}} \approx \hat{H}_0 + \sum_{n=1}^{\infty} \left\{ \frac{\left[\hat{H}_n, \hat{H}_{-n}\right]}{n\hbar\omega} - \frac{\left[\hat{H}_n, \hat{H}_0\right]}{e^{-i2\pi n\alpha}n\hbar\omega} + \frac{\left[\hat{H}_{-n}, \hat{H}_0\right]}{e^{i2\pi n\alpha}n\hbar\omega} \right\} + \dots . \qquad (7.83)$$

Here the terms in $\{\dots\}$ keeps $1/\omega$ order. The first term has a very clear physical meaning. From the expansion Eq. 7.82, it is clear that $\hat{H}_n$ and $\hat{H}_{-n}$ can be viewed as processes absorbing n-"photons" and emitting n-"photons," respectively, with energy change being $n\hbar\omega$. Thus, both $\hat{H}_n\hat{H}_{-n}/(n\hbar\omega)$ and $\hat{H}_{-n}\hat{H}_n/(-n\hbar\omega)$ are two second-order perturbation processes, and they together give the first term in $\{\dots\}$.

The last two terms in $\{\dots\}$ of Eq. 7.83 depend on α. Considering two different α_1 and α_2, it means that when the two sets of observations are shifted by $(\alpha_1 - \alpha_2)T$, it leads to different conclusions of the effective Hamiltonian. Note that $(\alpha_1 - \alpha_2)T$ is a microscopic time-scale, these two terms is therefore called the *micromotion* term. Whether the effect of such terms exists depends on experimental realization. Usually the experimental measurements are averaged over many runs. If every time the experimental realization and measurement are performed with a fixed α, then the effect of such terms exists. If every time α is chosen randomly, then the effect of these terms are averaged out.

In the single-band model discussed so far, two of the most important energy scales are the kinetic energy characterized by the band-width and the on-site interaction energy, both of which are smaller than the interband transition energy. Here we will consider two different cases, as shown in Figure 7.14. In the nonresonant case shown in Figure 7.14, the driving frequency is much larger than both the bandwidth and the interaction energy, and is not resonant with the interband transition energy. In this case we can straightforwardly apply the $1/\omega$ expansion for the single-band Hamiltonian. In the resonant case shown in

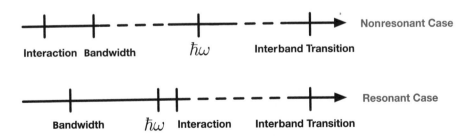

Figure 7.14 Two parameter regimes of periodical driven optical lattices. Four of the most important energy scales are compared in order, which are the on-site interaction energy, the bandwidth, the shaking frequency, and the interband transition energy. The nonresonant case and the resonant case are compared. The dashed line connects two regimes where one end is much larger than the other end. A color version of this figure can be found in the resources tab for this book at cambridge.org/zhai.

Figure 7.14, we consider the situation that the interaction energy is much larger than the kinetic energy, but the driving frequency is resonant with the interaction energy. In this case, $\hbar\omega$ is not the largest energy scale in the problem. Hence, in order to apply the Floquet theory, we need to first perform a rotating wave transformation to eliminate the largest energy scales of the problem, and then obtain the effective Floquet Hamiltonian for the time-periodical Hamiltonian after rotation. We will show below that these two cases yield different kind of physics.

Nonresonant Case. For simplicity, let us first consider a one-dimensional case. Considering $A(t) = -m\omega b\cos(\omega t)$, the single-particle Hamiltonian Eq. 7.77 can be written as

$$\hat{H} = -J\sum_{\langle ij\rangle}(e^{if_0\cos(\omega t)}\hat{b}_i^\dagger\hat{b}_j + \text{h.c.}),\tag{7.84}$$

where $f_0 = m\omega ba/\hbar$ is a constant and a is the lattice spacing. Using the identity for the Bessel function expansion, we can obtain that

$$\hat{H} = -J\sum_{\langle ij\rangle}\sum_n(i^n\mathcal{B}_n(f_0)\hat{b}_i^\dagger\hat{b}_j e^{in\omega t} + \text{h.c.}),\tag{7.85}$$

where $\mathcal{B}_n(f_0)$ is the nth order Bessel function. There are at least two direct applications of this simple case:

- To the lowest order of the $1/\omega$ expansion, we only keep $\hat{H}_0$ term, then

$$\hat{H}_0 = -J\mathcal{B}_0(f_0)\sum_{\langle ij\rangle}(\hat{b}_i^\dagger\hat{b}_j + \text{h.c.}).\tag{7.86}$$

$\mathcal{B}_0(f_0)$ oscillates with the increasing of f_0. In certain regime of f_0, $\mathcal{B}_0(f_0)$ can be negative and the sign of hopping term is inverted. This shifts the band minimum from momentum zero to momentum π. A straightforward generalization of this scheme can shift the band minimum from zero to any momentum in the Brillouin zone, which is equivalent to introducing a constant gauge field [169]. This scheme can also be generalized to a two-dimensional lattice [168]. For instance, in a triangular lattice, each triangular plaquette is surrounded by three nearest neighboring bonds. If the hopping sign is inverted only for one of the bonds, it is equivalent to introducing a π flux into each triangular plaquette. In such a situation, the single-particle band dispersion displays degenerate minima [168]. When we consider Bose condensation in such a band, bosons feel frustration, which is a similar as the spin-orbit coupled bosons discussed in Section 4.5.
- If we consider the next order terms in the $1/\omega$ expansion, it is straightforward to show that the $[H_n, H_{-n}]$ term can give rise to the next nearest neighbor hopping terms. We can use this method to generate the next nearest hopping in the Haldane model introduced in Section 7.3. Similar as the one-dimensional case, let us start with the nearest neighbor hopping only model and turn on the shaking, we can also generate the next nearest neighbor hopping. Then, the question is what kind of shaking can generate the specific type of next nearest hopping required by the Haldane model, which acquires the special phase factor in order to realize topological nontrivial bands. The answer can actually be found out by symmetry analysis. As we have discussed in Section 7.2 and Section 7.3,

the next nearest neighbor hopping term in the Haldane model respects the inversion symmetry but breaks the time-reversal symmetry. In the discussion above, we have build up the connection between atoms in shaking lattices and a charged particle in a laser field. It is easy to see that a linear polarized light always breaks the inversion symmetry but preserves the time-reversal symmetry, and a circular polarized light preserves the inversion symmetry but breaks the time-reversal symmetry, as the time-reversal operation brings the left circular polarized light to the right circular polarized one. Therefore, we should choose a circular shaking with $\phi = \pi/2$ to generate the next nearest hopping term in the Haldane model because their symmetries are compatible. This can be further verified by deriving the effective Hamiltonian H_{eff} following Eq. 7.83 and compute the topological invariant for H_{eff} [194]. Such a scheme has indeed been implemented to realize the Haldane model experimentally [85].

Resonant Case. Let us consider atoms of two different species, whose creation operators are denoted by $\hat{a}_i^\dagger$ and $\hat{b}_i^\dagger$, respectively, and both two species are subjected to a shaking with the same frequency ω. With the same tight-binding approximation described above, the model is given by

$$\hat{H} = -J^a \sum_{\langle ij \rangle}(e^{if\cos(\omega t)}\hat{a}_i^\dagger\hat{a}_j + \text{h.c.}) - J^b \sum_{\langle ij \rangle}(e^{if\cos(\omega t)}\hat{b}_i^\dagger\hat{b}_j + \text{h.c.}) + U\sum_i \hat{n}_i^a\hat{n}_i^b, \quad (7.87)$$

where J^a and J^b are the nearest neighbor hopping amplitudes for two different species, respectively, and U is the on-site interaction strength between two species with $\hat{n}_i^a = \hat{a}_i^\dagger\hat{a}_i$ and $\hat{n}_i^b = \hat{b}_i^\dagger\hat{b}_i$. Here we consider the situation that ω is very close to U. In this case we cannot directly apply the $1/\omega$ expansion, instead, we need to first apply a unitary rotation

$$\hat{R} = e^{-i\sum_i \omega t\hat{n}_i^a\hat{n}_i^b}. \quad (7.88)$$

According to the discussion in Section 1.3, the Hamiltonian after rotation is given by $\hat{R}^\dagger\hat{H}\hat{R} + i\hbar(\partial_t\hat{R}^\dagger)\hat{R}$. First, $i\hbar(\partial_t\hat{R}^\dagger)\hat{R}$ generates a term $-\hbar\omega\sum_i\hat{n}_i^a\hat{n}_i^b$, and this term can be combined with the on-site interaction term, which changes U to $\tilde{U} = U - \hbar\omega \ll U, \hbar\omega$ such that $1/\omega$ expansion can be safely applied to the Hamiltonian after rotation. Second, the hopping term does not commute with $\hat{R}$, and it is easy to see that

$$\hat{R}^\dagger\hat{H}\hat{R} = -J^a\sum_{\langle ij \rangle}(e^{if\cos(\omega t)}e^{i\omega t(\hat{n}_i^b-\hat{n}_j^b)}\hat{a}_i^\dagger\hat{a}_j + \text{h.c.})$$
$$-J^b\sum_{\langle ij \rangle}(e^{if\cos(\omega t)}e^{i\omega t(\hat{n}_i^a-\hat{n}_j^a)}\hat{b}_i^\dagger\hat{b}_j + \text{h.c.}). \quad (7.89)$$

Hence, with the Bessel function expansion, we obtain

$$e^{if\cos(\omega t)}e^{i\omega t(\hat{n}_i^b-\hat{n}_j^b)} = \sum_n i^n \mathcal{B}_n(f)e^{i\omega t(n+\hat{n}_i^b-\hat{n}_j^b)}, \quad (7.90)$$

and a similar result for $e^{if\cos(\omega t)}e^{i\omega t(\hat{n}_i^a-\hat{n}_j^a)}$. To the leading order, we keep the zero-frequency component and obtain

$$\hat{H}_0 = -J^a\sum_{\langle ij \rangle}(\mathcal{B}_{\hat{X}_{ij}}(f)\hat{a}_i^\dagger\hat{a}_j + \text{h.c.}) - J^b\sum_{\langle ij \rangle}(\mathcal{B}_{\hat{Y}_{ij}}(f)\hat{b}_i^\dagger\hat{b}_j + \text{h.c.}) + \tilde{U}\sum_i\hat{n}_i^a\hat{n}_i^b, \quad (7.91)$$

where $\hat{X}_{ij} = \hat{n}_j^b - \hat{n}_i^b$ and $\hat{Y}_{ij} = \hat{n}_j^a - \hat{n}_i^a$ represent the density difference between two neighboring sites.

This leads to an intriguing situation that the hopping amplitude of one species depends on the density of the other species. For instance, when the eigenvalue of $\hat{X}_{ij}$ or $\hat{Y}_{ij}$ changes from $+1$ to -1, the sign of the hopping is inverted and it is equivalent to inserting a π phase in this link. Thus, the Hamiltonian Eq. 7.91 describes a density-dependent gauge field. Since density itself is a dynamical variable of this quantum system that can fluctuate in space and time, unlike the static gauge field discussed in Section 1.3, this gauge field acquires dynamics. A further generalization of this method with fine-tuned parameters can also be used to realize *dynamical gauge field*, whose Hamiltonian possesses local gauge symmetries [156, 65]. In Box 7.3, we summarize a timeline of simulating various kinds of gauge fields with ultracold atoms.

Finally, when the shaking frequency is tuned to be resonant with an interband transition, even weak shaking amplitude can strongly hybridize these two bands, which provides a new tool to engineer the band dispersion [136]. To treat this situation, one has to go beyond the Peierls substitution and to consider the inter-band transition. A generalization of the formalism discussed above can also be applied to treat such cases [194]. In particular, we also need to first apply a rotating wave transformation to remove the interband energy separation resonating with the shaking frequency $\hbar\omega$ before we apply the $1/\omega$ expansion [194].

Box 7.3   **Simulating Static and Dynamical Gauge Fields**

Here we summarize the timeline for simulating gauge fields with ultracold atoms. Gauge fields are classified by abelian gauge fields and non-abelian gauge fields. Here we focus on the $U(1)$ gauge field, which is an abelian gauge field. The Maxwell theory is an example of a $U(1)$ gauge field. Charged particles in such a gauge field are described by the Hamiltonian $\hat{H} = \frac{1}{2m}(\mathbf{p} - \mathbf{A})^2 + \dots$, where $\mathbf{A}(\mathbf{r}, t)$ is the gauge field. First, a *constant gauge field* has been simulated by using atom-light interaction, as discussed in Section 1.3, or by using shaking optical lattices, as discussed in this section. A constant gauge field has no physical effect because it can be gauged away. Then, with the idea discussed in Section 1.2, a gauge field as a function of spatial coordinate $\mathbf{r}$ and time t has been realized. This creates either a *synthetic magnetic field* $\mathbf{B} = \nabla \times \mathbf{A}$ or a *synthetic electric field* $\mathbf{E} = -\partial\mathbf{A}/\partial t$. However, in these cases, the function form of $\mathbf{A}(\mathbf{r}, t)$ is fixed by external classical sources, and the gauge field itself has no dynamics. In this section, we describe the idea of how to create a density-dependent gauge field. In a quantum system, density of particles is a dynamical variable, and the density fluctuation gives rise to quantum dynamics of the gauge field. Therefore, this simulates a *gauge field with dynamics*. Nevertheless, the term *dynamical gauge field* has a more specific meaning in high-energy physics. It not only requires the Hamiltonian to include the dynamical term for the gauge field but also requires the entire Hamiltonian to possess local gauge symmetry, that is to say, the entire Hamiltonian is invariant under local gauge transformations. For $U(1)$ gauge theory, the Maxwell term in the electromagnetic theory is such a dynamical term with local $U(1)$ gauge symmetry. A Z_2 version of the dynamical gauge theory can be realized by the resonantly driven optical lattices, as discussed in this section.

7.5 Lattice from Cavity

In the above discussions, the optical lattice potential is created by external classical light sources, and we are only interested in the quantum mechanical motion of atoms under such classical lattice potentials, and the back-action of the atom to the light fields are safely ignored. Here we will discuss a different situation that the lattice itself is also a quantum field. That is to say, the photon field should not be treated classically but has to be treated as a quantum field as well. This is the situation when ultracold atoms are placed inside an optical cavity. Inside the cavity, the electromagnetic field is strongly confined by the cavity boundaries, which leads to discrete modes of the light field. Normally we consider the situation that only one or a few discrete modes are relevant to the atom-light coupling. Under this situation, the back-action of atoms to the light field becomes significant, especially nearby the superradiant transition discussed below where the photon occupation of the cavity mode is very few. Since the dynamics of motion of atoms and the dynamics of the cavity photon field are coupled and influence each other, they need to be treated on equal footing [150].

Atom-Cavity Interaction. As a typical example, here we discuss an experimental setup as shown in Figure 7.15 [14]. A pair of counter-propagating pumping lasers with frequency

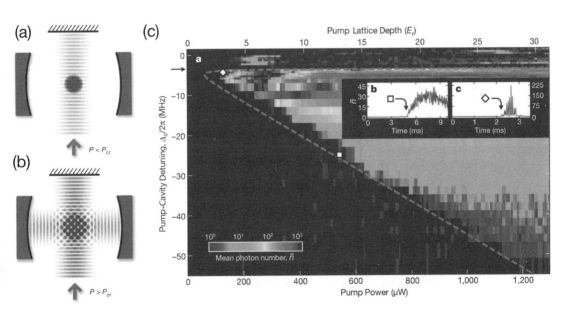

Figure 7.15 Bose–Einstein condensate inside a cavity. (a–b) Schematic of the phase before and after the superradiant transition. The vertical light is the pumping field, and the horizontal light is the cavity field. (c) The phase diagram in terms of the strength of the pumping field and the pumping-cavity detuning. The phase boundary separates the normal phase and the superradiant phase. Reprinted from Ref. [14]. A color version of this figure can be found in the resources tab for this book at cambridge.org/zhai.

ω_p, wave vector k_p and polarization along $\hat{y}$ is applied along the $\hat{z}$ direction, and they form a lattice. In addition, we consider a standing wave cavity field along $\hat{x}$ with the frequency ω_c, wave vector k_c and the polarization also along $\hat{y}$. Then, we can write the electric field as

$$\mathbf{E} = \left(E_p \cos(k_p z) \cos(\omega_p t) + E_c (\hat{a} e^{-i\omega_c t} + \hat{a}^\dagger e^{i\omega_c t}) \cos(k_c x) \right) \hat{y}. \qquad (7.92)$$

Here E_p denotes the strength of the electric field from the pumping laser, E_c is the electric field strength of each photon mode, and $\hat{a}$ and $\hat{a}^\dagger$ are the annihilation and creation operators of the standing wave cavity mode.

Following the same scheme discussed in the Section 1.3, we consider atoms interacting with such an electric field due to the dipole coupling $\hat{H}_d = \mathbf{d} \cdot \mathbf{E}$. By applying the unitary rotation $\mathcal{U}_{rot}(t) = e^{-i\omega_p t \mathcal{P}_e}$ and implementing the rotating wave approximation to ignore high frequency term with frequencies $2\omega_p$ and $\omega_p + \omega_c$, we reach the dipole coupling as

$$\hat{H}_d = d_y \left[\mathcal{P}_g \left(\frac{1}{2} E_p \cos(k_p z) + E_c \hat{a}^\dagger e^{-i\Delta_c t} \cos(k_c x) \right) \mathcal{P}_e + \text{h.c.} \right], \qquad (7.93)$$

where $\Delta_c = \omega_p - \omega_c$ is the detuning between the frequencies of the cavity field and the pumping field. Following the procedure described in Section 1.3, a second-order perturbation in terms of $\hat{H}_d$ leads to a scalar light shift as

$$\hat{V} = V_0 \cos^2(k_p z) + \eta(\hat{a} e^{i\Delta_c t} + \hat{a}^\dagger e^{-i\Delta_c t}) \cos(k_c x) \cos(k_p z) + U_0 \cos^2(k_c x) \hat{a}^\dagger \hat{a}. \qquad (7.94)$$

Here V_0 is the strength of lattice due to the pumping laser itself, and it is proportional to E_p^2. U_0 term describes a lattice potential from the cavity laser itself, and U_0 is fixed by the laser detuning and the ac polarization. η term is the interference term between the pumping laser and the cavity laser, with $\eta = \sqrt{V_0 U_0}$. Then, we can apply another unitary transformation $\hat{U} = e^{-i\Delta_c \hat{a}^\dagger \hat{a} t}$, under which $\hat{V}$ becomes time-independent as

$$\hat{V} = V_0 \cos^2(k_p z) + \eta(\hat{a} + \hat{a}^\dagger) \cos(k_c x) \cos(k_p z) + U_0 \cos^2(k_c x) \hat{a}^\dagger \hat{a}, \qquad (7.95)$$

and an extra term $-\hbar \Delta_c \hat{a}^\dagger \hat{a}$ is added into the Hamiltonian.

Langevin Force. In practice, there is certain probability that the cavity photon can leak out the system. On one hand, this leakage of photon allows nondestructive detection of the photon correlation without disturbing physics inside the cavity. On the other hand, it also makes the cavity system intrinsically nonequilibrium. To capture this effect, we usually add an imaginary part to the energy of the cavity photon, such that it becomes $-(\Delta_c + i\kappa)\hat{a}^\dagger \hat{a}$, where κ describes the leaking rate of cavity photons. However, there is a subtle issue after adding the κ term. Simply considering the cavity term along, the equation of motion for $\hat{a}(t)$ becomes

$$\hat{a}(t) = e^{-\kappa t + i\Delta_c t} \hat{a}(0), \qquad (7.96)$$

and therefore, the commutation relation $[\hat{a}(0), \hat{a}^\dagger(0)] = 1$ at $t = 0$ cannot be obeyed at any time t because

$$[\hat{a}(t), \hat{a}^\dagger(t)] = e^{-2\kappa t}. \qquad (7.97)$$

This problem can be eliminated by introducing the *Langevin force* $\hat{\xi}(t)$ [150], which modifies the Heisenberg equation as

$$i\hbar\frac{\partial\hat{a}}{\partial t} = [\hat{a},\hat{H}] + \hat{\xi}. \tag{7.98}$$

This is equivalent to adding a term $\hat{\xi}\hat{a}^\dagger + \hat{\xi}^\dagger\hat{a}$ into the Hamiltonian, where the Langevin force $\hat{\xi}(t)$ and $\hat{\xi}^\dagger(t)$ are random forces that should be averaged over. The solution of Eq. 7.98 can be written as

$$\hat{a}(t) = e^{-\kappa t + i\Delta_c t}\hat{a}(0) - \frac{i}{\hbar}\int_0^t \hat{\xi}(t')e^{-\kappa(t-t')}dt'. \tag{7.99}$$

By requiring $\langle\hat{\xi}(t)\rangle = \langle\hat{\xi}^\dagger(t)\rangle = 0$, $\langle\hat{\xi}(t)\hat{\xi}(t')\rangle = \langle\hat{\xi}^\dagger(t)\hat{\xi}^\dagger(t')\rangle = \langle\hat{\xi}^\dagger(t)\hat{\xi}(t')\rangle = 0$, and $\langle\hat{\xi}(t)\hat{\xi}^\dagger(t')\rangle = 2\kappa\hbar^2\delta(t-t')$, it can be shown that, by averaging over the Langevin noises, we can ensure that $[\hat{a}(t),\hat{a}^\dagger(t)] = 1$ for all time t. In fact, such a Langevin force term is required for all non-Hermitian quantum systems. This is because any non-Hermitian quantum system can always be viewed as a system coupled to an environment, and the entire system including environment is still a Hermitian one. When the environment is traced out to obtain an effective non-Hermitian description of the system, such a Langevin term is automatically generated when the Markovian coupling between system and environment is assumed.

Superradiant Transition. Let us first consider noninteracting bosons in such a lattice, the full Hamiltonian for the atom and cavity system can be written as

$$\hat{H} = \int d^3\mathbf{r}\hat{\Psi}^\dagger(\mathbf{r})\left(-\frac{\hbar^2\nabla^2}{2m} + \hat{V}\right)\hat{\Psi}(\mathbf{r}) - (\Delta_c + i\kappa)\hat{a}^\dagger\hat{a}, \tag{7.100}$$

where $\hat{V}$ is given by Eq. 7.95. This model contains two different phases. The expectation value of the cavity photon field $\alpha = \langle\hat{a}\rangle$ is the order parameter to characterize these two phases. In the normal phase, $\alpha = 0$ and atoms only experience the one-dimensional lattice from the pumping lattice along $\hat{z}$. In the superradiant phase, $\alpha \neq 0$ and atoms experience a two-dimensional lattice from both the pumping lattice and the cavity field. Note that the Hamiltonian is invariant under a transformation $\hat{a} \to -\hat{a}$ and $x \to x + \pi/k_c$ and it possesses the Z_2 symmetry. Thus, in the superradiant phase when $\alpha \neq 0$, the system breaks this Z_2 symmetry and the transition is described as the Z_2 symmetry-breaking transition. In this case, the transition between these two phases are controlled by parameters such as pumping laser strength V_0 and the detuning Δ_c. Since this is a transition happens at the zero-temperature, it is a quantum phase transition. More detailed discussion of the quantum phase transition will be given in Section 8.1.

To study this phase transition, we implement the mean-field theory by replacing $\hat{a}$ by $\alpha = \langle\hat{a}\rangle$. We denote $\Theta_x = \langle\cos^2(k_c x)\rangle$, $\Theta_z = \langle\cos^2(k_p z)\rangle$ and $\Theta_{xz} = \langle\cos(k_c x)\cos(k_p z)\rangle$, where the averages are taken over the wave function $\Psi(\mathbf{r})$ of atoms. Then $\mathcal{E} = \langle\hat{H}\rangle$ is given by

$$\mathcal{E} = \int d^3\mathbf{r} \left[\Psi^*(\mathbf{r}) \left(-\frac{\hbar^2 \nabla^2}{2m} \right) \Psi(\mathbf{r}) + V_0 \Theta_z + 2\eta \Theta_{xz} \mathrm{Re}\alpha + U_0 \Theta_x |\alpha|^2 \right] - (\Delta_c + i\kappa)|\alpha|^2.$$

$$(7.101)$$

Below we will consider real Θ_x, Θ_z and Θ_{xz} because the wave function of atoms does not break time-reversal symmetry.

Here we should also note that because this system is essentially an open system due to the photon leakage, when we talk about "phase" or "phase transition" in this case, we implicitly assume that there exists a steady state, in which the average photon number does not change with time. That is,

$$i \frac{\partial \alpha}{\partial t} = \eta \Theta_{xz} - (\Delta_c + i\kappa - U_0 \Theta_x)\alpha = 0, \quad (7.102)$$

which gives

$$\alpha = \frac{\eta \Theta_{xz}}{\Delta_c + i\kappa - U_0 \Theta_x}. \quad (7.103)$$

Eq. 7.103 locks the relation between the cavity field and the atom field. It says that in order for α to be nonzero, Θ_{xz} has to be nonzero. The proper procedure to obtain the stable steady state is to minimize the real part of Eq. 7.101 with respect to $\Psi(\mathbf{r})$ under the constraint Eq. 7.103.

As far as the phase boundary is concerned, we can derive a Landau-type theory for the phase transition. Assuming that the phase transition is a second-order one, following the same stratagem discussed in Section 6.2, we start from the normal phase with $\alpha = 0$ and derive the Landau theory by expanding energy in orders of α. At the zeroth order, $\alpha = 0$ and no lattice potential is applied along the $\hat{x}$ direction, therefore, both Θ_x and Θ_{xz} vanish because there exists continuous translational symmetry along $\hat{x}$. By the perturbation theory, the energy up to the second order of $|\alpha|^2$ is given by

$$\mathcal{E} = -4\eta^2 f (\mathrm{Re}\alpha)^2 - \Delta_c |\alpha|^2, \quad (7.104)$$

and

$$f = \sum_i \frac{|\langle \Psi_i | \cos(k_p z) \cos(k_c x) | \Psi_0 \rangle|^2}{E_i - E_0} \quad (7.105)$$

where $|\Psi_0\rangle$ and $|\Psi_i\rangle$ are the ground state and the excited state of the atom Hamiltonian in the absence of the cavity field, and E_0 and E_i are their corresponding eigenenergies, respectively. Moreover, to the leading order of α, Eq. 7.103 becomes

$$\alpha = \frac{\eta \Theta_{xz}}{\Delta_c + i\kappa}. \quad (7.106)$$

Substituting Eq. 7.106 into Eq. 7.104, Eq. 7.104 can be expressed in terms of Θ_{xz} as

$$\mathcal{E} = \left(-4 \frac{\eta^4 \Delta_c^2 f}{(\Delta_c^2 + \kappa^2)^2} - \frac{\Delta_c \eta^2}{\Delta_c^2 + \kappa^2} \right) \Theta_{xz}^2. \quad (7.107)$$

Therefore, the phase transition occurs when the coefficient in front of Θ_{xz} changes sign, and the phase boundary is determined by [14]

$$\eta_c = 2\sqrt{\frac{(\Delta_c^2 + \kappa^2)f}{-\Delta_c}}. \tag{7.108}$$

Here we always consider the situation that Δ_c is negative, and $\eta = \sqrt{V_0 U_0}$ is controlled by the strength of the pumping laser. Eq. 7.108 determines the phase boundary controlled by Δ_c and V_0, and this phase boundary has been observed experimentally, as shown in Figure 7.15(c).

The key information of Eq. 7.108 is the relation between the critical value for superradiant transition and the f-function. Note that Θ_{xz} is in fact a check-board density wave order in the xz plane, f is nothing but the susceptibility of the check-board order of the normal state. f is an intrinsically property of the normal state that characterizes the tendency toward forming the check-board density wave. Since the presence or the absence of the cavity field is equivalent to the presence or the absence of this check-board density wave of the atom field, the stronger this tendency, the easier for the superradiant transition to occur.

This discussion can also be generated to ultracold Fermi gases in cavity [31, 89, 142]. For fermions, the Fermi statistics and the Fermi surface geometry can play an important role in determining the check-board density wave susceptibility. Since the $\cos(k_p z)\cos(k_c x)$ scatters a fermion from momentum $\mathbf{p}$ to another momentum $\mathbf{p} + \mathbf{p}_0$, with $\mathbf{p}_0 = (\pm k_c, 0, \pm k_p)$, there are two features related to the Fermi statistics and the Fermi surface.

- In order for this scattering to happen, it requires momentum $\mathbf{p}$ state is occupied and momentum $\mathbf{p} + \mathbf{p}_0$ is empty. If for a sizable fraction of occupied $\mathbf{p}$ state, $\mathbf{p} + \mathbf{p}_0$ state is also occupied, then the check-board density wave susceptibility is strongly suppressed by the Pauli exclusion principle. This suppresses the superradiant transition.
- If for most $\mathbf{p}$ at the Fermi surface, $\mathbf{p} + \mathbf{p}_0$ also locates at or nearby the Fermi surface, these scattering processes cause little change of the kinetic energy. Such situation is known as the *Fermi surface nesting* as we will discuss again in Section 8.2. Due to the small kinetic energy cost, the check-board density wave susceptibility is strongly enhanced by the Fermi surface nesting. This enhances the superadiant transition [31, 89, 142].

Exercises

7.1 Solve the one-dimensional lattice model Eq. 7.10, and plot the band gap between the ground and the first excited bands as a function of V_x. Discuss the physical explanation of this function behavior for a small V_x regime and a large V_x regime, respectively.

7.2 Considering a two-dimensional square lattice or a three-dimensional cubic lattice, estimate the lattice potential when the bottom of the first excited band becomes higher than the top of the ground band and a real band gap with vanishing density-of-state appears.

7.3 Considering the one-dimensional lattice model Eq. 7.10, discuss when $V_x \to 0$ and, for a given quasi-momentum k and band index m, to which plane wave state it is adiabatically connected.

7.4 Considering the one-dimensional lattice model Eq. 7.10 in the tight-binding limit and considering the lowest band, compute the ratio of the next nearest hopping strength to the nearest hopping strength and the ratio of interaction strength between two nearest neighboring sites to the on-site interaction strength. Discuss how these two ratios change with increasing lattice potential.

7.5 Adding a J_3 term (as shown by t_3 in Figure 7.5(b)) into the tight-binding Hamiltonian Eq. 7.31, compute the band structure and the Dirac point. Discuss how these two Dirac points merge and annihilate when J_3 is large enough. Compare the band structure for $J_3 = J_1$ with the band structure of a two-dimensional square lattice.

7.6 Compute the Chern number using Eq. 7.59 for the Haldane model.

7.7 Diagonalize a finite size SSH chain as shown in Figure 7.8(b). Show the existence of the zero-energy mode in the spectrum and plot its wave function.

7.8 Deduce the effective Hamiltonian of the periodical-driven honeycomb lattice, and compute the topological invariant of the band structure of this effective Hamiltonian.

7.9 Derive Eq. 7.93 from $\hat{H}_d = \mathbf{d} \cdot \mathbf{E}$ with $\mathbf{E}$ given by Eq. 7.92, by using the rotating wave approximation introduced in Section 1.3.

8   The Hubbard Model

Learning Objectives

- Introduce quantum phase transition in the Bose–Hubbard model.
- Emphasize vanishing energy scales and universality at the quantum critical point.
- Introduce microscopic theories to describe the quantum phase transition, and explicitly show the vanishing of energy scales and the critical exponents.
- Illustrate the emergent Lorentz symmetry and the Higgs mode at the quantum critical point of the Bose–Hubbard model with the particle-hole symmetry.
- Discuss experimental probe of the superfluid to the Mott insulator phase transition.
- Show that the repulsive and the attractive Fermi–Hubbard models are related by the particle-hole symmetry.
- Discuss the origin of antiferromagnetic order in the repulsive Fermi–Hubbard model at half-filling.
- Introduce the enlarged $SO(4)$ symmetry of the Fermi–Hubbard model at half-filling and zero spin imbalance and its physical consequence.
- Introduce important unsolved challenge issues in the Fermi–Hubbard model.
- Introduce the concept of the eigenstate thermalization hypothesis and many-body localization as opposite to thermalization.
- Introduce a few metrics to characterize the many-body localization.
- Emphasize the role of entanglement entropy in characterizing quantum thermalization, and discuss how to measure entanglement entropy in ultracold atom systems.

8.1 Bose–Hubbard Model

The Hamiltonian for the Bose–Hubbard model (BHM) has been derived in Section 7.1, which is given by

$$\hat{H}_{\mathrm{BH}} = -J \sum_{\langle ij \rangle} (\hat{b}_i^\dagger \hat{b}_j + \mathrm{h.c.}) + \frac{U}{2} \sum_i \hat{n}_i(\hat{n}_i - 1) - \mu \sum_i \hat{n}_i. \tag{8.1}$$

In this section, we will focus on discussing the quantum phase transition in the BHM in two and three dimensions.[1] Let us first analyze two limits of the BHM. In the limit $U \to 0$, the ground state is a Bose condensate on the single-particle ground state, and the lowest

---

[1] The physics of one-dimensional BHM is somewhat different and will not be discussed here.

energy Bloch state with zero quasi-momentum is macroscopically occupied. With Eq. 7.7, a Bloch wave function with quasi-momentum zero is an equal weight superposition of all wannier wave functions with same phase, thus

$$\hat{b}^{\dagger}{}_{\mathbf{k}=0} = \frac{1}{\sqrt{N_s}} \sum_{i=1}^{N_s} \hat{b}_i^{\dagger},$$ (8.2)

where N_s is the total number of sites, and the wave function of this Bose condensate is given by

$$|SF\rangle = \frac{1}{\sqrt{N!}} \left(\frac{1}{\sqrt{N_s}} \sum_{i=1}^{N_s} \hat{b}_i^{\dagger} \right)^N |0\rangle,$$ (8.3)

where N is the total number of atoms. This state has following properties:

• The particle number at each site i approximately obeys a Poisson distribution,

$$P[n_i] = e^{-\bar{n}} \frac{\bar{n}^{n_i}}{n_i!},$$ (8.4)

where $\bar{n} = \langle \hat{n}_i \rangle$. Therefore atom number at each site has a large fluctuation as $\langle \delta n_i^2 \rangle = \bar{n}$.

• This state has a long-range order $\langle \hat{b}_i^{\dagger} \hat{b}_j \rangle \to C$ where C is a nonzero constant as $|i - j| \to \infty$. Following Eq. 3.10, this means the existence of ODLRO.

• Because this is a Bose condensed phase, as we discussed in Section 3.2, the excitation has a gapless phonon mode with linear dispersion when weak interactions between atoms are turned on, and the linear dispersion leads to superfluidity.

In the opposite limit $J \to 0$, each site becomes independent. n_i at each site is a good quantum number whose value is denoted by n_0. The Hamiltonian at each site becomes

$$\hat{H}_i = \frac{U}{2} n_0(n_0 - 1) - \mu n_0.$$ (8.5)

By minimizing the energy, we obtain $n_0 = [\mu/U] + 1$, as shown in Figure 8.1, where $[\mu/U]$ denotes the largest integer smaller than μ/U. The ground state is

$$|MI\rangle = \prod_i \frac{1}{\sqrt{n_0!}} (\hat{b}_i^{\dagger})^{n_0} |0\rangle.$$ (8.6)

This state has very different properties compared with the states at $U \to 0$ limit for the following reasons:

• At each site the state is a Fock state and the number fluctuation vanishes.

• $\langle \hat{b}_i^{\dagger} \hat{b}_j \rangle = 0$, which means the absence of ODLRO.

• The excitation is either adding a particle with the excitation energy

$$\Delta E = \left[\frac{U}{2} (n_0 + 1) n_0 - \mu(n_0 + 1) \right] - \left[\frac{U}{2} n_0(n_0 - 1) - \mu n_0 \right]$$

$$= U n_0 - \mu,$$ (8.7)

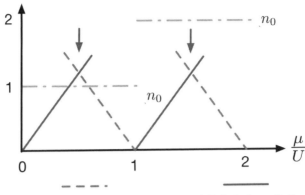

Figure 8.1 Mott insulator at $J = 0$ limit. The occupation number at each site (dash-dotted line), the particle excitation energy (dashed line), and the hole excitation energy (solid line), as a function of μ/U. The arrows indicate the particle-hole symmetric points. A color version of this figure can be found in the resources tab for this book at cambridge.org/zhai.

or taking a particle way with the excitation energy

$$\Delta E = \left[\frac{U}{2}(n_0 - 1)(n_0 - 2) - \mu(n_0 - 1) \right] - \left[\frac{U}{2}n_0(n_0 - 1) - \mu n_0 \right]$$

$$= \mu + U - U n_0. \tag{8.8}$$

They are respectively known as the particle excitation and the hole excitation. As shown in Figure 8.1, the charge excitations ΔE are gapped except for a discrete set of point with $\mu = nU$ with n being integers. According to what we have discussed in Box 7.1, the system is an insulator. Because the charge gap is now caused by the interactions between particles, such an insulator is called a *Mott insulator* (MI). Note that for each MI with a given n_0, there always exists a μ/U at which ΔE for the particle excitation equals to ΔE for the hole excitation, which is called the *particle-hole symmetric point*, as marked by arrows in Figure 8.1. We will emphasize its important role later.

Quantum Phase Transition. Therefore, the ground state of the BHM Hamiltonian at least contains two different phases, which are the SF phase and the MI phase. They are characterized by the presence or the absence of the superfluid order parameter, or characterized by the absence or the presence of the finite excitation gap. For the simplest situation, there is a single phase transition point that separates these two phases, and in fact it is indeed the case for the BHM. In general, a quantum phase transition refers to the situation that the ground state energy of an infinite system displays a discontinuity in its derivative with respect to continuously changing a parameter say, g, or a few parameters of a Hamiltonian. On one hand, similar as the thermal phase transition, these two states at two sides of the transition point should be fundamentally different, which are distinguished either by the order parameter or by topology. On the other hand, unlike the thermal phase transition, here the transition is driven by a parameter in the Hamiltonian instead of temperature, and we consider the discontinuity of the ground state energy instead of the free energy. In the

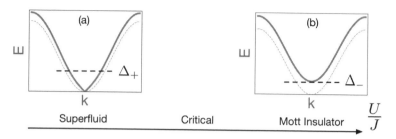

Schematic of dispersions for the BHM. (a) In the superfluid phase, low-energy excitation is a linear gapless mode, undergoing a crossover to the free-particle-like dispersion at high energy, and the crossover energy scale is denoted by Δ_+. (b) In the Mott insulator phase, the low-energy excitation is gapped, and the energy gap is denoted by Δ_-. The dashed lines show free-particle dispersion as a reference. A color version of this figure can be found in the resources tab for this book at cambridge.org/zhai.

case of the BHM, g is J/U. When $g < g_c$, it is the MI phase, and when $g > g_c$, it is the SF phase.

Here we specifically consider the quantum phase transitions that are second order. We can introduce Δ as a characteristic energy scale for the low-energy spectral of the Hamiltonian, and Δ_+ and Δ_- represent the characteristic energy scales for two different sides of the critical point g_c. In the BHM, when $g < g_c$ the system is an MI and the charge excitation is gapped. We introduce Δ_- as the charge gap, as shown in Figure 8.2. Across the transition the system is no longer an insulator, therefore, the charge gap Δ_- must vanish at the critical point. When $g > g_c$ the system is a SF and the low-energy excitation displays a linear gapless mode. As we discussed in Section 3.2, in the nonrelativistic case, the slope of the gapless mode determines the superfluid critical velocity, and there exists a characteristic energy scale at which the excitation spectrum undergoes the crossover from the linear dispersion at low-energy to the free-particle-like dispersion at high-energy. We denote this energy scale by Δ_+, as shown in Figure 8.2. If the slope of the linear dispersion is small, the linear regime is also small and therefore Δ_+ is small. At the critical point, the system loses superfluidity, hence, Δ_+ also must vanish at the critical point. Hence, when g approaches g_c, we have

$$\Delta \sim |g - g_c|^{z\nu}. \tag{8.9}$$

Here Δ represents either Δ_+ or Δ_-. Here the nontrivial point of the quantum criticality is that both two Δ share the same exponent $z\nu$, which is called the *critical exponent*.

Accompanying with the vanishing energy scales, there also exist divergent length scales ξ associated with a second-order phase transition. In the BHM, on the MI side, ξ_- determines the characteristic length scale of the exponential decay of the equal-time correlation function. And on the SF side, ξ_+ is the analogy of the healing length of a Bose superfluid discussed in Section 3.3. As approaching g_c, both two ξ diverge as

$$\xi^{-1} = |g - g_c|^\nu, \tag{8.10}$$

where ν is also a critical exponent. The ratio between Δ and ξ is $\Delta \sim \xi^{-z}$, where z is called the *dynamical critical exponent*. Normally $z = 2$ for a nonrelativistic system, and

$z = 1$ for a relativistic system. Below we will show that the BHM model can exhibit both relativistic and the nonrelativistic critical behaviors, depending on whether the system lies at the particle-hole symmetric point or not.

Because of the divergent length scales at the quantum critical point, the characteristic length scale is much larger than lattice spacing at the vicinity of the critical point. Hence, it is conceivable that there exists a long-wave length effective field theory to describe this phase transition, and both sides of the quantum phase transition are determined by the same field theory. Hence, when approaching the quantum phase transition from two different sides, it shares the same critical exponent ν determined by this field theory. Hence, ν is universal constant. By "universal," it means that the value is independent of microscopic details. For examples, if we consider the BHM in different lattice geometries, say, triangular, rectangular or square, or we can include hopping beyond the nearest neighbors hopping, the exact value of the critical point and the proportional constant in Eq. 8.9 will be changed, however, the critical exponent remains the same. This effective theory also determines the low-energy physics at the vicinity of the quantum critical point, and therefore, also determines the low-temperature physics above the quantum critical point. As we stated above, the quantum phase transition is defined as a zero-temperature transition. Nevertheless, we know that zero-temperature can never be reached. Therefore, the practical effects of a quantum critical point relies on the fact that this low-energy effective theory determines the low-temperature physics above the critical point.

Here we shall also emphasize that, although both energy scales shown in Figure 8.2 vanish at the quantum critical point, it does not mean that the quasi-particle dispersion becomes the same as the noninteracting one. In fact, in the quantum critical regime, the lifetimes of the low-energy quasi-particles are so short such that the quasi-particles are no longer well defined. Moreover, the divergence of all relevant low-energy length scales leads to the emergence of the scaling symmetry, and even more, the emergence of the conformal symmetry. Such a system with no well-defined quasi-particles and emergent conformal symmetry is reminiscent of the unitary Fermi gas discussed in Section 6.2. As we discussed in Box 6.2, on one hand, the absence of energy scales and the absence of quasi-particle description make such a quantum theory extremely difficult to deal with. And on the other hand, the concept of holographic duality discussed in Box 6.2 may be quite helpful in studying such field theory.

Hence, one can see that the vanishing of energy scales plays a crucial role in the quantum phase transition. Below we will concretely illustrate the vanishing of energy scales from microscopic theories of the BHM. We will present two different microscopic theories, which approach the quantum critical point from the Mott insulator side and the superfluid side, respectively.

Strongly Coupling Approach. As mentioned above, the excitation of a MI is fully gapped, and transition from a MI to a SF can be determined by the closing of the charge gap. In the limit $J \to 0$, suppose we add one extra particle at site-i on top of the ground state, the wave function for this particle excitation is written as

$$|p\rangle_i = \hat{b}_i^\dagger |MI\rangle, \tag{8.11}$$

where $|MI\rangle$ is given by Eq. 8.6, and this state causes a finite energy $\Delta_p = Un_0 - \mu$. This extra particle can be added in any site and the excitation energy is degenerate. When J is finite, hopping can connect all $|p\rangle_i$ with different site index i, and the matrix element between $|p\rangle_i$ and $|p\rangle_j$ is $-J(n_0 + 1)$ when i and j are neighboring sites. Here the factor $n_0 + 1$ comes from the Bose enhancement factor. Hence, due to the coupling, the gap for particle excitation becomes

$$\Delta_p = -ZJ(n_0 + 1) + Un_0 - \mu, \tag{8.12}$$

where Z denotes the coordination number. For instance, $Z = 4$ for a two-dimensional square lattice and $Z = 6$ for a three-dimensional cubic lattice. Similarly, in the limit $J \to 0$, we can take one particle away at site-i on top of the ground state, and the wave function for this hole excitation is given by

$$|h\rangle_i = \hat{b}_i|MI\rangle. \tag{8.13}$$

The excitation energy is $\Delta_h = \mu + U - Un_0$, and the energy is also degenerate for excitation at different sites. When J is finite, hopping can also couple $|h\rangle_i$ and $|h\rangle_j$ when i and j are neighboring sites, with the coupling strength given by $-Jn_0$. Hence, the excitation energy for hole excitation becomes

$$\Delta_h = -ZJn_0 + \mu + U - Un_0. \tag{8.14}$$

We can systematically improve the expression for Δ_p and Δ_h by considering multiple particle-hole excitations.

For a given μ/U, the SF-MI transition takes place at a critical $(J/U)_c$ at which either $\Delta_p = 0$ or $\Delta_h = 0$ is reached, and the charge excitation becomes gapless. Expanding around $(J/U)_c$, Eq. 8.12 and Eq. 8.14 also show that the gap Δ_- vanishes as $\Delta_- \sim |J/U - (J/U)_c|$ near the critical point.

This condition $\Delta_p = 0$ or $\Delta_h = 0$ gives rise to a phase boundary similar as shown in Figure 8.3, although Figure 8.3 is obtained by the mean-field theory described below. Several key features in this phase diagram can actually be easily understood by the strong coupling approach.

- As discussed above, for $n_0 - 1 < \mu/U < n_0$, the MI state has n_0 number of particle at each site. As shown in Figure 8.1, for $\mu/U = n_0$, either the particle or the hole excitation is gapless even when $J = 0$. Therefore, for these discrete set of chemical potentials, the system becomes a SF for infinitesimal small J.
- For each Mott lobe, when μ/U lies in the middle regime between $n_0 - 1$ and n_0, the particle or hole excitation gap is the largest, therefore, the critical $(J/U)_c$ is also the largest there. In the upper part of the Mott lobe, the transition is driven by the particle excitation becoming gapless. In the lower part of the Mott lobe, the transition is driven by the hole excitation becoming gapless.
- Due to the Bose enhancement factor, the hopping effect is enhanced by n_0 or $n_0 + 1$, the critical $(J/U)_c$ is suppressed for larger n_0. That is to say, the larger n_0, the smaller Mott lobe.

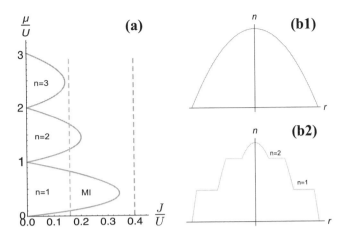

Figure 8.3 Phase diagram of the BHM. (a) The phase diagram of the Bose–Hubbard model, in terms of J/U and μ/U. (b) Schematic of the density profile inside a harmonic trap obtained from the local density approximation, for two different J/U, and the corresponding chemical potential trajectories indicated in (a). A color version of this figure can be found in the resources tab for this book at cambridge.org/zhai.

Mean-Field Approach. The mean-field theory for the BHM follows the standard mean-field approach summarized in Box 6.1, which allows us to approach the quantum critical point from the SF side. Here we decouple the hopping term as

$$-J\hat{b}_i^\dagger \hat{b}_j = -J\langle\hat{b}_i^\dagger\rangle\hat{b}_j - J\hat{b}_i^\dagger\langle\hat{b}_j\rangle + \left(-J\hat{b}_i^\dagger\hat{b}_j + J\langle\hat{b}_i^\dagger\rangle\hat{b}_j + J\hat{b}_i^\dagger\langle\hat{b}_j\rangle\right)$$
$$\approx -J\langle\hat{b}_i^\dagger\rangle\hat{b}_j - J\hat{b}_i^\dagger\langle\hat{b}_j\rangle + J\langle\hat{b}_i^\dagger\rangle\langle\hat{b}_j\rangle. \tag{8.15}$$

Now for simplicity, let us consider a uniform solution and assume all $\langle\hat{b}_i^\dagger\rangle$ are equal,[2] we introduce an order parameter ϕ as

$$\phi = ZJ\langle\hat{b}\rangle. \tag{8.16}$$

Then, the mean-field Hamiltonian becomes site independent as

$$\hat{H}_{\mathrm{MF}} = \left(-\phi\hat{b}^\dagger - \phi^*\hat{b} + \frac{U}{2}\hat{n}(\hat{n}-1) - \mu\hat{n} + \frac{|\phi|^2}{ZJ}\right). \tag{8.17}$$

This mean-field Hamiltonian should be solved self-consistently. The ground state wave function of the mean-field Hamiltonian Eq. 8.17 is a function of ϕ, and therefore, when we compute the expectation value of $\hat{b}$ under this wave function, the l.h.s. of Eq. 8.16 is a function of ϕ. Thus, Eq. 8.16 becomes a self-consistent equation for ϕ. If the solution is $\phi = 0$, effectively each site is disconnected from its neighboring sites, and the $\hat{H}_{\mathrm{MF}}$ is exactly the same as the BHM at $J = 0$, which gives rise to a MI phase. If the solution is $\phi \neq 0$, it is easy to see that the mean-field Hamiltonian Eq. 8.17 allows on-site number fluctuations. Moreover, because for each site $\langle\hat{b}_i\rangle = \phi$, the correlation $\langle\hat{b}_i^\dagger\hat{b}_j\rangle = |\phi|^2 \neq 0$,

[2] This assumption is not always correct, and in many circumstances of generalized BHM, one needs to consider nonuniform solutions.

which is independent of the relative distance between two sites. Hence the condensate fraction is given by

$$\frac{N_0}{N} = \frac{1}{N}\langle \hat{b}^\dagger_{\mathbf{k}=0}\hat{b}_{\mathbf{k}=0}\rangle = \frac{1}{NN_s}\sum_{ij}\langle \hat{b}^\dagger_i\hat{b}_j\rangle = \frac{N_s}{N}|\phi|^2 = \frac{|\phi|^2}{n} \neq 0. \qquad (8.18)$$

This shows that the state displays macroscopic occupation in the zero quasi-momentum mode and therefore this state is a SF state. In this way, we can determine the phase diagram by solving the self-consistent Eq. 8.16.

Although the procedure for carrying out the mean-field theory is quite standard, the success of a mean-field theory crucially depends on whether one can select out the correct term to perform the mean-field decoupling. Here it is worth comparing the BCS mean-field theory discussed in Section 6.1 and the mean-field theory for BHM discussed here.

- In the BCS case, the interactions between fermions scatter pairs of fermions from $(\mathbf{k}\uparrow, -\mathbf{k}\downarrow)$ to another momentum $(\mathbf{k}'\uparrow, -\mathbf{k}'\downarrow)$, and the BCS state can gain energy from this pair scattering. In the BHM case, the hopping of bosons transfer bosons from i-site to j-site, and the SF state can gain energy from this hopping.

- In the BCS case, because of the pair scattering, $(\mathbf{k}\uparrow, -\mathbf{k}\downarrow)$ should be either fully empty or both occupied, and this is the pair fluctuations. In the BHM model, because of hopping, on-site boson number fluctuates.

- In the mean-field theory, for the BCS case, the pair fluctuation is captured by coupling order parameter Δ to the fermion pair operator as $\Delta\hat{c}^\dagger_{\mathbf{k}\uparrow}\hat{c}^\dagger_{-\mathbf{k}\downarrow}$ + h.c. In other words, Δ describes the influence of all other momentum modes to the $(\mathbf{k}\uparrow, -\mathbf{k}\downarrow)$ mode due to the pair hopping. For the BHM case, the on-site boson number fluctuation is captured by coupling order parameter ϕ to the boson operator as $\phi\hat{b}^\dagger_i$ + h.c. In other words, ϕ describes the influence of other sites to the i-site due to the boson hopping.

- For the BCS case, the advantage of the mean-field theory is that the Hamiltonian becomes diagonal in momentum space. And because the Hamiltonian is quadratic for each fixed $\mathbf{k}$, the mean-field Hamiltonian can be solved exactly. For the BHM case, the advantage of the mean-field theory is that the Hamiltonian becomes diagonal in real space. Although in this case, the Hamiltonian is not quadratic for each site-i and cannot be solved analytically, it can be easily solved numerically.

- In the BCS case, the self-consistent condition leads to the BCS gap equation as shown in Eq. 6.24. In the BHM case, the self-consistent condition leads to Eq. 8.16 for ϕ as discussed above.

- In the BCS case, whether Δ is zero determines the transition from the normal state to the BCS superfluid. In the BHM case, whether ϕ is zero determines the transition from the MI to the SF.

In the BCS theory, we have discussed in Section 6.2 that we can treat Δ as a small parameter to obtain an effective theory on the vicinity of the phase transition. Similarly, considering the MI to SF transition as a second-order one, we can also use ϕ as a small parameter to treat the mean-field Hamiltonian Eq. 8.17 perturbatively. In this way, we can obtain

$$\mathcal{E} = -a|\phi|^2 + b|\phi|^4. \qquad (8.19)$$

Using the second-order perturbation theory, it is straightforward to show that

$$a = \frac{n_0 + 1}{n_0 U - \mu} + \frac{n_0}{\mu - U(n_0 - 1)} - \frac{1}{ZJ}, \tag{8.20}$$

where the first and the second terms in Eq. 8.20 come from the contributions of the particle excitation and the hole excitation being the intermediate states, respectively. As we have discussed in Section 6.2, for the Landau theory, the superfluid order parameter ϕ is nonzero and the system is a SF when a is positive, and the order parameter ϕ is zero when a is negative. Therefore, $a = 0$ determines the critical condition, that is, J_c/J is a function of μ/U given by

$$\frac{J_c}{U} = \frac{\left(n_0 - \frac{\mu}{U}\right)\left(\frac{\mu}{U} - (n_0 - 1)\right)}{Z\left(\frac{\mu}{U} + 1\right)}. \tag{8.21}$$

Nearby the critical point, a scales as $|(J/U)-(J/U)_c|$. Furthermore, because $|\phi|^2$ is proportional to a nearby the critical point, and therefore, the condensate fraction is proportional a. Moreover, because the low-energy scale Δ_+ is proportional to the condensate fraction, Δ_+ vanishes as $|(J/U) - (J/U)_c|$ as approaching the quantum critical point. Hence, we have shown that both Δ_+ and Δ_- vanish in the same way as approaching the quantum critical point, that gives $z\nu = 1$. When $z = 2$, it gives $\nu = 1/2$.

Emergent Lorentz Symmetry. Nearby the critical point, the effective theory should be determined by the spatial and temporal fluctuations of the order parameter ϕ. To describe such fluctuations, we should introduce the path integral formalism to write the partition function as

$$\mathcal{Z} = \int \prod_i \mathcal{D}\phi_i^* \mathcal{D}\phi_i(\tau) e^{-S[\phi^*,\phi]}, \tag{8.22}$$

where the action S is given by

$$S[\phi^*, \phi] = \int_0^\beta d\tau \int d^3\mathbf{r}[u\phi^*\partial_\tau\phi + v|\partial_\tau\phi|^2 + w|\nabla\phi|^2 - a|\phi|^2 + b|\phi|^4 + \dots]. \tag{8.23}$$

Here the first two terms in Eq. 8.23 describe the temporal fluctuation of ϕ and the third term in Eq. 8.23 describes the spatial fluctuation of ϕ.

Now we introduce a method to compute u using the gauge symmetry. Note that before making the small ϕ expansion, the dynamics of the BHM should be described by the partition function

$$\mathcal{Z} = \int \prod_i \mathcal{D}b_i^*(\tau)\mathcal{D}b_i(\tau) \exp\left\{\int_0^\beta d\tau \left[\sum_i -b_i^*\partial_\tau b_i - H_{\mathrm{MF}}(\{b_i^*, b_i\})\right]\right\}. \tag{8.24}$$

Here we have implemented the coherent state path integral representation, and $H_{\mathrm{MF}}(\{b_i^*, b_i\})$ is a functional of b_i^* and b_i obtained by replacing normal ordered Hamiltonian $\hat{H}_{\mathrm{MF}}$ Eq. 8.17 with $\hat{b}_i \rightarrow b_i$ and $\hat{b}_i^\dagger \rightarrow b_i^*$. Note that the original Lagrangian Eq. 8.24 has a global $U(1)$ symmetry, that is,

$$b(\tau) \rightarrow b(\tau)e^{i\theta(\tau)}, \quad b^*(\tau) \rightarrow b^*(\tau)e^{-i\theta(\tau)} \tag{8.25}$$

and

$$\phi(\tau) \rightarrow \phi(\tau)e^{i\theta(\tau)}, \quad \phi^*(\tau) \rightarrow \phi^*(\tau)e^{-i\theta(\tau)}, \quad \mu \rightarrow \mu + i\partial_\tau\theta. \tag{8.26}$$

Here "global" means that θ can only depend on τ but cannot depend on the spatial coordinate.

Therefore, we shall also require that the effective action Eq. 8.23 also obeys this gauge symmetry. Note that with this transformation, $u\phi^*\partial_\tau\phi$ term acquires an extra term as $iu(\partial_\tau\theta)|\phi|^2$, and $-a|\phi|^2$ term acquires an extra term as $-(\partial a)/(\partial\mu)(i\partial_\tau\theta)|\phi|^2$. They are only two extra terms proportional to $i(\partial_\tau\theta)|\phi|^2$, therefore, to keep the action invariant, these two extra terms have to cancel each other, which leads to

$$\frac{\partial a}{\partial\mu} - u = 0. \tag{8.27}$$

Thus we obtain

$$u = \frac{n_0 + 1}{(n_0 U - \mu)^2} - \frac{n_0}{(\mu - U(n_0 - 1))^2}. \tag{8.28}$$

Hence, for each n_0, there always exists a μ/U between $n_0 - 1$ and n_0, such that $u = 0$. For sufficiently large n_0, if we can ignore the difference between $n_0 + 1$ and n_0, $u = 0$ when $n_0 U - \mu = \mu - U(n_0 - 1)$. That is to say, the particle excitation energy equals the hole excitation energy, and therefore, this particular value of μ/U is the particle-hole symmetric point. When the first order temporal derivative term vanishes at the particle-hole symmetric point, the second-order temporal derivative becomes dominative, and the effective action looks like

$$S[\phi^*, \phi] = \int_0^\beta d\tau \int d^3\mathbf{r}[v|\partial_\tau\phi|^2 + w|\nabla\phi|^2 - a|\phi|^2 + b|\phi|^4 + \ldots]. \tag{8.29}$$

In the action Eq. 8.29, both the spatial fluctuation and the temporal fluctuations are determined by the second-order derivative terms. Therefore, after a proper scaling of the space and time, that is, $\tau \rightarrow \tau/\sqrt{v}$ and $\mathbf{r} \rightarrow \mathbf{r}/\sqrt{w}$, the effective field theory becomes

$$S[\phi^*, \phi] = \int_0^\beta d\tau \int d^3\mathbf{r}[|\partial_\tau\phi|^2 + |\nabla\phi|^2 - a|\phi|^2 + b|\phi|^4 + \ldots], \tag{8.30}$$

which possesses the Lorentz symmetry. In a Lorentz theory, the energy and the length scale the same way, and thus $z = 1$. The critical theory along the particle-hole symmetric line is therefore different from the critical theory away from this line. Since the original Hamiltonian is an nonrelativistic theory, the Lorentz symmetry displayed by the low-energy theory is larger than the symmetry of the original Hamiltonian. This phenomenon is known as the *emergent symmetry*, which refers to the phenomenon that the symmetry of low-energy physics is larger than the symmetry of the original Hamiltonian.

Let us consider the effective action Eq. 8.29 with the Lorentz symmetry. For $a < 0$, the ground state has $\phi = 0$ and the system is a MI. Considering the fluctuation around $\phi = 0$, this field theory gives a gapped excitation $\omega = \sqrt{k^2 + |a|}$, and the gap scales with $\sqrt{a} \sim (1 - J/J_c)^{1/2}$. Note that this gap is different from both the particle excitation energy Eq. 8.12 and the hole excitation energy Eq. 8.14. It is because in this particle-hole symmetric case, one needs to consider the coupling between the particle excitation and the hole excitation since they are nearly degenerate.

For $a > 0$, the system is in the symmetry-breaking phase, and we should consider both the phase and the amplitude around a nonzero ϕ. In this case, it can be shown that the phase and amplitude fluctuation are decoupled at the lowest order in low-energy. We can obtain two branches of excitation spectrum as

$$\omega = k \tag{8.31}$$

$$\omega = \sqrt{k^2 + 2a}. \tag{8.32}$$

The first one is a gapless Goldstone mode for the phase fluctuation, and the second is for density mode called the *Higgs mode*, which is always gapped. If the Goldstone mode is coupled to another gauge field, it can give rise to a mass for the gauge field, which is known as the *Higgs mechanism*. We discuss the Higgs model and the Higgs mechanism in Box 8.1. The gap of Higgs mode Δ scales as $\Delta \sim \sqrt{a} \sim \sqrt{J/J_c - 1}$. Thus, we show that for the particle-hole symmetric case, the critical behavior is different, and the characteristic energy scales in both sides vanish as approaching the critical point with $z = 1$ and $\nu = 1/2$.

Experimental Signatures. Now we will discuss several experimental measurements that can distinguish the SF and the MI phases and determine the quantum phase transition.

- Time-of-Flight: In the SF phase, bosons are condensed in the Bloch state with zero quasi-momentum. In Section 7.1, we have discussed two types of time-of-flight measurements, which is called momentum mapping and band mapping, respectively. In the momentum mapping, the time-of-flight measurement of a SF phase shows peaks at both the zero momentum and all reciprocal lattice vectors. In the MI phase, bosons are all localized, and therefore the momentum distribution will become a broad featureless

---

Box 8.1   **Higgs Mode and Higgs Mechanism**

The discussion here shows that the existence of the Higgs mode in a $U(1)$ symmetry-breaking phase relies on the emergent Lorentz symmetry. Another system that can exhibit the Higgs mode is a BCS type of fermion superfluid, where the effective theory for pairing order parameter displays the Lorentz symmetry because of the underlying particle-hole symmetry of the fermionic excitation spectrum, as we discussed in Chapter 6. Nevertheless, when the particle-hole symmetry gradually disappears in the strong pairing BEC regime, the system loses its Lorentz symmetry and returns to a nonrelativistic boson theory. Hence, the Higgs mode can be identified in the BCS regime and gradually disappears in the BEC regime. Here we should distinguish the concept of the Higgs mode and the Higgs mechanism. If the fermions are charged and are coupled to an external $U(1)$ gauge field, as in the case of a superconductor, the coupling between the Goldstone mode of the superconductor and the gauge field can make the gauge field gapped. In this way, the gauge particle can acquire a mass, which is now known as the Higgs mechanism.

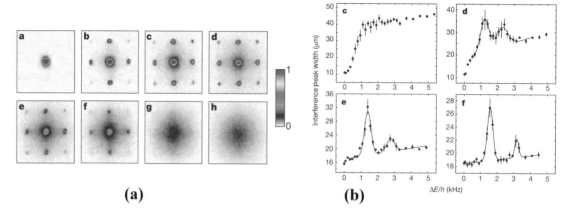

(a)   **(b)**

Figure 8.4   Experimental evidence of quantum phase transition in the BHM. (a) The momentum distribution measured by the time of flight for different lattice depths. (b) Response to a potential gradient ΔE for different lattice depths. Reprinted from Ref. [63]. A color version of this figure can be found in the resources tab for this book at cambridge.org/zhai.

distribution. Such experimental measurements provided the first evidence for the quantum phase transition in the BHM [63]. As shown in Figure 8.4(a), one finds that the momentum distribution smears out as the interaction strength increases.

- Transport: As an insulator, there absents charge transport for a small potential gradient in the MI phase. And because the density mode of an SF is gapless, a SF can transport charge with any small potential gradient. This feature has also been explored experimentally [63]. It has been found the system responds to infinitesimal gradient for weak interaction, but it requires a critical gradient for the system to respond for strong interaction, as shown in Figure 8.4(b).

- In-situ Density Profile: The in-situ density profile inside a harmonic trap can be determined by the local density approximation. One can define a local chemical potential $\mu(\mathbf{r}) = \mu_0 - V(\mathbf{r})$, where μ_0 is the chemical potential at the center of the harmonic trap. Suppose we know $n(\mu)$ for a uniform system, by replacing μ with $\mu_0 - V(\mathbf{r})$, we can map out a real space density distribution. When the trajectory $\mu(\mathbf{r})$ lies inside the MI phase, the density does not change with the changing of r, and therefore, the density distribution shows a so-called wedding cake structure, as shown in Figure 8.3(b). The reason is in fact the same as the transport measurement because the density does not respond to small potential gradient due to the charge gap. By measuring the in-situ density profile $n(\mathbf{r})$ one can in fact read out the local compressibility using the relation

$$\kappa = \frac{\partial n}{\partial \mu} = \frac{\partial n}{\partial \mathbf{r}} \frac{\partial \mathbf{r}}{\partial \mu} = -\frac{\partial n}{\partial \mathbf{r}} \frac{\partial \mathbf{r}}{\partial V(\mathbf{r})}, \tag{8.33}$$

where $\partial n / \partial \mathbf{r}$ can be deduced from the measurements. A MI can be identified from $\kappa \approx 0$. In addition, by repeating the density measurements under the same condition, one can not only determine the mean value $\bar{n} = \langle \hat{n} \rangle$ but also measure the variance $\sigma^2 = \langle (\hat{n} - \bar{n})^2 \rangle$. The variance σ^2 and the compressibility κ are related, and σ^2 vanishes when κ

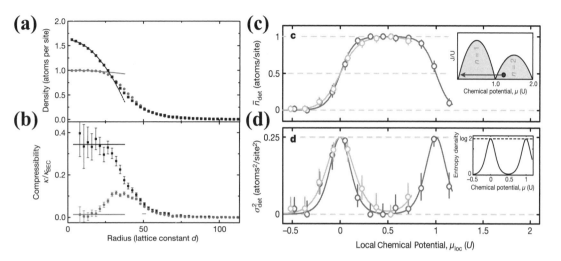

Figure 8.5 Experimental measurements of density profile of the BHM. (a): Density distribution as a function of position. (b) The compressibility κ deduced from (a). (c) Density as a function of the local chemical potential. (d) The particle number variance as a function of the local chemical potential. (a–b) are measured with high-resolution in situ imaging, and (c–d) are measured with single-site in situ imaging. (a–b) are reprinted from Ref. [58], and (c–d) are reprinted from Ref. [158]. A color version of this figure can be found in the resources tab for this book at cambridge.org/zhai.

vanishes. Such measurements have been carried out [58, 158, 11] and the representative results are shown in Figure 8.5. One can see that when the density shows a plateau in the MI regime, where both the compressibility and the particle number variance vanish. Precisely speaking, since the experiments are always performed at finite temperature, when the temperature is much smaller compared with the charge gap in the MI phase, both the compressibility and the particle number variance are exponentially small.

Below we further discuss some experimental consequences related to the quantum critical point. First, the Higgs mode has been observed in this system [53]. A periodic modulation of lattice depth with a frequency ν_{mod} has been applied for certain duration, after which the lattice is ramped to very deep limit to measure the increases of temperature. This modulation is coupled to the density excitations, and the increasing of temperature is used as a probe of the density response excited by the lattice modulation. Figure 8.6(b) shows a gapped response, from which the onset frequency ν_0 of the response can be extracted. As the coupling J approaches the critical value J_c, it is clear that the onset frequency moves to lower frequencies. Figure 8.6(a) displays ν_0 as a function of J/J_c. In the MI side it is consistent with a charge gap which scale with $\sqrt{1 - J/J_c}$, and in the SF side it is consistent with a Higgs gap which scales with $\sqrt{J/J_c - 1}$.

Second, we consider the question that whether the phase transition also exists at finite temperature. Generally, there are also two different scenarios for a quantum phase transition. One scenario is that the singularity only occurs at zero temperature and all thermodynamic quantities behave smoothly at finite temperature. The other scenarios is that the singularity also occurs at finite temperature. The BHM belongs to the latter, as shown

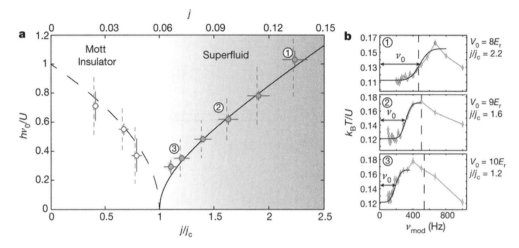

Figure 8.6 Observation of the Higgs model in the BHM model. (a) Circles are the fitted gap values (in units of U), which show that both the Higgs gap (solid line) and the Mott gap (dashed line) vanish at the quantum critical point, and both scale as $\sqrt{|1 - J/J_c|}$. j/j_c in the label of horizontal axis is J/J_c in the text. (b) Temperature response to lattice modulation (circles) and the fitting function (solid line) for the three different points labeled in (a). Reprinted from Ref. [53]. A color version of this figure can be found in the resources tab for this book at cambridge.org/zhai.

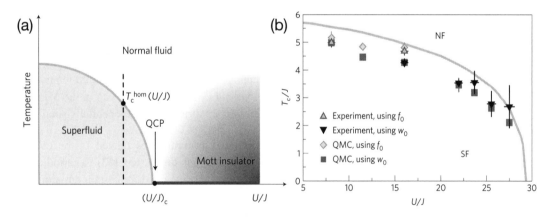

Figure 8.7 Finite temperature phase diagram of the BHM. (a) A schematic plot of the finite temperature phase diagram of the BHM in terms of temperature and U/J. (b) Experimental measurement of the SF transition at the finite temperature for different U/J. The experimental measurements are compared with the Monte Carlo simulation. Reprinted from Ref. [175]. A color version of this figure can be found in the resources tab for this book at cambridge.org/zhai.

in Figure 8.7(a). This is because at the SF side, there should always be a finite temperature transition to the normal gas. From the single-particle picture, this transition is determined by the effective mass, or the hopping J. As long as the mass is not infinite, or equivalent to say, as long as J does not vanish, the transition temperature should retain finite. However, because of the presence of the quantum critical point, the superfluid transition temperature is suppressed to zero at the quantum critical regime. This is a phenomenon that cannot be explained by the single-particle picture and it is a strong manifestation of the quantum critical point. In the BHM, the suppression of the superfluid transition temperature has been measured and compared very well with the quantum Monte Carlo simulation [175]. The results are shown in Figure 8.7(b).

8.2 Fermi–Hubbard Model

The Hamiltonian of the Fermi–Hubbard model (FHM) has been derived in Eq. 7.24. Here we slightly rewrite the Hamiltonian as

$$\hat{H}_{\text{FH}} = - J \sum_{\langle ij \rangle, \sigma} \hat{c}_{i\sigma}^\dagger \hat{c}_{j\sigma} + U \sum_i \left(\hat{n}_{i\uparrow} - \frac{1}{2} \right) \left(\hat{n}_{i\downarrow} - \frac{1}{2} \right)$$

$$- \mu \sum_i (\hat{n}_{i\uparrow} + \hat{n}_{i\downarrow}) - h \sum_i (\hat{n}_{i\uparrow} - \hat{n}_{i\downarrow}). \tag{8.34}$$

Here we first rewrite the interaction term, which is equivalent to shifting the chemical potential by $U/2$. The advantage of writing the interaction term in this way will be clear shortly. We also anticipate that the chemical potentials can be different for different spin components, which is equivalent to introducing a Zeeman field h coupled to spin polarization along $\hat{z}$. Alternatively, we can also work in the canonical ensemble, and write the Hamiltonian as

$$\hat{H}_{\text{FH}} = - J \sum_{\langle ij \rangle, \sigma} \hat{c}_{i\sigma}^\dagger \hat{c}_{j\sigma} + U \sum_i \left(\hat{n}_{i\uparrow} - \frac{1}{2} \right) \left(\hat{n}_{i\downarrow} - \frac{1}{2} \right). \tag{8.35}$$

In the canonical ensemble, we need to impose constraints for two conserved quantities, which are $\delta = N_\uparrow + N_\downarrow - N_s$ and $m = N_\uparrow - N_\downarrow$, respectively. Here $N_\sigma = \sum \langle \hat{c}_{i\sigma}^\dagger \hat{c}_{i\sigma} \rangle$. N_s is the total number of site. The reason we write the first conserved quantity as $N_\uparrow + N_\downarrow - N_s$ instead of $N_\uparrow + N_\downarrow$ will also be made clear below. Both two conserved quantities can take any integer between $-N_s$ to N_s. For spin-1/2 fermions, because each site can be maximally occupied by two particles, we have $0 \leqslant N_\uparrow + N_\downarrow \leqslant 2N_s$. Hence, $N_\uparrow + N_\downarrow = N_s$ is called *half-filling*, which corresponds to $\delta = 0$. In condensed matter literatures, a nonzero δ is called *doping*. m represents the spin polarization, which is also called the spin imbalance in many ultracold atom literatures.

Spin Rotational Symmetry. Here we first discuss the spin rotational symmetry of the FHM. First of all, by using the completeness relation of $SU(2)$ Pauli matrix

$$\sum_{i=1}^3 \sigma_{\alpha\beta}^i \sigma_{\gamma\delta}^i = 2\delta_{\alpha\delta}\delta_{\beta\gamma} - \delta_{\alpha\beta}\delta_{\delta\gamma}, \tag{8.36}$$

we have

$$\frac{1}{2}\hat{n}_i^2 = \hat{n}_i - \frac{2}{3}\hat{\mathbf{S}}_i \cdot \hat{\mathbf{S}}_i, \tag{8.37}$$

where $\hat{n}_i = \hat{n}_{i\uparrow} + \hat{n}_{i\downarrow}$ and $\hat{\mathbf{S}}_i = (1/2)\hat{c}_{i\alpha}^\dagger \boldsymbol{\sigma}_{\alpha\beta}\hat{c}_{i\beta}$. Thus, the FHM can be rewritten as

$$\hat{H}_{\text{FH}} = -J\sum_{\langle ij \rangle, \sigma} \hat{c}_{i\sigma}^\dagger \hat{c}_{j\sigma} - \frac{2U}{3}\sum_i \hat{\mathbf{S}}_i^2 - \mu\sum_i \hat{n}_i - 2h\sum_i \hat{S}_i^z. \qquad (8.38)$$

It can be verified that when $h = 0$, both the interaction term and the chemical potential term are invariant under the local $SU(2)$ spin rotations, but the kinetic energy term is only invariant under the global $SU(2)$ spin rotation, which are generated by following three total spin operators $\hat{\mathbf{S}}$ defined as

$$\hat{S}^x = \frac{1}{2}\sum_i (\hat{c}_{i\uparrow}^\dagger \hat{c}_{i\downarrow} + \hat{c}_{i\downarrow}^\dagger \hat{c}_{i\uparrow}), \qquad (8.39)$$

$$\hat{S}^y = \frac{i}{2}\sum_i (\hat{c}_{i\uparrow}^\dagger \hat{c}_{i\downarrow} - \hat{c}_{i\downarrow}^\dagger \hat{c}_{i\uparrow}), \qquad (8.40)$$

$$\hat{S}^z = \frac{1}{2}\sum_i (\hat{c}_{i\uparrow}^\dagger \hat{c}_{i\uparrow} - \hat{c}_{i\downarrow}^\dagger \hat{c}_{i\downarrow}). \qquad (8.41)$$

When h is nonzero, the magnetic field along $\hat{z}$ direction breaks the full $SU(2)$ spin rotational symmetry, and the system is only invariant under the spin rotation along $\hat{z}$.

Particle-Hole Transformation. Let us first introduce a general situation that two different systems are related by a symmetry. Considering $\hat{H}$ and $\hat{H}'$ as Hamiltonians of two systems, if $\hat{H}' = \hat{U}^\dagger \hat{H}\hat{U}$, then we say that these two systems are related by the symmetry operation $\hat{U}$. In this case, if Ψ is the ground state of $\hat{H}$, then $\hat{U}^\dagger \Psi$ is the ground state of $\hat{H}'$. That is to say, for two systems related by a symmetry, we can deduce the property of one system with the knowledge of the other system. Here we consider two systems that are the FHM with repulsive interaction and attractive interaction, respectively, and the symmetry operation is the particle-hole transformation.

First, let us consider the following particle-hole transformation which acts on spin-down particle only, that is to say, we keep $\hat{c}_{i\uparrow}$ and $\hat{c}_{i\uparrow}^\dagger$ unchanged but we change $\hat{c}_{i\downarrow} \to \hat{c}_{i\downarrow}^\dagger$ and $\hat{c}_{i\downarrow}^\dagger \to \hat{c}_{i\downarrow}$. It can be further shown that under this particle-hole transformation, $\hat{n}_{i\downarrow}$ becomes $1 - \hat{n}_{i\downarrow}$. Thus, the U-term changes sign, and the h- and $\mu-$ term exchange with each other, that is,

$$U\left(\hat{n}_{i\uparrow} - \frac{1}{2}\right)\left(\hat{n}_{i\downarrow} - \frac{1}{2}\right) \to -U\left(\hat{n}_{i\uparrow} - \frac{1}{2}\right)\left(\hat{n}_{i\downarrow} - \frac{1}{2}\right), \qquad (8.42)$$

$$\mu(\hat{n}_{i\uparrow} + \hat{n}_{i\downarrow}) \to \mu(\hat{n}_{i\uparrow} - \hat{n}_{i\downarrow} + 1), \qquad (8.43)$$

$$h(\hat{n}_{i\uparrow} - \hat{n}_{i\downarrow}) \to h(\hat{n}_{i\uparrow} + \hat{n}_{i\downarrow}). \qquad (8.44)$$

If we work on the canonical ensemble, it is easy to see that the two conserved quantities also exchange with each other, that is,

$$\delta = N_\uparrow + N_\downarrow - N_s \to N_\uparrow - N_\downarrow = m \qquad (8.45)$$

$$m = N_\uparrow - N_\downarrow \to N_\uparrow + N_\downarrow - N_s = \delta. \qquad (8.46)$$

Now we consider the kinetic energy. The kinetic energy for spin-up fermions is unchanged, but the kinetic energy for spin-down fermions changes sign because of the particle-hole transformation. However, we would like to compare two FHM with the same kinetic energy. To this end, we introduce a concept called the *bipartite lattice*. The bipartite lattice means that we can divide all the lattice sites into two groups, denoted by A sublattice and B sublattice, and hopping only takes places between A sublattices and B sublattices and cannot happen within A sublattices or within B sublattices. For instance, if we only consider the nearest neighbor hopping, a two-dimensional square lattice or honeycomb lattice, and a three-dimensional cubic lattice are all bipartite lattices, but a two-dimensional triangular lattice is not a bipartite lattice. If one includes the next nearest neighbor hopping, even the two-dimensional square lattice or honeycomb lattice is not a bipartite lattice.

To keep the kinetic energy term invariant for both spin components, we modify the particle-hole transformation by introducing a minus sign for fermion operator in one sublattice but not the other sublattice. To be concrete, let us consider a two-dimensional square lattice with site-i labeled by $R_i = (i_x, i_y)$. Let us now modify the particle-hole transformation for spin-down particles as $\hat{c}_{i\downarrow} \rightarrow (-1)^{i_x+i_y}\hat{c}_{i\downarrow}^\dagger$ and $\hat{c}_{i\downarrow}^\dagger \rightarrow (-1)^{i_x+i_y}\hat{c}_{i\downarrow}$. Since the hopping only occurs between two sublattices, this extra minus sign cancels out the minus sign introduced by the particle-hole transformation. Hence, the kinetic energy is invariant for hopping on a bipartite lattice. Note that

$$(-1)^{i_x+i_y} = e^{i(\pi i_x + \pi i_y)} = e^{iQ\cdot R_i}, \quad \text{or,} \quad (-1)^{i_x+i_y} = e^{-i(\pi i_x + \pi i_y)} = e^{-iQ\cdot R_i}, \quad (8.47)$$

where $Q = (\pi, \pi)$. Hence, when we consider the operator in momentum space, say, $\hat{c}_k$, we have

$$\hat{c}_k = \sum_i e^{ik\cdot R_i}\hat{c}_i \rightarrow \sum_i e^{i(k-Q)\cdot R_i}\hat{c}_i^\dagger = \hat{c}_{-k+Q}^\dagger, \quad (8.48)$$

and

$$\hat{c}_k^\dagger = \sum_i e^{-ik\cdot R_i}\hat{c}_i^\dagger \rightarrow \sum_i e^{i(-k+Q)\cdot R_i}\hat{c}_i = \hat{c}_{-k+Q}. \quad (8.49)$$

For the nearest hopping, the single-particle dispersion is given by

$$\epsilon(k) = -2J(\cos(k_x) + \cos(k_y)), \quad (8.50)$$

and it is easy to see that $\epsilon(k) = -\epsilon(-k+Q)$. Generally speaking, if there exists a momentum Q such that $\epsilon(k) = -\epsilon(-k+Q)$ for all k, the kinetic energy can be invariant under the particle-hole transformation as

$$\sum_k \epsilon(k)\hat{c}_k^\dagger\hat{c}_k \rightarrow \sum_k \epsilon(-k+Q)\hat{c}_{-k+Q}^\dagger\hat{c}_{-k+Q}. \quad (8.51)$$

Hence, we show that this particle-hole transformation relates to a FHM with repulsive interaction $U_0 > 0$ to a FHM with attractive interaction with $-U_0 < 0$, and meanwhile, it exchanges h- and μ- terms. As shown in Figure 8.8, a FHM with $(U = U_0, \mu = x, h = y)$ is mapped to a FHM with $(U = -U_0, \mu = y, h = x)$. Or equivalently, a FHM with $(U = U_0, \delta = x, m = y)$ is mapped to a FHM with $(U = -U_0, \delta = y, m = x)$.

Pairing and Antiferromagnetism. Let us first consider a FHM with equal spin population $N_\uparrow = N_\downarrow$ and attractive interaction, whose parameters are $(-U_0, \mu = x, h = 0)$ $(U_0 > 0)$. Based on the discussion in Section 6.1, the Cooper instability due to the

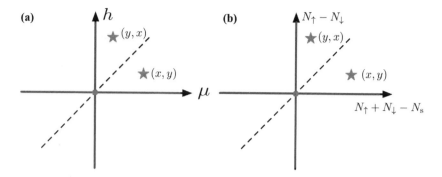

Figure 8.8 Repulsive and attractive FHM related by the particle-hole symmetry. In each plot, two stars label two parameters where repulsive FHM in one place can be mapped to attractive FHM in another place with equal interaction strength. Especially, the vertical axes with repulsive (attractive) interaction are mapped to horizontal axes with attractive (repulsive) interaction. The central point is self-dual between repulsive and attractive models. (a) and (b) correspond to the grand-canonical and canonical ensembles, respectively. A color version of this figure can be found in the resources tab for this book at cambridge.org/zhai.

attractive interaction can lead to pairing of fermions, which gives rise to pairing order $\Delta = \sum_{\mathbf{k}} \langle \hat{c}_{-\mathbf{k}\downarrow} \hat{c}_{\mathbf{k}\uparrow} \rangle \neq 0$ or $\Delta^* = \sum_{\mathbf{k}} \langle \hat{c}^{\dagger}_{\mathbf{k}\uparrow} \hat{c}^{\dagger}_{-\mathbf{k}\downarrow} \rangle \neq 0$ in the ground state. In three dimensions, there is a finite temperature phase transition to the ordered phase with fermion pairing order, as shown in Figure 8.9.

The particle-hole transformation transfers the attractive FHM to the repulsive FHM at half-filling with $N_\uparrow + N_\downarrow = N_s$, whose parameters are $(U_0, \mu = 0, h = x)$. The same particle-hole transformation transfers pairing order Δ into $\sum_{\mathbf{k}} \langle \hat{c}^{\dagger}_{\mathbf{k}+\mathbf{Q}\downarrow} \hat{c}_{\mathbf{k}\uparrow} \rangle$ and Δ^* into $\sum_{\mathbf{k}} \langle \hat{c}^{\dagger}_{\mathbf{k}\uparrow} \hat{c}_{\mathbf{k}+\mathbf{Q}\downarrow} \rangle$. These two new order parameters are nothing but the in-plane spin orders S^- and S^+ with momentum $\mathbf{Q}$. Such spin orders with nonzero momentum are usually called *spin density wave*. Because the momentum $\mathbf{Q} = (\pi, \pi)$, the order parameter takes opposite sign between A sublattices and neighboring B sublattices, and therefore, it is also called *antiferromagnetic spin order*. Note that the fermion pairing order in the attractive FHM does not have to be in half-filling or $\mu = 0$, the corresponding antiferromagnetic spin order in the repulsive FHM does not have to be spin balanced or $h = 0$. The presence of nonzero h breaks the full $SU(2)$ spin rotational symmetry and only retains the in-plane spin rotational symmetry. Note that in the fermion pairing case, the pairing order parameter can take arbitrary phase and the energy is degenerate. This phase translates into an arbitrary phase in S^+ or S^-, which corresponds to the azimuthal spin angle along the xy plane, which reflects the spin rotational symmetry along $\hat{z}$. That is to say, the ground state has a $U(1)$ degeneracy.

In Figure 8.9, we show the correspondence between the attractive and the repulsive FHM in three dimensions. There are two notable features:

• The antiferromagnetism occurs in the repulsive side even with infinitesimal small interaction strength, this is equivalent to the fact that the pairing order always occurs for the attractive FHM with equal spin population even with infinitesimal small interaction.

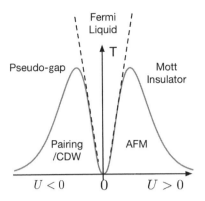

Schematic of the phase diagram of the FHM in three dimension. Here we consider a half-filled ($N_\uparrow + N_\downarrow = N_s$) and spin-balanced ($N_\uparrow = N_\downarrow$) FHM. At low temperature there exists a phase transition to the antiferromagnetic ordered phase (AFM) for positive U and a transition to pairing or the charge-density-wave (CDW) phase for negative U. The dashed line indicates a crossover temperature to Mott insulator phase for positive U and a pseudo-gap phase for negative U. The high-temperature regime is a Fermi liquid. This phase diagram at the positive U side can be one-to-one mapped to that at the negative U side by the particle-hole transformation. A color version of this figure can be found in the resources tab for this book at cambridge.org/zhai.

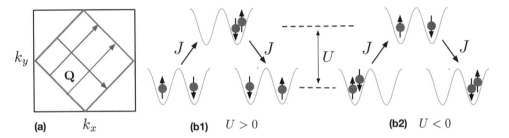

Mechanism for the origin of the antiferromagnetic order. (a) Illustration of the Fermi surface nesting at half-filling. (b) Schematic of the second-order perturbation processes in terms of hopping in the strongly interacting regime. (b1) the repulsive FHM with $U > 0$. The ground state favors the singly occupied state, and the second-order perturbation leads to superexchange of spins. (b2) The attractive FHM with $U < 0$. The ground state favors either a doubly occupied or empty state, and the second-order perturbation leads to hopping of pairs. The excitation energies in both cases are given by $|U|$. (b1) and (b2) are related by the particle-hole transformation. A color version of this figure can be found in the resources tab for this book at cambridge.org/zhai.

This phenomenon can also be understood using the concept of the *Fermi surface nesting*. A perfect Fermi surface nesting means that for any $\mathbf{k}$ point at Fermi surface, there exists a wave vector $\mathbf{Q}$ such that $\mathbf{k} + \mathbf{Q}$ also locate at the Fermi surface. $\mathbf{Q}$ is called the nesting momentum. The Fermi surface of a half-filling band has a diamond shape, as shown in Figure 8.10(a), which is an example of Fermi surface nesting and the nesting momentum $\mathbf{Q} = (\pi, \pi)$. In the presence of Fermi surface nesting, it is easy to see that a particle-hole excitation $c_{\mathbf{k}\sigma}^{\dagger} c_{\mathbf{k}+\mathbf{Q}\sigma'}$ causes zero energy. Here $\sigma \neq \sigma'$ corresponds to the spin wave order and $\sigma = \sigma'$ corresponds to the density wave order. Fermi surface nesting plays an important role in the weakly interacting regime, because the reconstruction of quantum many-body state mainly takes play at the Fermi surface for weak interactions. For example, when we discuss the BCS pairing, we mainly concern a pair of states at the Fermi surface, where pairing fluctuation causes zero energy. Because these excitations cause zero energy in the presence of the Fermi surface nesting, the corresponding spin wave order, or the density wave order, can also occur with infinitesimal small interaction.

- The transition temperature first increases and then decreases as the interaction strength further increases. In the weakly interacting regime, it is quite natural that the transition temperature increases as the interaction strength increases. In the strongly interacting regime, we can take the interaction term as the most dominate part of the Hamiltonian, and treat the hopping term as perturbation. For the half-filling case with average one particle per site, let us consider a simple two-site case as an example, and the total number of possible states are

$$| \uparrow, \uparrow \rangle; \; | \downarrow, \downarrow \rangle; \; | \uparrow, \downarrow \rangle; \; | \downarrow, \uparrow \rangle; \; | \uparrow\downarrow, 0 \rangle; \; | 0, \uparrow\downarrow \rangle. \tag{8.52}$$

For the first four states, both sites are singly occupied, whose energy are degenerate at the zeroth order when $J = 0$. For the last two states, one of the site is doubly occupied and the other site is empty, and therefore, their energies are higher by the interaction strength U for the repulsive interaction. Hence, the charge excitation is gapped, which prevents charge transfer. This state is a Mott insulator as we have discussed in the BHM.

Now turning on the hopping J as perturbation, and we will see how the hopping lifts the degeneracy between the first four states. It is easy to see that the hopping is prohibited for the first two states due to the Pauli exclusion principle, and therefore, their energies remain unchanged. As we show in Figure 8.10(b1), hopping can couple $| \uparrow, \downarrow \rangle$ and $| \downarrow, \uparrow \rangle$, and the coupling is made through a second-order perturbation. The intermediate states of this second-order process are the last two states in Eq. 8.52, whose energies are higher by U. Therefore, the coupling is proportional to J^2/U. Because of the coupling, the energy is lowered by forming a superposition of these two states. This intuitively explains why the antiferromagnetic spin configurations is energetically favorable. This second-order process is called the *superexchange*.

For the case of many sites, we can formally write down the effective Hamiltonian following the second-order perturbation as

$$H_{\text{eff}} = -\frac{J^2}{U} \sum_{\langle ij \rangle ss'} (c_{is}^\dagger c_{js} c_{js'}^\dagger c_{is'} + \text{h.c.}). \tag{8.53}$$

Using the relation

$$\sum_{ss'} c_{is}^\dagger c_{js'}^\dagger c_{is'} c_{js} = -2\mathbf{S}_i \cdot \mathbf{S}_j + \frac{1}{2} n_i n_j, \tag{8.54}$$

one can obtain an effective spin Hamiltonian as the Heisenberg model

$$H_{\text{eff}} = J_{\text{ex}} \sum_{\langle ij \rangle} \mathbf{S}_i \cdot \mathbf{S}_j, \tag{8.55}$$

with $J_{\text{ex}} = 4J^2/U$, where the number four counts total possible paths of this second-order processes. Since $J_{\text{ex}} > 0$, it favors an antiferromagnetic ground state, that is, spins at two neighboring sites are antiparallel to each other. This also tells us that in the strongly interacting regime, the antiferromagnetic order occurs only when $T \sim J_{\text{ex}} = 4J^2/U$, which decreases with the increasing of interaction strength. We shall also note that, because the charge gap is $\sim U$, the Mott insulator physics occurs when $T \lesssim U$. However, the antiferromagnetic order appears at a temperature $\sim J_{\text{ex}} \sim (J/U)^2 U \ll U$ when $U \gg J$. Therefore, there will be an intermediate temperature regime where the system is already an insulator but the spin order has not been formed, as shown in Figure 8.9.

Similar scenario can be applied to the attractive interaction side. Now at the zeroth order of J, the ground state is either doubly occupied or empty, as the last two state in Eq. 8.52, and the excited states are that two sites are singly occupied. As shown in Figure 8.10(b2), starting from the state both two atoms initially occupied the left site, a single-particle hopping process can connect this initial state to an excited state with the excitation energy $|U|$. Then through another single-particle hopping process, two atoms can be both in the right site. In this way the pair hops from one site to another, and thus the effective hopping strength for the pairs is also $4J^2/|U|$, where the number four also counts total possible paths of this hopping processes. Therefore, as the absolute value of U increases, this effective hopping actually decreases, which means the effective mass of these pairs becomes larger. Therefore, the condensation temperature decreases. Here in the attractive case, the energy to break a pair is also $\sim |U|$ in the strongly interacting regime. Therefore, fermion pairs form at the temperature $\sim |U|$. This temperature is also much larger than the pair condensation temperature $\sim J^2/U$. Hence, there is also an intermediate temperature regime, where pairs form but they are not condensed. This is exactly the pseudo-gap regime discussed in Section 6.2.

Hence, as shown in Figure 8.9, upon the particle-hole transformation, the pair formation temperature in the attractive side is mapped to the Mott insulator formation temperature in the repulsive side, and the pair condensation temperature in the attractive side is mapped to the antiferromagnetic order formation temperature in the repulsive side. In both cases, there exists a temperature window where either pairing gap or charge gap has formed but orders have not yet formed.

The short-range antiferromagnetic correlation has been observed in the FHM in a three-dimensional cubic lattice by the Bragg spectroscopy [68]. Later it has also been observed by using the Fermi gas microscope for two-dimensional FHM. The experimental observation of a quasi-long-range antiferromagnetic correlation has been observed in a system with about 10×10 two-dimensional lattice sites [9]. As shown in Figure 8.11, although in this case the spin correlator is always exponentially decayed, the correlation length increases as

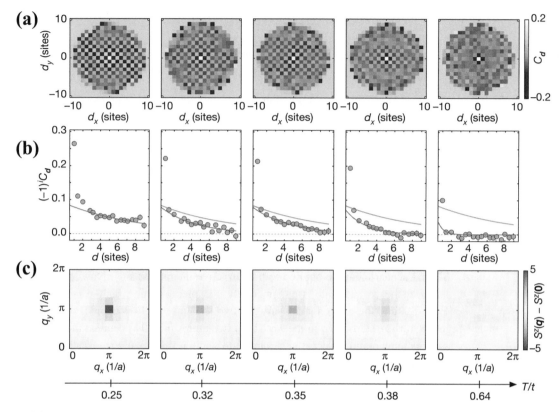

Figure 8.11 Observation of antiferromagnetic long-range correlation. (a) The spin correlation function $C_\mathbf{d}$ is defined as $C_\mathbf{d} = \frac{1}{N_\mathbf{d}} \frac{1}{S^2} \sum_{\mathbf{r_1},\mathbf{r_2},\mathbf{d}=\mathbf{r_1}-\mathbf{r_2}} \langle \hat{S}^z_{\mathbf{r_1}} \hat{S}^z_{\mathbf{r_2}} \rangle - \langle \hat{S}^z_{\mathbf{r_1}} \rangle \langle \hat{S}^z_{\mathbf{r_2}} \rangle$, where $N_\mathbf{d}$ is the total number of different two point correlators with displacement $\mathbf{d}$ between $\mathbf{r_1}$ and $\mathbf{r_2}$, and $S = 1/2$ in this case. This correlator is shown for an about 10×10 sites Fermi–Hubbard model with U/J fixed at 7.2. (b) The amplitude of this antiferromagnetic correlator $(-1)^i C_\mathbf{d}$ after an azimuthal average. (c) The measured spin structure factor $S^z(\mathbf{q}) - S^z(\mathbf{0})$ obtained from the Fourier transformation of single images, where $S^z(\mathbf{q})$ is defined as $S^z(\mathbf{q}) = \frac{1}{N} \sum_{\mathbf{r_1},\mathbf{r_2}} \frac{1}{S^2} \langle \hat{S}^z_{\mathbf{r_1}} \hat{S}^z_{\mathbf{r_2}} \rangle e^{i\mathbf{q}(\mathbf{r_1}-\mathbf{r_2})}$ and N is the total number of sites in this area. All the results are shown for different temperature T/J. Reprinted from Ref. [9]. A color version of this figure can be found in the resources tab for this book at cambridge.org/zhai.

the temperature decreases, and at the lowest temperature, the correlation length is already comparable to or exceeds the system size.

SO$_4$ Symmetry. Considering the FHM with $h = 0$, it possesses the $SU(2)$ spin rotational symmetry generated by Eq. 8.39-8.41. Now consider the FHM with $\mu = 0$, it can be mapped to another FHM with $h = 0$, and the latter possesses the $SU(2)$ spin rotational symmetry generated by Eq. 8.39-8.41. Now under the particle-hole transformation, Eq. 8.39-8.41 becomes

$$\hat{S}^x \rightarrow \hat{L}^x = \frac{1}{2} \sum_i (-1)^{i_x+i_y}(\hat{c}^\dagger_{i\uparrow}\hat{c}^\dagger_{i\downarrow} + \hat{c}_{i\downarrow}\hat{c}_{i\uparrow}), \tag{8.56}$$

$$\hat{S}^y \to \hat{L}^y = \frac{i}{2}\sum_i (-1)^{i_x+i_y}(\hat{c}_{i\uparrow}^\dagger \hat{c}_{i\downarrow}^\dagger - \hat{c}_{i\downarrow}\hat{c}_{i\uparrow}), \qquad (8.57)$$

$$\hat{S}^z \to \hat{L}^z = \frac{1}{2}\left(\sum_i (\hat{c}_{i\uparrow}^\dagger \hat{c}_{i\uparrow} + \hat{c}_{i\downarrow}^\dagger \hat{c}_{i\downarrow}) - N_s\right). \qquad (8.58)$$

Hence, it is straightforward to verify that the Hamiltonian of the FHM with $\mu = 0$ commute with $\hat{L}^x$, $\hat{L}^y$ and $\hat{L}^z$, which generates another $SU(2)$ rotational symmetry. By combining these two symmetries, we can conclude that when both μ and h are zero, the FHM possesses an $SU(2) \times SU(2)$ symmetry. Furthermore, we note that $S_z + L_z = N_\uparrow - N_s/2$, and for given N_s, $S_z + L_z$ is either an integer or a half integer. Thus, not all representations of the $SU(2) \times SU(2)$ symmetry group are allowed. In fact, only half of them are allowed. Therefore, the actual symmetry group is $SU(2) \times SU(2)/Z_2 \sim SO(4)$ [185].

The enlarged symmetry at $\mu = h = 0$ has dramatical physical consequence in two dimensions. First, let us consider the repulsive FHM with $\mu = 0$. If $h \neq 0$, as we have discussed above, the degenerate space of the antiferromagnetic spin order is the $U(1)$ azimuthal angle. As we have discussed in Section 4.2, in two dimensions, there exists a finite temperature Kosterlitz–Thouless (KT) transition to the in-plane anti-ferromagnetic spin ordered phase, as shown in Figure 8.12(b). This KT transition is driven by the proliferation of topological defect of $U(1)$ order parameter known as vortex. If one further considers the case that h is also zero, the full spin rotational symmetry is restored. In this case, the spin order can be taken along any direction and all these states are degenerate. For instance, the antiferromagnetic spin order can be taken along the $\hat{z}$ direction, that is, $\sum_k \langle \hat{c}_{k\uparrow}^\dagger \hat{c}_{k+Q\uparrow} - \hat{c}_{k\downarrow}^\dagger \hat{c}_{k+Q\downarrow}\rangle \neq 0$, and its energy is degenerate with states whose the

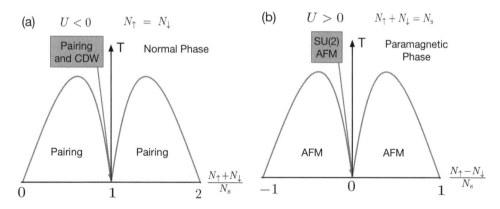

Figure 8.12   Schematic of the phase diagram for a two-dimensional FHM. (a) $U < 0$ and equal spin population with $N_\uparrow = N_\downarrow$. Here we plot the Kosterlitz–Thouless transition temperature to a fermion paired superfluid phase as a function of total filling number $(N_\uparrow + N_\downarrow)/N_s$. (b) $U > 0$ and half-filling with $N_\uparrow + N_\downarrow = N_s$. Here we plot the Kosterlitz–Thouless transition temperature to an in-plane antiferromagnetic (AFM) phase as a function of spin imbalance $(N_\uparrow - N_\downarrow)/N_s$; "CDW" denotes the charge-density-wave. These two phase diagrams are related to each other by the particle-hole transformation. A color version of this figure can be found in the resources tab for this book at cambridge.org/zhai.

antiferromagnetic order is taken in the xy plane, that is, $\sum_{\mathbf{k}} \langle \hat{c}_{\mathbf{k}+\mathbf{Q}\downarrow} \hat{c}^{\dagger}_{\mathbf{k}\uparrow} \rangle \neq 0$. Therefore, the manifold of the antiferromagnetic spin order is no longer $U(1)$ but an S^2 sphere. As we have discussed in Section 4.2, because an S^2 degenerate manifold does not support topological defect in two dimensions, it does not support a finite temperature KT transition. Thus, as shown in Figure 8.12(b), the KT transition temperature vanishes at $h = 0$ or when $N_{\uparrow} = N_{\downarrow}$.

Then we turn to consider the attractive FHM. When $\mu = h = 0$, the attractive FHM is mapped to the repulsive FHM with same interaction strength. As we have discussed above, for $h = \mu = 0$ and repulsive interaction, all antiferromagnetic states with spin order taken along any direction are degenerate. Now taking the particle-hole transformation, the in-plane antiferromagnetic order is mapped to pairing order with zero momentum, but the antiferromagnetic order along $\hat{z}$ is mapped to the charge density wave order with momentum $\mathbf{Q}$, that is, $\sum_{\mathbf{k}} \langle \hat{c}^{\dagger}_{\mathbf{k}\uparrow} \hat{c}_{\mathbf{k}+\mathbf{Q}\uparrow} + \hat{c}^{\dagger}_{\mathbf{k}\downarrow} \hat{c}_{\mathbf{k}+\mathbf{Q}\downarrow} \rangle \neq 0$. That is to say, for the attractive FHM at half-filling and with equal spin population, the charge density wave state is always energetically degenerate with the fermion pairing state, which is in fact ensured by the $SO(4)$ symmetry discussed above. Similar as the discussion above in the repulsive FHM, when $\mu \neq 0$, the degenerate space of the order parameter is the $U(1)$ phase of the paring order. But when $\mu = 0$, the degenerate space includes the charge density wave order and becomes S^2. Hence, the topological defect no long exists in two dimensions and the KT transition temperature is suppressed to zero, as shown in Figure 8.12(a).

In Figure 8.13 we show the measurements of the short-range spin correlations for the repulsive FHM at half-filling [23] and the short-range density correlation for the attractive FHM with equal spin population, as well as the calculated short-range fermion pairing correlation for the latter model [121]. They are related by the particle-hole transformation and the measurements presented in Figure 8.13 does show that they behave similarly by equaling $1 - n$ in Figure 8.13 (b) with p^{s} in Figure 8.13(a). One can see that in Figure 8.13(a), only for zero spin polarization, the spin correlations along $\hat{z}$ nearly coincide with the in-plane spin correlations. For any finite spin polarization, the spin correlation along $\hat{z}$ is always weaker. Similarly, in Figure 8.13(b), only for half-filling the density correlations become nearly coincide with the pairing correlations. Away from half-filling, the density correlations are always weaker.

Challenging Issues. Above, we have discussed some established results of the FHM. However, for most regimes of the FHM in two and three dimensions, the physics is actually unknown.[3] For repulsive interaction, we only know unambiguously the results for the half-filled case. Once the density is away from the half-filling, there is little consensus. Here we will highlight two unsolved issues. One is away but nearby the half-filling with $n \sim 1$ in two dimensions, where $n = (N_{\uparrow} + N_{\downarrow})/N_{\mathrm{s}}$ is the density. The other is at low-density regime with $n \ll 1$. By the particle-hole transformation, they are mapped to the attractive FHM. The former is the regime with small spin imbalance regime in two dimensions, and the latter is the regime with nearly fully spin polarization. Below we will discuss them respectively. We hope quantum simulation with ultracold atoms in optical lattice can offer new promises and provide new insights for solving these problems.

---

[3] Except in one dimension, the Luttinger liquid theory is expected to work well.

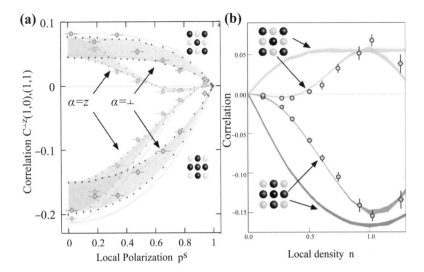

Observation of correlations in the FHM. (a) Spin correlations as a function of spin imbalance for the repulsive FHM at half-filling. U/J is fixed at 8.0. Large circles and squares denote the nearest neighbor and the diagonal neighbor spin correlations, respectively. The spin correlation is defined as $C^\alpha(\mathbf{d}) = 4(\langle \hat{S}_{\mathbf{i}}^\alpha \hat{S}_{\mathbf{i+d}}^\alpha \rangle - \langle \hat{S}_{\mathbf{i}}^\alpha \rangle \langle \hat{S}_{\mathbf{i+d}}^\alpha \rangle)$. Both $\alpha = z$ and $\alpha = \perp$ components of the spin correlations are plotted. (b) Density correlation as a function of averaged density for the attractive FHM with equal spin population. U/J is fixed at -5.7. Circles represent measurements of the doublon correlations for the nearest neighbor and the diagonal neighbor. Solid lines are the Monte Carlo simulation of the nearest and the diagonal neighbor pairing correlation. These two measurements are also related by the particle-hole transformation. By this mapping, $p^s = 1$ in (a) corresponds to $n = 0$ in (b), and $p^s = 0$ in (a) corresponds to $n = 1$ in (b). $\hat{S}^z$ correlation in (a) corresponds to the doublon correlation in (b); and $\hat{S}^\perp$ correlation in (a) corresponds to the pairing correlation in (b). Reprinted from Ref. [23] and [121] with modifications. A color version of this figure can be found in the resources tab for this book at cambridge.org/zhai.

The first issue is about the d-wave superconductivity in the high-Tc superconductor. The studies of physical mechanism behind the high-Tc cuprate superconductor have been an important topic in the condensed matter physics for centuries. Here we should first stress that the material structures of the cuprates are quite complicated, and it is hard to say whether the physics can be well captured by the FHM or not. However, the parent compounds of the cuprates are always antiferromagnetic Mott insulators with a half-filled band, which share the same feature as the FHM at half-filling. When the density is tuned slightly away from the half-filling by doping, there exists a range of doping where the low-temperature state displays superconductivity. There were also discussions whether this superconductivity also comes from the effective attractive interaction due to the electron-phonon interaction, as in the case of the conventional BCS superconductors, although it is commonly believed that the attractive interaction originated from the electron-phonon interaction is normally too weak to give rise to such a high transition temperature. If this is not originated from the electron-phonon interaction, it is quite mysterious that how superconductivity with such a high transition temperature can emerge from the repulsive interaction between electrons. It is an interesting question to ask whether similar d-wave pairing of fermion can emerge for the FHM with the repulsive interaction, when it is slightly doped away from the half-filling. There is no theoretical consensus on this problem so far. Note that the d-wave pairing order parameter is given by

$$\Delta^* = \sum_{\mathbf{k}} f(\mathbf{k}) \langle \hat{c}_{\mathbf{k}\uparrow}^\dagger \hat{c}_{-\mathbf{k}\downarrow}^\dagger \rangle, \tag{8.59}$$

where $f(\mathbf{k}) \propto k_x^2 - k_y^2$. Upon the particle-hole transformation, it is equivalent to a d-wave antiferromagnetic order given by

$$\sum_{\mathbf{k}} f(\mathbf{k}) \langle \hat{c}_{\mathbf{k}\uparrow}^\dagger \hat{c}_{\mathbf{k}+\mathbf{Q}\downarrow} \rangle. \tag{8.60}$$

In other words, if a d-wave fermion pairing state can exist for the repulsive FHM slightly away from the half-filling, a d-wave antiferromagnetic order can also exist for the attractive FHM with slight spin imbalance. It is possible to answer these questions by using ultracold fermions in optical lattices to simulate the FHM. But so far the temperatures of ultracold atom systems are still too high to reach the transition temperature, even if such a phase exists. Here we should recall that ultracold atom systems and condensed matter systems have very different energy scales. Even the absolute temperatures for superconductivity in the cuprates are quite high, they are still small in unit of the Fermi energy. Replacing the Fermi temperature with the typical Fermi temperature of ultracold atomic gases, the transition temperature becomes too low and it is quite challenge to reach in practice.

The second issue is about itinerant ferromagnetism. Human has observed the phenomenon of ferromagnetism for thousands of years but its mechanism is still not crystal clear. Here we shall focus on the question whether the ground state of the repulsive FHM can display ferromagnetism in the low-density regime with $n \ll 1$. In this regime, if we apply the Hatree–Fock mean-field approximation to the interaction energy and consider uniform solutions, we can obtain interaction energy as

$$E_{\text{int}} = U n_\uparrow n_\downarrow = U \left(\frac{n^2}{4} - S_z^2 \right), \tag{8.61}$$

where $S_z = (n_\uparrow - n_\downarrow)/2$. Thus, the interaction energy favors a state that can maximize the absolute value of S_z, that is, a ferromagnetic state. If $S_z = n/2$ or $S_z = -n/2$, the interaction energy vanishes. It is easy to understand because there will be no doubly occupied sites when the system is fully spin polarized. On the other hand, we should also look at the kinetic energy. In the low-density limit, we can approximate the dispersion as $\epsilon_{\mathbf{k}} = \mathbf{k}^2/(2m^*)$. In a three-dimensional system, given $n_\uparrow = n/2 + S_z$ and $n_\downarrow = n/2 - S_z$ the kinetic energy is given by

$$E_{\text{kin}} = \frac{(6\pi^2)^{5/3}}{20m^*\pi^2} \left[\left(\frac{n}{2} + S_z \right)^{5/3} + \left(\frac{n}{2} - S_z \right)^{5/3} \right], \tag{8.62}$$

which favors a minimum located at $S_z = 0$. Therefore, it was proposed that the competition between the interaction energy and the kinetic energy can lead a transition at critical U_c. The system is a nonmagnetic state of $S_z = 0$ with $U < U_c$, and it undergoes a transition to a ferromagnetic state of $S_z \neq 0$ with $U > U_c$, which eventually saturates to a fully polarized state with $S_z = n/2$ or $S_z = -n/2$ at very large U. This is known as the *Stoner mechanism of ferromagnetism*. In reality, if the total number of fermions in each component is conserved, ferromagnetism manifests itself as phase separation. The larger the effective mass m^*, the smaller the required critical interaction U_c. Especially, when the band is strictly flat, the effective mass diverges and the Stoner ferromagnetism can occur with infinitesimal small interaction strength.

However, there is debate about this Stoner mechanism for ferromagnetism when the band is not strictly flat. One can image the mean-field ferromagnetic state is not the only candidate that can avoid the interaction energy. In fact, Gutzwiller has proposed another type of correlated state as [64]

$$|\Psi\rangle = \prod_i (1 - \alpha n_{i\uparrow} n_{i\downarrow})|\text{FS}\rangle, \tag{8.63}$$

where $|\text{FS}\rangle$ is a free fermion Fermi sea with equal number of spin-up and spin-down particles. The projection operator $\mathcal{P} = \prod_i (1 - \alpha n_{i\uparrow} n_{i\downarrow})$ does not change the particle numbers of each spin component, and therefore this state is still a nonmagnetic state. However, it can suppress the interaction energy by projecting out the doubly occupied sites. α is taken as a variational parameter. When $\alpha = 0$, it recovers the free Fermi gas, and when $\alpha = 1$, all the doubly occupied states are projected out and the interaction energy also vanishes. On the other hand, when α deviates from zero, the projection operator distorts the free Fermi sea and increases the kinetic energy. By minimizing the energy with respect to α, one can also find that α increases with the increasing of U, and eventually α reaches unity for very large U.

Both the Gutzwiller projection state and the ferromagnetic state pay the price of increasing the kinetic energy in order to reduce the interaction energy. The difference is that the Gutzwiller state has zero polarization but strong local correlations. So in reality, which state is be the true ground state depends on the energy comparison. In other words, it depends on which state costs less kinetic energy when reducing the same amount of interaction energy. However, given the fact that the Gutzwiller wave function is a variational wave function, one can always come up with more complicated wave function that hopefully can do a better job in reducing interaction energy and costing less kinetic energy. On the other hand, the ferromagnetic state considered here is a noninteracting one. One can also further optimize the energy of the ferromagnetic state by including short-range correlations. Hence, it is still an open issue that whether the system will become a ferromagnetic state or a strongly correlated nonmagnetic state with strong repulsive interaction. By the particle-hole transformation, the ferromagnetic state is mapped to a state with density inhomogeneity, and therefore, it is equivalent to ask whether the attractive FHM will develop density inhomogeneity when very low density spin-up fermions are added into a nearly fully filled band of spin-down fermions.

8.3 Thermalization and Entanglement

Thermalization in Quantum System. Thermalization lies at the center of statistical mechanics. Thermalization in a closed many-body quantum system means that when the system evolves from an arbitrary initial state, it can eventually reach thermal equilibrium.

It looks like there is an apparent paradox of quantum thermalization. On the one hand, we know that a system in thermal equilibrium is fully characterized by few number of parameters such as temperature and chemical potential. That is to say, when a system thermalizes, all detailed information about the initial state has been erased. On the other hand, the quantum evolution governed by a Hamiltonian is a unitary transformation, and it is known that a unitary evolution cannot erase information. Therefore, all the information of the initial state has to be preserved during the evolution. It is also interesting to note that this paradox bears similarity with the black hole information paradox. In the black hole information paradox, we consider a material throwing into the black hole and emitting the Hawking radiation. Suppose the material is described by a pure state, and because the Hawking radiation only depends on temperature, all the information of the initial state are lost. However, assuming that the black hole is also a quantum system and the evolution is also a unitary transformation, it conserves all the information. Hence, the black hole information paradox is quite similar as the quantum thermalization paradox.

The resolution to this apparent paradox is that all the local information of the initial state has been spread into the entire system, such that it cannot be retrieved by local unitary measurements. In this sense, the local information has been "erased." That is the reason why the two concepts of the quantum thermalization and the quantum information scrambling are tied together. The lesson from this discussion is that, for quantum thermalization, we should not focus on the entire system but should focus on local observables. Hence, we consider a small fraction of the system called region $\mathcal{A}$ and all the remainder part is called region $\mathcal{B}$. Thermalization means that the subregion $\mathcal{A}$ reaches thermal equilibrium in contact with the rest part of the system $\mathcal{B}$, which serves as a reservoir.

We consider wave function $|\Psi(t)\rangle$ evolved from the initial state $|\Psi(0)\rangle$. By expanding $|\Psi(0)\rangle$ over a complete set of bases given by the many-body eigenstates $|\alpha\rangle$ with energy E_α, that is,

$$|\Psi(0)\rangle = \sum_\alpha a_\alpha |\alpha\rangle, \tag{8.64}$$

we have

$$|\Psi(t)\rangle = \sum_\alpha a_\alpha e^{-iE_\alpha t} |\alpha\rangle. \tag{8.65}$$

We consider local observable $\hat{O}$ in the region $\mathcal{A}$, and for simplicity, we consider an infinite-time average of this physical observable as[4]

$$\langle \hat{O} \rangle_\infty = \lim_{T \to \infty} \frac{1}{T} \int_0^T \langle \Psi(t)|\hat{O}|\Psi(t)\rangle dt = \sum_\alpha |a_\alpha|^2 \langle \alpha|\hat{O}|\alpha\rangle. \tag{8.66}$$

Here we should emphasize a key difference between thermalization in a classical system and in a quantum system. Thermalization in the classical system is based on the *Ergodicity Hypothesis*, which states that all microscopic states of the system can be accessed with equal probability as long as the evolution time is sufficiently long. However, this ergodicity

---

[4] With a refined study, the long time average is actually not necessary. Here we consider the long time average just for the simplicity of our discussion.

cannot be applied to a quantum system. It is clear from Eq. 8.66 that the occupations of different many-body eigenstates are fixed by the choice of the initial state, and the occupations do not change as time evolves. Therefore, in order for any generic initial state to thermalize, a natural assumption is that all generic eigenstates have to thermalize. That is to say, for a quantum system, the *Ergodicity Hypothesis* should be replaced by *Eigenstate Thermalization Hypothesis*, short noted as ETH. Roughly speaking, ETH states that for a generic many-body eigenstate $|\alpha\rangle$ with energy E_α, the expectation value of a local operator $\hat{O}$ is identical to the measured value of this observable in a micro-canonical ensemble with mean energy E_α,[5] that is,

$$\langle\alpha|\hat{O}|\alpha\rangle = \langle\rho_{\mathrm{mc}}(E_\alpha)\hat{O}\rangle, \tag{8.67}$$

where $\rho_{\mathrm{mc}}(E_\alpha)$ is the density matrix of the micro-canonical ensemble for the entire system with energy E_α.

Here we consider the initial state whose energy fluctuation $(\delta E)^2$ is small enough, and here $(\delta E)^2$ is defined as

$$(\delta E)^2 = \langle\Psi_0|\hat{H}^2|\Psi_0\rangle - (\langle\Psi_0|\hat{H}|\Psi_0\rangle)^2. \tag{8.68}$$

That is to say, for all states $|\alpha\rangle$ in Eq. 8.64 whose $|a_\alpha|^2$ is not negligible, $E_\alpha \approx E = \langle\Psi_0|\hat{H}|\Psi_0\rangle$. For such initial states, when Eq. 8.67 is satisfied, it naturally leads to

$$\langle\hat{O}\rangle_\infty = \langle\rho_{\mathrm{mc}}(E)\hat{O}\rangle \sum_\alpha |a_\alpha|^2 = \langle\rho_{\mathrm{mc}}(E)\hat{O}\rangle. \tag{8.69}$$

This means that based on all local measurements within the subregion $\mathcal{A}$, one cannot tell whether the entire system is in a pure quantum state, or in a thermal equilibrium state. In this sense, we state that this initial state thermalizes after sufficient long evolution time.

A strong version of the ETH can be stated in terms of the density matrix of an eigenstate. To be more precise, for a many-body eigenstate $|\alpha\rangle$ of the entire system, we consider the pure state density matrix $|\alpha\rangle\langle\alpha|$, and by tracing out the subregion $\mathcal{B}$, one obtains $\rho_\mathcal{A}^\alpha = \mathrm{Tr}_\mathcal{B}|\alpha\rangle\langle\alpha|$. On the other hand, we can choose a temperature T such that $E_\alpha = \langle\rho_{\mathrm{eq}}(T)\hat{H}\rangle$, where $\rho_{\mathrm{eq}}(T)$ is the thermal equilibrium density matrix given by

$$\rho_{\mathrm{eq}}(T) = \frac{1}{Z}e^{-\hat{H}/(k_{\mathrm{B}}T)}, \tag{8.70}$$

and Z is the partition function. We can obtain a reduced density matrix at thermal equilibrium as $\rho_\mathcal{A}^{\mathrm{eq}} = \mathrm{Tr}_\mathcal{B}\rho_{\mathrm{eq}}(T)$. A strong version of the ETH states that

$$\rho_\mathcal{A}^\alpha = \rho_\mathcal{A}^{\mathrm{eq}}. \tag{8.71}$$

This means that, when the entire system is prepared in a many-body eigenstate, any subregion experiences the remainder region as a heat bath and looks like thermal.

Eq. 8.71 has a direct consequence on the entanglement entropy. For the left-hand side of Eq. 8.71, one can compute the entanglement entropy of the quantum state $|\alpha\rangle$ between subregion $\mathcal{A}$ and $\mathcal{B}$. And for the right-hand side of Eq. 8.71, one can compute the thermal entropy of the subregion $\mathcal{A}$. Hence, Eq. 8.71 says that these two entropies have to be equal.

---

[5] A more rigorous definition can be found in Ref. [1, 42].

Because the thermal entropy is an extensive quantity that should be proportional to the volume of $\mathcal{A}$. Therefore, the entanglement entropy of state $|\alpha\rangle$ should also be proportional to the volume of $\mathcal{A}$. This is called the *volume law of the entanglement entropy* and it is a strong evidence to support whether a system obeys ETH.

Many-Body Localization. ETH is a hypothesis. It has been tested numerically for a number of models. Nevertheless, the numerical tests always suffer from the finite size effect. Therefore, another route to understand ETH is to study examples that violate ETH. First of all, the exactly integrable systems violate ETH. Examples of such models include noninteracting bosons and fermions, and one-dimensional models that are exactly solvable by the Bethe–Ansatz method, as we have discussed in Section 3.4. However, we should emphasize that the violation of ETH in these models requires fine tuning of model parameters. For instance, for the non-interacting bosons and fermions, any finite interactions between particles can lead to thermalization when disorder is absent. In order for the one-dimensional model to be exactly solvable by the Bethe–Ansatz method, these one-dimensional models require specific form of interaction potentials or interaction parameters. Any deviation from such interactions can break the integrability and lead to thermalization.

It is now known that among all the models violating ETH, there exists another class of models aside from the exactly integrable models. In contrast to these exactly integrable models, the violation of ETH in these models is stable against small perturbation to the Hamiltonian. These models are called the *many-body localizations* (MBL), because so far most of these models always include disorders. Nevertheless, whether there exists other possibilities is still an open question. We summarize the relation between ETH, MBL and exact integrable models in Figure 8.14. More details about MBL and ETH can be found in Ref. [1, 42, 124].

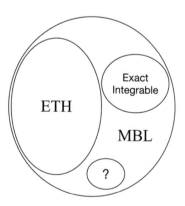

Figure 8.14   Schematic of different types of models. The largest circle includes all possible physical models. One circle labeled by "ETH" includes models obeying eigenstate thermalization hypothesis (ETH). Models outside this circle do not obey ETH, among which a smaller circle includes all exactly integrable systems. Currently, models excluded by these two circles are systems understood in terms of the many-body localization system (MBL), with other open possibilities denoted by the circle with a question mark.

There are a few metrics to characterize the breakdown of the ETH [1, 42, 124]:

- Observable-Energy Relation: Note that the right-hand side of Eq. 8.67 only depends on energy. Thus, it says that for many-body eigenstates, the expectation value of any local observable under a many-body eigenstate should be a smooth function of its eigenenergy only. In other words, if we consider different many-body eigenstates whose energies are very close, but the expectation value of local operators under these wave functions are very different, it indicates that the ETH fails. This criterion can be tested in numerical calculation.
- Initial Local Information: As discussed above, if a system thermalizes, information of initial state will be scrambled into the entire system and cannot be observed in local observables. However, if a system does not thermalize, at least part of the local information of the initial state information can be maintained in the local measurements even the system has evolved for sufficiently long time. This criterion has been used for experimental probe of MBL.[6]
- Entanglement Entropy: As discussed above, if a system obeys ETH, the entanglement entropy between a subregion $\mathcal{A}$ and the remaining region $\mathcal{B}$ obeys the volume law. In contrast, for MBL, this entanglement entropy is proportional to the area of the interface between $\mathcal{A}$ and $\mathcal{B}$, which is known as the *area law*. We will discuss how to measure the entanglement entropy in ultracold atom system, and this criterion have also been used experimentally to distinguish a MBL from ETH.

There are also other metrics such as level statistics that involves the random matrix theory description [42], the out-of-time-ordered correlation function that describes how fast the information scrambles in the system [54], and the entanglement growth after quench that describes how the entanglement property of a system responds to an external probe [124].

Theoretically, two one-dimensional lattice models are often considered in literature for studying MBL [1, 42, 124]. One model is the spinless fermion in a disorder potential with the nearest neighbor interactions, and the Hamiltonian is written as

$$\hat{H} = -J \sum_i (\hat{c}_i^\dagger \hat{c}_{i+1} + \text{h.c.}) + V \sum_i \hat{n}_i \hat{n}_{i+1} + \sum_i \epsilon_i \hat{n}_i, \qquad (8.72)$$

where J is the hopping amplitude, V is the nearest neighbor interaction strength and ϵ_i is the on-site disorder potential ranging from $[-W, W]$. The other model is the Heisenberg model in a random Zeeman field, whose Hamiltonian is given by

$$\hat{H} = J_\perp \sum_i (\hat{\sigma}_i^x \hat{\sigma}_{i+1}^x + \hat{\sigma}_i^y \hat{\sigma}_{i+1}^y) + J_z \sum_i \hat{\sigma}_i^z \hat{\sigma}_{i+1}^z + \sum_i h_i \hat{\sigma}_i^z, \qquad (8.73)$$

where $\hat{\sigma}_{x,y,z}$ are spin-1/2 operators, $J_\perp$ and J_z are two spin coupling strengths, and h_i is a random Zeeman field ranging from $[-W/2, W/2]$.

---

[6] Although in practices, because of the finite lifetime of the system, it is always hard to distinguish MBL from ETH with very long thermalization time.

These two lattice models can be mapped to each other through the Jordan–Wigner transformation. The Jordan–Wigner transformation is defined as

$$\sigma_i^+ = \frac{1}{2}(\sigma_i^x + i\sigma_i^y) = e^{-i\pi \sum_{j=1}^{i-1} \hat{c}_j^\dagger \hat{c}_j} \hat{c}_i^\dagger,$$ (8.74)

$$\sigma_i^- = \frac{1}{2}(\sigma_i^x - i\sigma_i^y) = e^{i\pi \sum_{j=1}^{i-1} \hat{c}_j^\dagger \hat{c}_j} \hat{c}_i,$$ (8.75)

$$\sigma_i^z = 2\hat{c}_i^\dagger \hat{c}_i - 1,$$ (8.76)

where the phase factor is introduced to ensure the fermion statistics between different sites. Here it is important to recognize that the local spin operators are mapped to non-local fermionic operators. Under this mapping, the $J_\perp$ and the J_z terms in Eq. 8.73 are mapped to the hopping term and the nearest neighbor interaction terms in Eq. 8.72, respectively. The random Zeeman field in Eq. 8.73 is mapped to the on-site disorder potential in Eq. 8.72. Thus, these two models share the same phase diagram.

Extensive numerical studies have been performed in these models using various metrics mentioned above. Here we just briefly mention the results [1]. In the fermion model, when $V = 0$, the noninteracting model displays the single-particle Anderson localization in one dimension for any weak disorder. When the interaction strength is finite, the many-body system remains localized until the interaction strength is beyond a threshold. By the Jordan–Wigner transformation, it means that in the spin model, there is also a transition from MBL to ETH at a critical J_z^*. And if the interaction strength is fixed, it requires a finite disorder strength W in order for all the many-body eigenstates are localized. These two models, as well as the phases therein, are schematically shown in Figure 8.15.

Experimentally, one of the results are shown in Figure 8.16. Here two-component fermions are loaded in a one-dimensional optical lattice. Because the nearest neighboring interaction is weak in optical lattices, here the on-site interaction between two components are used. Initially, atoms only populate all the even sites and no atoms populate

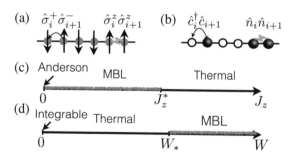

Figure 8.15 Schematic of models and phase diagram for MBL. (a) Model Eq. 8.73 of the Heisenberg-type spin model in a random Zeeman field. (b) Model Eq. 8.72 of fermions in disorder potential with the nearest neighbor interactions. (c–d) The phase diagram of the spin model Eq. 8.73. Reprinted from Ref. [1]. A color version of this figure can be found in the resources tab for this book at cambridge.org/zhai.

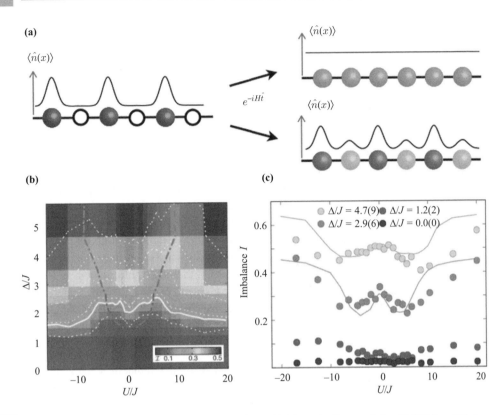

Figure 8.16 Memory of the initial state information as an indicator of MBL. (a) Schematic of how to distinguish MBL from ETH. For the upper one, the initial information of density order has been erased, which is the ETH case. For the lower one, the initial information of density order is still partially maintained, which is the MBL case. (b–c) The density imbalance $\mathcal{I}$ between even and odd sites taken as a measure of the initial state information, and measurements of $\mathcal{I}$ after sufficiently long time plotted as a function of the interaction strength and the disorder strength. (b) is a contour plot of $\mathcal{I}$, and the solid line indicates the phase boundary between the MBL regime and the ETH regime. (c) is a plot of $\mathcal{I}$ as a function of U for different disorder strengths. In the figure, Δ is the disorder strength, and U is the interaction strength. Reprinted from Ref. [154]. A color version of this figure can be found in the resources tab for this book at cambridge.org/zhai.

the odd sites, which corresponds to a charge density wave order state characterized by an imbalance $\mathcal{I}$ defined as

$$\mathcal{I} = \frac{N_{\mathrm{e}} - N_{\mathrm{o}}}{N_{\mathrm{e}} + N_{\mathrm{o}}}, \tag{8.77}$$

where N_{e} and N_{o} are the total number of atoms in the even and the odd sites, respectively. Another counterpropagating laser beam with incommensurate wave length creates the second lattice potential. Together with the original lattice potential, it generates a quasi-random one-dimensional lattice, which is known as the Aubry-André model in absence of interactions. Here we should note that there is a difference between this quasi-randomness and the true random potential in the noninteracting case. In one dimension, for this quasi-random potential, it requires the disorder strength to be large enough compared with the hopping strength in order for all single-particle states to be localized. But for the true disorder potential, all states are localized for arbitrary weak random potential. In this experiment, the quasi-random potential is tuned to the single-particle localization regime, and this experiment studies whether the system enters a MBL regime or still obeys ETH when the interaction strength is finite. The criterion implemented in the experiment is whether the initial density imbalance information $\mathcal{I}$ can be maintained after sufficiently long evolving time, as illustrated in Figure 8.16(a) [154]. It is found that, after sufficiently long evolution time, $\mathcal{I}$ reaches a stationary value and this stationary value is plotted in Figure 8.16(b) and (c). When this stationary value of $\mathcal{I}$ remains finite, it means the breakdown of ETH in this regime. In this way, a boundary between ETH and MBL is drawn in Figure 8.16(b).

Finally we should remark that the nature of the MBL to ETH transition is still an open question. All the other phase transitions discussed in this book is either for the ground state, or low-energy steady state, or for a thermal equilibrium state. But the MBL to ETH transition is actually about whether a system evolving from a generic initial state can or cannot reach thermal equilibrium. Since a generic initial state is composited with superpositions of many excited states, the MBL to ETH transition concerns all the excited states. Therefore, it calls for developing new stratagem to describe such a transition.

Entanglement Entropy. Above we have discussed that the entanglement entropy plays an important role in characterizing whether a quantum system can thermalize or not. Here we will discuss how the entanglement entropy can be measured in experiment. For measuring the entanglement entropy, we need to turn the entanglement entropy into a physical observable. Below we will show that two ingredients are important to fulfill this goal. First, we consider the Rényi entropy instead of the von Neumann entropy, and second, we need to duplicate the system into two identical copies.

The Rényi entropy is defined as

$$S_n(\rho) = \frac{1}{1-n} \log \text{Tr}(\rho^n).$$ (8.78)

Here we show evidences that S_n is indeed a proper definition of entropy. Taking S_2 as an example, first, the entropy for a pure state is zero. It is easy to see that the density matrix of pure state satisfies $\text{Tr}(\rho^2) = 1$, and therefore, $S_2 = 0$. Second, the entropy for a mixed state is always positive. It is also easy to see that $\text{Tr}\rho^2 < 1$ for a mixed state and thus $S_2 > 0$. For this reason, $\text{Tr}(\rho^2)$ is also called the *purity*. The von Neumann entropy S_{vN} can be viewed as the $n \to 1$ limit of the Rényi entropy, that is,

$$S_{vN} = -\text{Tr}(\rho \log \rho) = \lim_{n \to 1} S_n(\rho).$$ (8.79)

Furthermore, it can be shown that $dS_n(\rho)/dn \leqslant 0$, and as a result,

$$S_{vN}(\rho) \geqslant S_2(\rho).$$ (8.80)

In other words, the second Rényi entropy gives a lower bound for the von Neumann entropy.

We consider two systems α and β whose Hilbert spaces are identical and are spanned by a set of bases $\{|n\rangle\}$, thus the total Hilbert space of the joint system $\alpha \otimes \beta$ is spanned by the bases $|n\rangle_\alpha \otimes |n'\rangle_\beta$. We define a swap operator $\hat{V}$ which acts on the base as

$$\hat{V}|n\rangle_\alpha \otimes |n'\rangle_\beta = |n'\rangle_\alpha \otimes |n\rangle_\beta. \tag{8.81}$$

Below we can prove the identity that [43, 80]

$$\mathrm{Tr}(\rho_\alpha \rho_\beta) = \mathrm{Tr}(\hat{V}\rho_\alpha \otimes \rho_\beta). \tag{8.82}$$

Let us start from the r.h.s. of Eq. 8.82; it can be explicitly written as

$$\hat{V}\rho_\alpha \otimes \rho_\beta = \sum_{nmn'm'} \rho^\alpha_{nm}\rho^\beta_{n'm'} \hat{V}|n\rangle_\alpha \otimes |n'\rangle_\beta \langle m|_\alpha \otimes \langle m'|_\beta$$

$$= \sum_{nmn'm'} \rho^\alpha_{nm}\rho^\beta_{n'm'} |n'\rangle_\alpha \otimes |n\rangle_\beta \langle m|_\alpha \otimes \langle m'|_\beta. \tag{8.83}$$

By taking the trace, it gives $n' = m$ and $n = m'$, and therefore,

$$\mathrm{Tr}(\hat{V}\rho_\alpha \otimes \rho_\beta) = \sum_{nm} \rho^\alpha_{nm}\rho^\beta_{mn}|m\rangle_\alpha \otimes |n\rangle_\beta \langle m|_\alpha \otimes \langle n|_\beta = \mathrm{Tr}(\rho_\alpha \rho_\beta). \tag{8.84}$$

With this identity, if we take $\rho_\alpha = \rho_\beta = \rho$, we obtain that

$$\mathrm{Tr}(\rho^2) = \mathrm{Tr}(\hat{V}\rho \otimes \rho). \tag{8.85}$$

Similar identity can also be generalized for the nth Rényi entropy. This identity is crucial for measuring the Rényi entropy, because it turns the second Rényi entropy in a single system into the expectation of an observable in a double system [43, 80]. The trade-off is that we need to enlarge the system to include two identical copies of the original system, and the observable is the swap operator in the enlarged system.

Furthermore, we will discuss how to measure the expectation value of the swap operator in an optical lattice experiment [43, 80]. Let us take two copies of one-dimensional chain BHM as an example. The Hamiltonian is written as

$$\hat{H} = \sum_{\sigma=\alpha,\beta} \left[-J \sum_{\langle ij \rangle} \hat{a}^\dagger_{i\sigma}\hat{a}_{j\sigma} + \frac{U}{2} \sum_i \hat{n}_{i\sigma}(\hat{n}_{i\sigma} - 1) \right], \tag{8.86}$$

where $\hat{a}^\dagger_{i\alpha}$ and $\hat{a}^\dagger_{i\beta}$ are creation operators at site-i, and α and β label two different copies. For simplicity, we first ignore the site index and consider one site in each copy, and under the swap operator,

$$\hat{V}\hat{a}^\dagger_\alpha \hat{V}^\dagger = \hat{a}^\dagger_\beta, \quad \hat{V}\hat{a}^\dagger_\beta \hat{V}^\dagger = \hat{a}^\dagger_\alpha. \tag{8.87}$$

Therefore, if we define the symmetric and antisymmetric operators as

$$\hat{a}^\dagger_s = \frac{\hat{a}^\dagger_\alpha + \hat{a}^\dagger_\beta}{\sqrt{2}} \tag{8.88}$$

$$\hat{a}^\dagger_t = \frac{\hat{a}^\dagger_\alpha - \hat{a}^\dagger_\beta}{\sqrt{2}}, \tag{8.89}$$

we have

$$\hat{V}\hat{a}_s^\dagger\hat{V}^\dagger = \hat{a}_s^\dagger; \quad \hat{V}\hat{a}_t^\dagger\hat{V}^\dagger = -\hat{a}_t^\dagger. \tag{8.90}$$

We introduce

$$|n_s\rangle = \frac{\hat{a}_s^{\dagger n_s}}{\sqrt{n_s!}}|0\rangle, \quad |n_t\rangle = \frac{\hat{a}_t^{\dagger n_t}}{\sqrt{n_t!}}|0\rangle \tag{8.91}$$

as new bases, and $\hat{V}$ is diagonal in these bases. By using Eq. 8.90, we can obtain that $\langle n_s|\hat{V}|n_s\rangle = 1$ and $\langle n_t|\hat{V}|n_t\rangle = (-1)^{n_t}$. Thus we reach the conclusion that

$$\mathrm{Tr}(\rho^2) = \mathrm{Tr}(\hat{V}\rho \otimes \rho) = \langle(-1)^{n_t}\rangle. \tag{8.92}$$

Adding the site index back, it is straightforward to generalize this formula as

$$\mathrm{Tr}(\rho^2) = \mathrm{Tr}(\hat{V}\rho \otimes \rho) = \Big\langle \prod_i (-1)^{n_{it}} \Big\rangle. \tag{8.93}$$

Eq. 8.92 tells us that the second Rényi entropy depends on the population on the antisymmetric state bases of each site. For a pure state, in order for $\mathrm{Tr}(\rho^2) = 1$, there must be even number of site whose n_{it} is odd.

To measure the population on the antisymmetric bases, we first freeze the tunneling by increasing the barrier of optical lattice potential such that $J \to 0$. This projects the many-body wave function into the Fock state bases of each site, and different sites become disconnected. Then, we turn on the tunneling between two systems, which is written as

$$\hat{T} = -J_\perp \sum_i (\hat{a}_{i\alpha}^\dagger\hat{a}_{i\beta} + \hat{a}_{i\beta}^\dagger\hat{a}_{i\alpha}). \tag{8.94}$$

The evolution under $\hat{T}$ is a periodical oscillation. Considering the time evolution for $1/4$ of a single period, the unitary evolution is equivalent to changing the bases as

$$\hat{a}_{is} \leftrightarrow \hat{a}_{i\alpha}, \quad \hat{a}_{it} \leftrightarrow \hat{a}_{i\beta}. \tag{8.95}$$

Hence, by measuring the total number of atoms at site-i of the β-system after evolution of $1/4$ period, we can determine the occupation in n_{it}, and subsequently determine S_2 through Eq. 8.93. The entire procedure is schematically shown in Figure 8.17(a).

Figure 8.17(b) shows the measurements of the entanglement entropy for a low-temperature equilibrium phase of one-dimensional BHM with four sites [80]. Ideally, the results should be as what are shown in Figure 8.17(a). For the MI phase, since the many-body wave function is a product state of different sites as Eq. 8.6, there is no entanglement between any of two subsystems, no matter how the entire system is divided. Thus, the occupations of n_{it} are always even for all sites. For the SF state, two subsystems are entangled. As shown in the lower part of Figure 8.17(a), in $\mathcal{A}$ or $\mathcal{B}$ subsystem, there is odd number of sites whose n_{it} is odd. But since the total system is always a pure state, in the total system there is always even number of sites whose n_{it} is odd. In real experimental data, the measurement of entropy for the entire system is nearly a constant independent

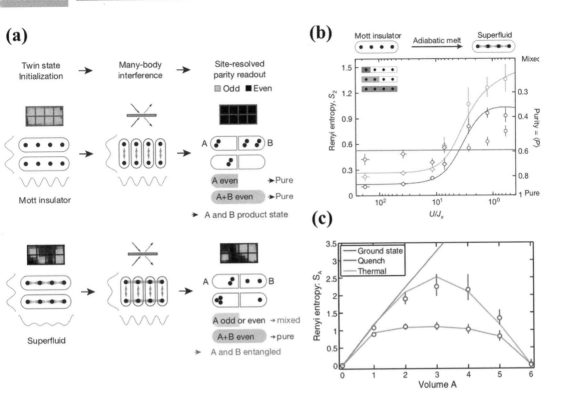

Figure 8.17 Measuring the entanglement entropy. (a) Experimental procedure of measuring the second Rényi entropy S_2 of a BHM and ideal experimental results for the MI regime and the SF regime. (b) Experimental results of S_2 as a function of U/J for the low-temperature equilibrium phase. (c) Experimental measurement of S_2 of a quantum state under a long time evolution after quench. Reprinted from Ref. [80] and [88]. A color version of this figure can be found in the resources tab for this book at cambridge.org/zhai.

of U/J, although it is not absolutely zero because of the finite temperature effect and the imperfection in measurement. Two sets of data correspond to two different ways of dividing the total system into two subregions $\mathcal{A}$ and $\mathcal{B}$. Clearly, in both cases, the entanglement entropy is smaller in the MI regime and larger in the SF regime.

Figure 8.17(c) shows the entanglement entropy after a quench in a one-dimensional BHM with six sites [88]. The initial state is a Fock state with fixed number of atoms at each site. Then, the hopping J term along the chain is turned on and the system evolves for sufficiently long time. During the evolution the entanglement entropy continuously grows. The entanglement entropy is measured when the system has evolved for sufficiently long time, as shown in Figure 8.17(c). The entanglement entropy is plotted as a function of the size of the subregion $\mathcal{A}$. As we have discussed above, for a system obeying ETH, the entanglement entropy will obey the volume law. The data points shown in Figure 8.17(c) indeed obey the volume law and can be compared with the thermal entropy shown by the solid line. The discrepancy is larger when $\mathcal{A}$ subsystem has three sites, which is already half of the system, and the subregion and the bath are of equal sizes. As we have emphasized above, ETH is a feature about a generic excited state. Usually the entanglement entropy of the ground state does not possess the volume law. This can also be seen in Figure 8.17(c).

Exercises

8.1 Show that for the Bose condensation wave function Eq. 8.3, the density fluctuation at each site approximately obeys the Poisson distribution Eq. 8.4. Discussion the conditions when the approximation is good.

8.2 Solve the BHM Hamiltonian with the Bogoliubov theory introduced in Section 3.3. Compute the condensation fraction as a function of J/U, and compare the results with the condensation fraction obtained by the mean-field theory introduced in this chapter.

8.3 Starting from the mean-field Hamiltonian, use the second-order perturbation theory to derive the critical value for the phase transition

$$\frac{J_c}{U} = \frac{\left(n_0 - \frac{\mu}{U}\right)\left(\frac{\mu}{U} - (n_0 - 1)\right)}{Z\left(\frac{\mu}{U} + 1\right)}. \tag{8.96}$$

8.4 Considering a relativistic nonlinear equation

$$\frac{\hbar^2 \partial^2 \phi}{\partial t^2} = -\frac{\hbar^2 \nabla^2}{2m} \phi + U|\phi|^2 \phi, \tag{8.97}$$

study the small amplitude oscillations of the phase and the amplitude of wave function ϕ, and obtain a gapless Goldstone mode and a gapped Higgs mode from the real and imaginary parts of this equation.

8.5 Consider a two-dimensional FHM without interactions ($U = 0$), and for a nonmagnetic state ($n_\uparrow = n_\downarrow$), discuss how the Fermi surface evolves when the filling number n increases from zero to two.

8.6 Discuss the Hanbury–Brown–Twiss effect for noninteracting fermions and compare its difference with bosons.

8.7 Considering an attractively interacting FHM, and $|U| \gg J$, derive the effective Hamiltonian for hopping of pairs.

8.8 Verify that $\hat{L}_x$, $\hat{L}_y$ and $\hat{L}_z$ defined in Eq. 8.56–8.58 commute with the FHM with $\mu = 0$.

8.9 Show that the anticommutator relation $\{\hat{c}_i, \hat{c}_j\} = \delta_{ij}$ can be satisfied when $\hat{c}_i$ and $\hat{c}_i^\dagger$ are related to spin operators by the following Jordan–Wigner transformation:

$$\sigma_i^+ = \frac{1}{2}(\sigma_i^x + i\sigma_i^y) = e^{-i\pi \sum_{j=1}^{i-1} \hat{c}_j^\dagger \hat{c}_j} \hat{c}_i^\dagger, \tag{8.98}$$

$$\sigma_i^- = \frac{1}{2}(\sigma_i^x - i\sigma_i^y) = e^{i\pi \sum_{j=1}^{i-1} \hat{c}_j^\dagger \hat{c}_j} \hat{c}_i, \tag{8.99}$$

$$\sigma_i^z = 2\hat{c}_i^\dagger \hat{c}_i - 1. \tag{8.100}$$

8.10 Show that the fermion model Eq. 8.72 and the spin model Eq. 8.73 are equivalent by the Jordan–Wigner transformation.

References

[1] Abanin, Dmitry A., Altman, Ehud, Bloch, Immanuel, and Serbyn, Maksym. 2019. Colloquium: Many-Body Localization, Thermalization, and Entanglement. *Rev. Mod. Phys.*, **91**(2), 021001.

[2] Abo-Shaeer, J. R., Raman, C., Vogels, J. M., and Ketterle, W. 2001. Observation of Vortex Lattices in Bose–Einstein Condensates. *Science*, **292**(5516), 476–479.

[3] Adams, Allan, Carr, Lincoln D., Schäfer, Thomas, Steinberg, Peter, and Thomas, John E. 2012. Strongly Correlated Quantum Fluids: Ultracold Quantum Gases, Quantum Chromodynamic Plasmas and Holographic Duality. *New J. Phys.*, **14**(11), 115009.

[4] Albiez, Michael, Gati, Rudolf, Fölling, Jonas, Hunsmann, Stefan, Cristiani, Matteo, and Oberthaler, Markus K. 2005. Direct Observation of Tunneling and Nonlinear Self-Trapping in a Single Bosonic Josephson Junction. *Phys. Rev. Lett.*, **95**, 010402.

[5] Anderson, M. H., Ensher, J. R., Matthews, M. R., Wieman, C. E., and Cornell, E. A. 1995. Observation of Bose–Einstein Condensation in a Dilute Atomic Vapor. *Science*, **269**(5221), 198–201.

[6] Anderson, P. W. 1958. Random-Phase Approximation in the Theory of Superconductivity. *Phys. Rev.*, **112**, 1900–1916.

[7] Andrews, M. R., Townsend, C. G., Miesner, H.-J., Durfee, D. S., Kurn, D. M., and Ketterle, W. 1997a. Observation of Interference between Two Bose Condensates. *Science*, **275**(5300), 637–641.

[8] Andrews, M. R., Kurn, D. M., Miesner, H.-J., Durfee, D. S., Townsend, C. G., Inouye, S., and Ketterle, W. 1997b. Propagation of Sound in a Bose–Einstein Condensate. *Phys. Rev. Lett.*, **79**, 553–556.

[9] Mazurenko, Anton, Chiu, Christie S., Ji, Geoffrey, Parsons, Maxwell F., Kanász-Nagy, Mrton, Schmidt, Richard, Grusdt, Fabian, Demler, Eugene, Greif, Daniel, and Greiner, Markus. 2017. A Cold-Atom Fermi–Hubbard Antiferromagnet. *Nature*, **545**, 462.

[10] Bacry, H. 1974. Orbits of the Rotation Group on Spin States. *J. Math. Phys.*, **15**(10), 1686.

[11] Bakr, W. S., Peng, A., Tai, M. E., Ma, R., Simon, J., Gillen, J. I., Fölling, S., Pollet, L., and Greiner, M. 2010. Probing the Superfluid to Mott Insulator Transition at the Single-Atom Level. *Science*, **329**(5991), 547–550.

[12] Bardeen, J., Cooper, L. N., and Schrieffer, J. R. 1957. Theory of Superconductivity. *Phys. Rev.*, **108**(Dec), 1175–1204.

[13] Barnett, Ryan, Turner, Ari, and Demler, Eugene. 2006. Classifying Novel Phases of Spinor Atoms. *Phys. Rev. Lett.*, **97**, 180412.

[14] Baumann, Kristian, Guerlin, Christine, Brennecke, Ferdinand, and Esslinger, Tilman. 2010. Dicke Quantum Phase Transition with a Superfluid Gas in an Optical Cavity. *Nature*, **464**(7293), 1301–1306.

[15] Becker, Christoph, Stellmer, Simon, Soltan-Panahi, Parvis, Dörscher, Sören, Baumert, Mathis, Richter, Eva Maria, Kronjäger, Jochen, Bongs, Kai, and Sengstock, Klaus. 2008. Oscillations and Interactions of Dark and Dark-Bright Solitons in Bose-Einstein Condensates. *Nat. Phys.*, **4**(6), 496–501.

[16] Ben Dahan, Maxime, Peik, Ekkehard, Reichel, Jakob, Castin, Yvan, and Salomon, Christophe. 1996. Bloch Oscillations of Atoms in an Optical Potential. *Phys. Rev. Lett.*, **76**(Jun), 4508–4511.

[17] Bergeman, T., Moore, M. G., and Olshanii, M. 2003. Atom-Atom Scattering under Cylindrical Harmonic Confinement: Numerical and Analytic Studies of the Confinement Induced Resonance. *Phys. Rev. Lette.*, **91**(16), 163201.

[18] Bettermann, Oscar, Oppong, Nelson Darkwah, Pasqualetti, Giulio, Riegger, Luis, Bloch, Immanuel, and Fölling, Simon. 2020. *Clock-Line Photoassociation of Strongly Bound Dimers in a Magic-Wavelength Lattice*, arXiv number 2003.10599.

[19] Blackley, Caroline L., Julienne, Paul S., and Hutson, Jeremy M. 2014. Effective-Range Approximations for Resonant Scattering of Cold Atoms. *Phys. Rev. A*, **89**(Apr), 042701.

[20] Boyd, Martin M., Zelevinsky, Tanya, Ludlow, Andrew D., Blatt, Sebastian, Zanon-Willette, Thomas, Foreman, Seth M., and Ye, Jun. 2007. Nuclear Spin Effects in Optical Lattice Clocks. *Phys. Rev. A*, **76**(Aug), 022510.

[21] Braaten, Eric, and Hammer, H.-W. 2006. Universality in Few-Body Systems with Large Scattering Length. *Phys. Rep.*, **428**(5–6), 259–390.

[22] Braaten, Eric, and Hammer, H.-W. 2007. Efimov Physics in Cold Atoms. *Ann. Phys.*, **322**(1), 120–163.

[23] Brown, Peter T., Mitra, Debayan, Guardado-Sanchez, Elmer, Schauß, Peter, Kondov, Stanimir S., Khatami, Ehsan, Paiva, Thereza, Trivedi, Nandini, Huse, David A. and Bakr, Waseem S. 2017. Spin–Imbalance in a 2D Fermi-Hubbard System *Science*, **357**(6358), 1385–1388.

[24] Burdick, Nathaniel Q., Tang, Yijun, and Lev, Benjamin L. 2016. Long-Lived Spin Orbit-Coupled Degenerate Dipolar Fermi Gas. *Phys. Rev. X*, **6**(Aug), 031022.

[25] Campbell, S. L., Hutson, R. B., Marti, G. E., Goban, A., Darkwah Oppong, N. McNally, R. L., Sonderhouse, L., Robinson, J. M., Zhang, W., Bloom, B. J., and Ye, J. 2017. A Fermi-Degenerate Three-Dimensional Optical lattice Clock. *Science* **358**(6359), 90–94.

[26] Cao, C., Elliott, E., Joseph, J., Wu, H., Petricka, J., Schafer, T., and Thomas, J. E 2010. Universal Quantum Viscosity in a Unitary Fermi Gas. *Science*, **331**(6013) 58–61.

[27] Cappellini, G., Mancini, M., Pagano, G., Lombardi, P., Livi, L., Siciliani de Cumis M., Cancio, P., Pizzocaro, M., Calonico, D., Levi, F., Sias, C., Catani, J., Inguscio M., and Fallani, L. 2014. Direct Observation of Coherent Interorbital Spin-Exchange Dynamics. *Phys. Rev. Lett.*, **113**, 120402.

[28] Castin, Yvan, and Dalibard, Jean. 1997. Relative Phase of Two Bose-Einstein Condensates. *Phys. Rev. A*, **55**, 4330–4337.

[29] Chandrasekhar, B. S. 1962. A Note on the Maximum Critical Field of High Field Superconductor. *Appl. Phys. Lett.*, **1**(1), 7–8.

[30] Chang, Ming Shien, Qin, Qishu, Zhang, Wenxian, You, Li, and Chapman, Michael S. 2005. Coherent Spinor Dynamics in a Spin-1 Bose Condensate. *Nat. Phys.*, **1**(2), 111–116.

[31] Chen, Yu, Yu, Zhenhua, and Zhai, Hui. 2014. Superradiance of Degenerate Fermi Gases in a Cavity. *Phys. Rev. Lett.*, **112**(Apr), 143004.

[32] Cheuk, Lawrence W., Sommer, Ariel T., Hadzibabic, Zoran, Yefsah, Tarik, Bakr, Waseem S., and Zwierlein, Martin W. 2012. Spin-Injection Spectroscopy of a Spin-Orbit Coupled Fermi Gas. *Phys. Rev. Lett.*, **109**(Aug), 095302.

[33] Chevy, F. 2006. Universal Phase Diagram of a Strongly Interacting Fermi Gas with Unbalanced Spin Populations. *Phys. Rev. A*, **74**(6), 063628.

[34] Chin, Cheng, Grimm, Rudolf, Julienne, Paul, and Tiesinga, Eite. 2010. Feshbach Resonances in Ultracold Gases. *Rev. Mod. Phys.*, **82**, 1225–1286.

[35] Chiu, Ching-Kai, Teo, Jeffrey C. Y., Schnyder, Andreas P., and Ryu, Shinsei. 2016. Classification of Topological Quantum Matter with Symmetries. *Rev. Mod. Phys.*, **88**(3), 035005.

[36] Choi, Jae-yoon, Kang, Seji, Seo, Sang Won, Kwon, Woo Jin, and Shin, Yong-il. 2013. Observation of a Geometric Hall Effect in a Spinor Bose-Einstein Condensate with a Skyrmion Spin Texture. *Phys. Rev. Lett.*, **111**, 245301.

[37] Clogston, A. M. 1962. Upper Limit for the Critical Field in Hard Superconductors. *Phys. Rev. Lett.*, **9**(Sep), 266–267.

[38] Combescot, R., Recati, A., Lobo, C., and Chevy, F. 2007. Normal State of Highly Polarized Fermi Gases: Simple Many-Body Approaches. *Phys. Rev. Lett.*, **98**(18), 180402.

[39] Cooper, Leon N. 1956. Bound Electron Pairs in a Degenerate Fermi Gas. *Phys. Rev.*, **104**(Nov), 1189–1190.

[40] Cui, Xiaoling, and Zhai, Hui. 2010. Stability of a Fully Magnetized Ferromagnetic State in Repulsively Interacting Ultracold Fermi Gases. *Phys. Rev. A*, **81**, 041602.

[41] Cui, Xiaoling, Lian, Biao, Ho, Tin-Lun, Lev, Benjamin L., and Zhai, Hui. 2013. Synthetic Gauge Field with Highly Magnetic Lanthanide Atoms. *Phys. Rev. A*, **88**(Jul), 011601.

[42] D'Alessio, Luca, Kafri, Yariv, Polkovnikov, Anatoli, and Rigol, Marcos. 2016. From Quantum Chaos and Eigenstate Thermalization to Statistical Mechanics and Thermodynamics. *Adv. Phys.*, **65**(3), 239–362.

[43] Daley, A. J., Pichler, H., Schachenmayer, J., and Zoller, P. 2012. Measuring Entanglement Growth in Quench Dynamics of Bosons in an Optical Lattice. *Phys. Rev. Lett.*, **109**(Jul), 020505.

[44] Davis, K. B., Mewes, M. O., Andrews, M. R., van Druten, N. J., Durfee, D. S., Kurn, D. M., and Ketterle, W. 1995. Bose-Einstein Condensation in a Gas of Sodium Atoms. *Phys. Rev. Lett.*, **75**, 3969–3973.

[45] De Marco, Luigi, Valtolina, Giacomo, Matsuda, Kyle, Tobias, William G., Covey, Jacob P., and Ye, Jun. 2019. A Degenerate Fermi Gas of Polar Molecules. *Science*, **363**(6429), 853–856.

[46] DeMarco, B., and Jin, D. S. 1999. Onset of Fermi Degeneracy in a Trapped Atomic Gas. *Science*, **285**(5434), 1703–1706.

[47] DeMarco, B., Papp, S. B., and Jin, D. S. 2001. Pauli Blocking of Collisions in a Quantum Degenerate Atomic Fermi Gas. *Phys. Rev. Lett.*, **86**, 5409–5412.

[48] Deng, S., Shi, Z.-Y., Diao, P., Yu, Q., Zhai, H., Qi, R., and Wu, H. 2016. Observation of the Efimovian Expansion in Scale-Invariant Fermi Gases. *Science*, **353**(6297), 371–374.

[49] Desbuquois, Rémi, Chomaz, Lauriane, Yefsah, Tarik, Léonard, Julian, Beugnon, Jérôme, Weitenberg, Christof, and Dalibard, Jean. 2012. Superfluid Behaviour of a Two-Dimensional Bose Gas. *Nat. Phys.*, **8**(9), 645–648.

[50] Donadello, Simone, Serafini, Simone, Tylutki, Marek, Pitaevskii, Lev P., Dalfovo, Franco, Lamporesi, Giacomo, and Ferrari, Gabriele. 2014. Observation of Solitonic Vortices in Bose-Einstein Condensates. *Phys. Rev. Lett.*, **113**(6), 065302.

[51] Duca, L., Li, T., Reitter, M., Bloch, I., Schleier-Smith, M., and Schneider, U. 2015. An Aharonov-Bohm Interferometer for Determining Bloch Band Topology. *Science*, **347**(6219), 288–292.

[52] Eagles, D. M. 1969. Possible Pairing without Superconductivity at Low Carrier Concentrations in Bulk and Thin-Film Superconducting Semiconductors. *Phys. Rev.*, **186**(Oct), 456–463.

[53] Endres, Manuel, Fukuhara, Takeshi, Pekker, David, Cheneau, Marc, Schauß, Peter, Gross, Christian, Demler, Eugene, Kuhr, Stefan, and Bloch, Immanuel. 2012. The "Higgs" Amplitude Mode at the Two-Dimensional Superfluid/Mott Insulator Transition. *Nature*, **487**(7408), 454–458.

[54] Fan, Ruihua, Zhang, Pengfei, Shen, Huitao, and Zhai, Hui. 2017. Out-of-Time-Order Correlation for Many-Body Localization. *Sci. Bull.*, **62**(10), 707–711.

[55] Ferlaino, Francesca, and Grimm, Rudolf. 2010. Trend: Forty Years of Efimov Physics: How a Bizarre Prediction Turned into a Hot Topic. *Physics*, **3**(9), 9.

[56] Fölling, Simon, Gerbier, Fabrice, Widera, Artur, Mandel, Olaf, Gericke, Tatjana, and Bloch, Immanuel. 2005. Spatial Quantum Noise Interferometry in Expanding Ultracold Atom Clouds. *Nature*, **434**, 481.

[57] Fulde, Peter, and Ferrell, Richard A. 1964. Superconductivity in a Strong Spin-Exchange Field. *Phys. Rev.*, **135**(Aug), A550–A563.

[58] Gemelke, Nathan, Zhang, Xibo, Hung, Chen Lung, and Chin, Cheng. 2009. In Situ Observation of Incompressible Mott-Insulating Domains in Ultracold Atomic Gases. *Nature*, **460**(7258), 995–998.

[59] Giorgini, Stefano, Pitaevskii, Lev P., and Stringari, Sandro. 2008. Theory of Ultracold Atomic Fermi Gases. *Rev. Mod. Phys.*, **80**(4), 1215–1274.

[60] Girardeau, M. 1960. Relationship between Systems of Impenetrable Bosons and Fermions in One Dimension. *J. Math. Phys.*, **1**(6), 516–523.

[61] Goldman, N., Juzelianas, G., A-hberg, P., and Spielman, I. B. 2014. Light-Induced Gauge Fields for Ultracold Atoms. *Rep. Prog. Phys.*, **77**(12), 126401.

[62] Gorshkov, A. V., Hermele, M., Gurarie, V., Xu, C., Julienne, P. S., Ye, J., Zoller, P., Demler, E., Lukin, M. D., and Rey, A. M. 2010. Two-Orbital SU(N) Magnetism with Ultracold Alkaline-Earth Atoms. *Nat. Phys.*, **6**(4), 289–295.

[63] Greiner, Markus, Mandel, Olaf, Esslinger, Tilman, Hänsch, Theoder W., and Bloch, Immanuel. 2002. Quantum Phase Transition from a Superfluid to a Mott Insulator in a Gas of Utracold Atoms. *Nature*, **415**, 39.

[64] Gutzwiller, Martin C. 1963. Effect of Correlation on the Ferromagnetism of Transition Metals. *Phys. Rev. Lett.*, **10**, 159–162.

[65] Görg, Frederik, Sandholzer, Kilian, Minguzzi, Joaquín, Desbuquois, Rémi, Messer, Michael, and Esslinger, Tilman. 2019. Realization of Density-Dependent Peierls Phases to Engineer Quantized Gauge Fields Coupled to Ultracold Matter. *Nat. Phys.*, **15**(11), 1161–1167.

[66] Hadzibabic, Zoran, Krüger, Peter, Cheneau, Marc, Battelier, Baptiste, and Dalibard, Jean. 2006. Berezinskii-Kosterlitz-Thouless Crossover in a Trapped Atomic Gas. *Nature*, **441**(7097), 1118–1121.

[67] Haldane, F. D. M. 1988. Model for a Quantum Hall Effect without Landau Levels: Condensed-Matter Realization of the "Parity Anomaly." *Phys. Rev. Lett.*, **61**, 2015–2018.

[68] Hart, Russell A., Duarte, Pedro M., Yang, Tsung-Lin, Liu, Xinxing, Paiva, Thereza, Khatami, Ehsan, Scalettar, Richard T., Trivedi, Nandini, Huse, David A., and Hulet, Randall G. 2015. Observation of Antiferromagnetic Correlations in the Hubbard Model with Ultracold Atoms. *Nature*, **519**(7542), 211–214.

[69] Hartnoll, Sean A., Lucas, Andrew, and Sachdev, Subir. 2016. *Holographic Quantum Matter*. MIT Press.

[70] Ho, Tin-Lun. 1998. Spinor Bose Condensates in Optical Traps. *Phys. Rev. Lett.*, **81**, 742–745.

[71] Ho, Tin-Lun, and Ciobanu, C. V. 2000. *The Schrodinger Cat Family in Attractive Bose Gases and Their Interference*, arXiv number cond-mat/0011095.

[72] Ho, Tin-Lun, and Mueller, Erich J. 2004. High Temperature Expansion Applied to Fermions near Feshbach Resonance. *Phys. Rev. Lett.*, **92**(16), 160404.

[73] Ho, Tin-Lun, and Shenoy, V. B. 1996. Local Spin-Gauge Symmetry of the Bose-Einstein Condensates in Atomic Gases. *Phys. Rev. Lett.*, **77**(Sep), 2595–2599.

[74] Ho, Tin-Lun, and Zhang, Shizhong. 2011. Bose-Einstein Condensates with Spin-Orbit Interaction. *Phys. Rev. Lett.*, **107**(Oct), 150403.

[75] Höfer, M., Riegger, L., Scazza, F., Hofrichter, C., Fernandes, D. R., Parish, M. M., Levinsen, J., Bloch, I., and Fölling, S. 2015. Observation of an Orbital Interaction-Induced Feshbach Resonance in ^{173}Yb. *Phys. Rev. Lett.*, **115**, 265302.

[76] Hu, Jiazhong, Urvoy, Alban, Vendeiro, Zachary, Crépel, Valentin, Chen, Wenlan, and Vuleti, Vladan. 2017. Creation of a Bose-Condensed Gas of 87Rb by Laser Cooling. *Science*, **358**(6366), 1078–1080.

[77] Huang, Kerson, and Yang, C. N. 1957. Quantum-Mechanical Many-Body Problem with Hard-Sphere Interaction. *Phys. Rev.*, **105**(Feb), 767–775.

[78] Hugenholtz, N. M., and Pines, D. 1959. Ground-State Energy and Excitation Spectrum of a System of Interacting Bosons. *Phys. Rev.*, **116**(Nov), 489–506.

[79] Inouye, S., Andrews, M. R., Stenger, J., Miesner, H.-J., Stamper-Kurn, Dan M., and Ketterle, W. 1998. Observation of Feshbach Resonances in a Bose-Einstein Condensate. *Nature*, **392**(6672), 151–154.

[80] Islam, Rajibul, Ma, Ruichao, Preiss, Philipp M., Eric Tai, M., Lukin, Alexander, Rispoli, Matthew, and Greiner, Markus. 2015. Measuring Entanglement Entropy in a Quantum Many-Body System. *Nature*, **528**(7580), 77–83.

[81] Ji, Si-Cong, Zhang, Jin-Yi, Zhang, Long, Du, Zhi-Dong, Zheng, Wei, Deng, You-Jin, Zhai, Hui, Chen, Shuai, and Pan, Jian-Wei. 2014. Experimental Determination of the Finite-Temperature Phase Diagram of a Spin-Orbit Coupled Bose Gas. *Nat. Phys.*, **10**(4), 314–320.

[82] Ji, Si-Cong, Zhang, Long, Xu, Xiao-Tian, Wu, Zhan, Deng, Youjin, Chen, Shuai, and Pan, Jian-Wei. 2015. Softening of Roton and Phonon Modes in a Bose-Einstein Condensate with Spin-Orbit Coupling. *Phys. Rev. Lett.*, **114**(10), 105301.

[83] Jiang, Yuzhu, Qi, Ran, Shi, Zhe-Yu, and Zhai, Hui. 2017. Vortex Lattices in the Bose-Fermi Superfluid Mixture. *Phys. Rev. Lett.*, **118**(8), 080403.

[84] Jin, D. S., Matthews, M. R., Ensher, J. R., Wieman, C. E., and Cornell, E. A. 1997. Temperature-Dependent Damping and Frequency Shifts in Collective Excitations of a Dilute Bose-Einstein Condensate. *Phys. Rev. Lett.*, **78**, 764–767.

[85] Jotzu, Gregor, Messer, Michael, Desbuquois, Rémi, Lebrat, Martin, Uehlinger, Thomas, Greif, Daniel, and Esslinger, Tilman. 2014. Experimental Realization of the Topological Haldane Model with Ultracold Fermions. *Nature*, **515**(7526), 237–240.

[86] Huang, K. 1963. Chap. 10 of: *Statistical Mechanics*. Braun-Brumfield.

[87] Katz, N., Steinhauer, J., Ozeri, R., and Davidson, N. 2002. Beliaev Damping of Quasiparticles in a Bose-Einstein Condensate. *Phys. Rev. Lett.*, **89**, 220401.

[88] Kaufman, A. M., Tai, M. E., Lukin, A., Rispoli, M., Schittko, R., Preiss, P. M., and Greiner, M. 2016. Quantum Thermalization through Entanglement in an Isolated Many-Body System. *Science*, **353**(6301), 794–800.

[89] Keeling, J., Bhaseen, M. J., and Simons, B. D. 2014. Fermionic Superradiance in a Transversely Pumped Optical Cavity. *Phys. Rev. Lett.*, **112**(Apr), 143002.

[90] Kinoshita, Toshiya, Wenger, Trevor, and Weiss, David S. 2004. Observation of a One-Dimensional Tonks-Girardeau Gas. *Science*, **305**(5687), 1125–1128.

[91] Kohn, W. 1959. Analytic Properties of Bloch Waves and Wannier Functions. *Phys. Rev.*, **115**(Aug), 809–821.

[92] Kolkowitz, S., Pikovski, I., Langellier, N., Lukin, M. D., Walsworth, R. L., and Ye, J. 2016. Gravitational Wave Detection with Optical Lattice Atomic Clocks. *Phys. Rev. D*, **94**(Dec), 124043.

[93] Konotop, Vladimir V., and Pitaevskii, Lev. 2004. Landau Dynamics of a Grey Soliton in a Trapped Condensate. *Phys. Rev. Lett.*, **93**, 240403.

[94] Kraemer, T., Mark, M., Waldburger, P., Danzl, J. G., Chin, C., Engeser, B., Lange, A. D., Pilch, K., Jaakkola, A., Nägerl, H.-C., et al. 2006. Evidence for Efimov Quantum States in an Ultracold Gas of Caesium Atoms. *Nature*, **440**(7082), 315–318.

[95] Krinner, Sebastian, Stadler, David, Husmann, Dominik, Brantut, Jean-Philippe, and Esslinger, Tilman. 2014. Observation of Quantized Conductance in Neutral Matter. *Nature*, **517**(7532), 64–67.

[96] Krinner, Sebastian, Esslinger, Tilman, and Brantut, Jean-Philippe. 2017. Two-Terminal Transport Measurements with Cold Atoms. *J. Phys. Condensed Matter*, **29**(34), 343003.

[97] Ku, M. J. H., Sommer, A. T., Cheuk, L. W., and Zwierlein, M. W. 2012. Revealing the Superfluid Lambda Transition in the Universal Thermodynamics of a Unitary Fermi Gas. *Science*, **335**(6068), 563–567.

[98] Ku, Mark J. H., Ji, Wenjie, Mukherjee, Biswaroop, Guardado-Sanchez, Elmer, Cheuk, Lawrence W., Yefsah, Tarik, and Zwierlein, Martin W. 2014. Motion of a Solitonic Vortex in the BEC-BCS Crossover. *Phys. Rev. Lett.*, **113**(6), 065301.

[99] Köhl, Michael, Moritz, Henning, Stöferle, Thilo, Günter, Kenneth, and Esslinger, Tilman. 2005. Fermionic Atoms in a Three Dimensional Optical Lattice: Observing Fermi Surfaces, Dynamics, and Interactions. *Phys. Rev. Lett.*, **94**(8), 080403.

[100] Landau, L. D., and Lifshitz, E. M. 1958. Chap. 139 of: *Fluid Mechanics: Vol. 6 of Course of Theoretical Physics*. Butterworth-Heinemann.

[101] Landau, L. D., and Lifshitz, E. M. 1980. *Statistical Physics, Part I*. Butterworth-Heinemann.

[102] Landau, L. D., Lifshitz, E. M., Sykes, J. B., Bell, J. S., and Rose, M. E. 1958. Chap. 36 of: *Quantum Mechanics, Non-Relativistic Theory: Vol. 3 of Course of Theoretical Physics*. Butterworth-Heinemann.

[103] Larkin, A. I., and Ovchinnikov, Yu. N. 1965. Inhomogeneous State of Superconductors. *Sov. Phys. JETP*, **20**, 762.

[104] Law, C. K., Pu, H., and Bigelow, N. P. 1998. Quantum Spins Mixing in Spinor Bose-Einstein Condensates. *Phys. Rev. Lett.*, **81**(Dec), 5257–5261.

[105] Lee, T. D., Huang, Kerson, and Yang, C. N. 1957. Eigenvalues and Eigenfunctions of a Bose System of Hard Spheres and Its Low-Temperature Properties. *Phys. Rev.*, **106**(Jun), 1135–1145.

[106] Leggett, A. J. 1980. *Diatomic Molecules and Cooper Pairs*. Lecture Notes in Physics 115. Springer.

[107] Li, Jun-Ru, Lee, Jeongwon, Huang, Wujie, Burchesky, Sean, Shteynas, Boris, Top, Furkan Çar, Jamison, Alan O., and Ketterle, Wolfgang. 2017. A Stripe Phase with Supersolid Properties in Spin-Orbit-Coupled Bose-Einstein Condensates. *Nature*, **543**(7643), 91–94.

[108] Li, Xiaoke, Zhu, Bing, He, Xiaodong, Wang, Fudong, Guo, Mingyang, Xu, Zhi-Fang, Zhang, Shizhong, and Wang, Dajun. 2015. Coherent Heteronuclear Spin Dynamics in an Ultracold Spinor Mixture. *Phys. Rev. Lett.*, **114**, 255301.

[109] Li, Yun, Pitaevskii, Lev P., and Stringari, Sandro. 2012. Quantum Tricriticality and Phase Transitions in Spin-Orbit Coupled Bose-Einstein Condensates. *Phys. Rev. Lett.*, **108**(22), 225309.

[110] Lian, Biao, Ho, Tin-Lun, and Zhai, Hui. 2012. Searching for Non-Abelian Phases in the Bose-Einstein Condensate of Dysprosium. *Phys. Rev. A*, **85**(5), 051606.

[111] Lieb, Elliott H., and Liniger, Werner. 1963. Exact Analysis of an Interacting Bose Gas. I. The General Solution and the Ground State. *Phys. Rev.*, **130**(May), 1605–1616.

[112] Lin, Y.-J., Compton, R. L., Perry, A. R., Phillips, W. D., Porto, J. V., and Spielman, I. B. 2009a. Bose-Einstein Condensate in a Uniform Light-Induced Vector Potential. *Phys. Rev. Lett.*, **102**(Mar), 130401.

[113] Lin, Y. J., Compton, R. L., Jiménez-García, K., Porto, J. V., and Spielman, I. B. 2009b. Synthetic Magnetic Fields for Ultracold Neutral Atoms. *Nature*, **462**(7273), 628–632.

[114] Lin, Y. J., Jiménez-García, K., and Spielman, I. B. 2011a. Spin-Orbit-Coupled Bose–Einstein Condensates. *Nature*, **471**(03), 83.

[115] Lin, Y.-J., Compton, R. L., Jiménez-García, K., Phillips, W. D., Porto, J. V., and Spielman, I. B. 2011b. A Synthetic Electric Force Acting on Neutral Atoms. *Nat. Phys.*, **7**(7), 531–534.

[116] Madison, K. W., Chevy, F., Wohlleben, W., and Dalibard, J. 2000. Vortex Formation in a Stirred Bose-Einstein Condensate. *Phys. Rev. Lett.*, **84**, 806–809.

[117] Majorana, Ettore. 1932. Atomi orientati in campo magnetico variabile. *Il Nuovo Cimento (1924–1942)*, **9**(2), 43–50.

[118] Mermin, N. D. 1979. The Topological Theory of Defects in Ordered Media. *Rev. Mod. Phys.*, **51**, 591–648.

[119] Mermin, N. D., and Ho, Tin-Lun. 1976. Circulation and Angular Momentum in the *A* Phase of Superfluid Helium-3. *Phys. Rev. Lett.*, **36**(Mar), 594–597.

[120] Miller, D. E., Chin, J. K., Stan, C. A., Liu, Y., Setiawan, W., Sanner, C., and Ketterle, W. 2007. Critical Velocity for Superfluid Flow across the BEC-BCS Crossover. *Phys. Rev. Lett.*, **99**, 070402.

[121] Mitra, Debayan, Brown, Peter T., Guardado-Sanchez, Elmer, Kondov, Stanimir S., Devakul, Trithep, Huse, David A., Schauß, Peter, and Bakr, Waseem S. 2017. Quantum Gas Microscopy of an Attractive Fermi-Hubbard System. *Nat. Phys.*, **14**(2), 173–177.

[122] Mueller, Erich J., and Ho, Tin-Lun. 2002. Two-Component Bose-Einstein Condensates with a Large Number of Vortices. *Phys. Rev. Lett.*, **88**, 180403.

[123] Mueller, Erich J., Ho, Tin-Lun, Ueda, Masahito, and Baym, Gordon. 2006. Fragmentation of Bose-Einstein Condensates. *Phys. Rev. A*, **74**, 033612.

[124] Nandkishore, Rahul, and Huse, David A. 2015. Many-Body Localization and Thermalization in Quantum Statistical Mechanics. *Ann. Rev. Condensed Matter Phys.*, **6**(1), 15–38.

[125] Navon, N., Nascimbène, S., Chevy, F., and Salomon, C. 2010. The Equation of State of a Low-Temperature Fermi Gas with Tunable Interactions. *Science*, **328**(5979), 729–732.

[126] Navon, Nir, Piatecki, Swann, Günter, Kenneth, Rem, Benno, Nguyen, Trong Canh, Chevy, Frédéric, Krauth, Werner, and Salomon, Christophe. 2011. Dynamics and Thermodynamics of the Low-Temperature Strongly Interacting Bose Gas. *Phys. Rev. Lett.*, **107**(13), 135301.

[127] Ni, K. K., Ospelkaus, S., De Miranda, M. H.G., Pe'er, A., Neyenhuis, B., Zirbel, J. J., Kotochigova, S., Julienne, P. S., Jin, D. S., and Ye, J. 2008. A High Phase-Space-Density Gas of Polar Molecules. *Science*, **322**(5899), 231–235.

[128] Nozieres, P., and Schmitt-Rink, S. 1985. Bose Condensation in an Attractive Fermion Gas: From Weak to Strong Coupling Superconductivity. *J. Low. Temp. Phys.*, **59**, 195.

[129] O'Hara, K. M., Hemmer, S. L., Gehm, M. E., Granade, S. R., and Thomas, J. E. 2002. Observation of a Strongly Interacting Degenerate Fermi Gas of Atoms. *Science*, **298**(5601), 2179–2182.

[130] Ohmi, Tetsuo, and Machida, Kazushige. 1998. Bose-Einstein Condensation with Internal Degrees of Freedom in Alkali Atom Gases. *J. Phys. Soc. Japan*, **67**(6), 1822–1825.

[131] Olshanii, M. 1998. Atomic Scattering in the Presence of an External Confinement and a Gas of Impenetrable Bosons. *Phys. Rev. Lett.*, **81**, 938–941.

[132] Onofrio, R., Durfee, D. S., Raman, C., Köhl, M., Kuklewicz, C. E., and Ketterle, W. 2000. Surface Excitations of a Bose-Einstein Condensate. *Phys. Rev. Lett.*, **84**, 810–813.

[133] Pagano, G., Mancini, M., Cappellini, G., Livi, L., Sias, C., Catani, J., Inguscio, M., and Fallani, L. 2015. Strongly Interacting Gas of Two-Electron Fermions at an Orbital Feshbach Resonance. *Phys. Rev. Lett.*, **115**, 265301.

[134] Paredes, Belén, Widera, Artur, Murg, Valentin, Mandel, Olaf, Fölling, Simon, Cirac, Ignacio, Shlyapnikov, Gora V., Hänsch, Theodor W., and Bloch, Immanuel. 2004. Tonks–Girardeau Gas of Ultracold Atoms in an Optical Lattice. *Nature*, **429**(6989), 277–281.

[135] Parish, Meera M. 2014. The BCS-BEC Crossover. *Quantum Gas Experiments*, September, 179–197.

[136] Parker, Colin V., Ha, Li-Chung, and Chin, Cheng. 2013. Direct Observation of Effective Ferromagnetic Domains of Cold Atoms in a Shaken Optical Lattice. *Nature Physics*, **9**(12), 769–774.

[137] Pérez-García, Víctor M., Michinel, H., Cirac, J. I., Lewenstein, M., and Zoller, P. 1996. Low Energy Excitations of a Bose-Einstein Condensate: A Time-Dependent Variational Analysis. *Phys. Rev. Lett.*, **77**(Dec), 5320–5323.

[138] Pethick, C. J., and Smith, H. 2008. Chap. 4 of: *Bose-Einstein Condensation in Dilute Gases*. Cambridge University Press.

[139] Pethick, C. J., and Smith, H. 2008. Chap. 14 of: *Bose-Einstein Condensation in Dilute Gases*. Cambridge University Press.

[140] Petrich, Wolfgang, Anderson, Michael H., Ensher, Jason R., and Cornell, Eric A. 1995. Stable, Tightly Confining Magnetic Trap for Evaporative Cooling of Neutral Atoms. *Phys. Rev. Lett.*, **74**(Apr), 3352–3355.

[141] Petrov, D. S. 2012. *The Few-Atom Problem*, arXiv 1206.5752.

[142] Piazza, Francesco, and Strack, Philipp. 2014. Umklapp Superradiance with a Collisionless Quantum Degenerate Fermi Gas. *Phys. Rev. Lett.*, **112**(Apr), 143003.

[143] Pires, R., Ulmanis, J., Häfner, S., Repp, M., Arias, A., Kuhnle, E. D., and Wei-demüller, M. 2014. Observation of Efimov Resonances in a Mixture with Extreme Mass Imbalance. *Phys. Rev. Lett.*, **112**(25), 250404.

[144] Pitaevskii, L. P., and Stringari, S. 1997. Landau Damping in Dilute Bose Gases. *Phys. Lett. A*, **235**(4), 398–402.

[145] Qi, Xiao-Liang, and Zhang, Shou-Cheng. 2011. Topological Insulators and Super-conductors. *Rev. Mod. Phys.*, **83**(Oct), 1057–1110.

[146] Raman, C., Köhl, M., Onofrio, R., Durfee, D. S., Kuklewicz, C. E., Hadzibabic, Z., and Ketterle, W. 1999. Evidence for a Critical Velocity in a Bose-Einstein Condensed Gas. *Phys. Rev. Lett.*, **83**, 2502–2505.

[147] Ray, M. W., Ruokokoski, E., Kandel, S., Möttönen, M., and Hall, D. S. 2014. Observation of Dirac Monopoles in a Synthetic Magnetic Field. *Nature*, **505**(7485), 657–660.

[148] Regal, C. A., Greiner, M., and Jin, D. S. 2004. Observation of Resonance Conden-sation of Fermionic Atom Pairs. *Phys. Rev. Lett.*, **92**, 040403.

[149] Riegger, L., Darkwah Oppong, N., Höfer, M., Fernandes, D. R., Bloch, I., and Fölling, S. 2018. Localized Magnetic Moments with Tunable Spin Exchange in a Gas of Ultracold Fermions. *Phys. Rev. Lett.*, **120**(14), 143601.

[150] Ritsch, Helmut, Domokos, Peter, Brennecke, Ferdinand, and Esslinger, Tilman. 2013. Cold Atoms in Cavity-Generated Dynamical Optical Potentials. *Rev. Mod. Phys.*, **85**(2), 553–601.

[151] Sá de Melo, C. A. R., Randeria, Mohit, and Engelbrecht, Jan R. 1993. Crossover from BCS to Bose superconductivity: Transition Temperature and Time-Dependent Ginzburg-Landau Theory. *Phys. Rev. Lett.*, **71**(Nov), 3202–3205.

[152] Scazza, F., Hofrichter, C., Höfer, M., De Groot, P. C., Bloch, I., and Fölling, S. 2014. Observation of Two-Orbital Spin-Exchange Interactions with Ultracold SU(N)-Symmetric Fermions. *Nat. Phys.*, **10**(10), 779–784.

[153] Schirotzek, André, Wu, Cheng-Hsun, Sommer, Ariel, and Zwierlein, Martin W. 2009. Observation of Fermi Polarons in a Tunable Fermi Liquid of Ultracold Atoms. *Phys. Rev. Lett.*, **102**(Jun), 230402.

[154] Schreiber, M., Hodgman, S. S., Bordia, P., Luschen, H. P., Fischer, M. H., Vosk, R., Altman, E., Schneider, U., and Bloch, I. 2015. Observation of Many-Body Localiza-tion of Interacting Fermions in a Quasirandom Optical Lattice. *Science*, **349**(6250), 842–845.

[155] Schunck, Christian H., Shin, Yong Il, Schirotzek, André, and Ketterle, Wolf-gang. 2008. Determination of the Fermion Pair Size in a Resonantly Interacting Superfluid. *Nature*, **454**(7205), 739–743.

[156] Schweizer, Christian, Grusdt, Fabian, Berngruber, Moritz, Barbiero, Luca, Dem-ler, Eugene, Goldman, Nathan, Bloch, Immanuel, and Aidelsburger, Monika. 2019. Floquet Approach to Z2 Lattice Gauge Theories with Ultracold Atoms in Optical Lattices. *Nat. Phys.*, **15**(11), 1168–1173.

[157] Seaton, M. J. 1983. Quantum Defect Theory. *Rep. Progr. Phys.*, **46**(2), 167–257.

[158] Sherson, Jacob F., Weitenberg, Christof, Endres, Manuel, Cheneau, Marc, Bloch, Immanuel, and Kuhr, Stefan. 2010. Single-Atom-Resolved Fluorescence Imaging of an Atomic Mott Insulator. *Nature*, **467**(7311), 68–72.

[159] Shin, Yong-il, Schunck, Christian H., Schirotzek, André, and Ketterle, Wolfgang. 2008. Phase Diagram of a Two-Component Fermi Gas with Resonant Interactions. *Nature*, **451**(7179), 689–693.

[160] Solano, Pablo, Duan, Yiheng, Chen, Yu-Ting, Rudelis, Alyssa, Chin, Cheng, and Vuleti, Vladan. 2019. Strongly Correlated Quantum Gas Prepared by Direct Laser Cooling. *Phys. Rev. Lett.*, **123**(17), 173401.

[161] Son, D. T. 2007. Vanishing Bulk Viscosities and Conformal Invariance of the Unitary Fermi Gas. *Phys. Rev. Lett.*, **98**(2), 020406.

[162] Son, D. T. 2008. Toward an AdS/Cold Atoms Correspondence: A Geometric Realization of the Schrödinger Symmetry. *Phys. Rev. D*, **78**(Aug), 046003.

[163] Stamper-Kurn, D. M., Chikkatur, A. P., Görlitz, A., Inouye, S., Gupta, S., Pritchard, D. E., and Ketterle, W. 1999. Excitation of Phonons in a Bose-Einstein Condensate by Light Scattering. *Phys. Rev. Lett.*, **83**, 2876–2879.

[164] Stamper-Kurn, Dan M., and Ueda, Masahito. 2013. Spinor Bose Gases: Symmetries, Magnetism, and Quantum Dynamics. *Rev. Mod. Phys.*, **85**, 1191–1244.

[165] Steinhauer, J., Ozeri, R., Katz, N., and Davidson, N. 2002. Excitation Spectrum of a Bose-Einstein Condensate. *Phys. Rev. Lett.*, **88**, 120407.

[166] Stewart, J. T., Gaebler, J. P., and Jin, D. S. 2008. Using Photoemission Spectroscopy to Probe a Strongly Interacting Fermi Gas. *Nature*, **454**, 744.

[167] Stringari, S. 1996. Collective Excitations of a Trapped Bose-Condensed Gas. *Phys. Rev. Lett.*, **77**(Sep), 2360–2363.

[168] Struck, J., Olschlager, C., Le Targat, R., Soltan-Panahi, P., Eckardt, A., Lewenstein, M., Windpassinger, P., and Sengstock, K. 2011. Quantum Simulation of Frustrated Classical Magnetism in Triangular Optical Lattices. *Science*, **333**(6045), 996–999.

[169] Struck, J., Ölschläger, C., Weinberg, M., Hauke, P., Simonet, J., Eckardt, A., Lewenstein, M., Sengstock, K., and Windpassinger, P. 2012. Tunable Gauge Potential for Neutral and Spinless Particles in Driven Optical Lattices. *Physical Review Letters*, **108**(22), 225304.

[170] Su, W. P., Schrieffer, J. R., and Heeger, A. J. 1980. Soliton Excitations in Polyacetylene. *Phys. Rev. B*, **22**(Aug), 2099–2111.

[171] Tan, Shina. 2008. Energetics of a Strongly Correlated Fermi Gas. *Ann. Phys.*, **323**(12), 2952–2970.

[172] Tarnowski, Matthias, Ünal, F. Nur, Fläschner, Nick, Rem, Benno S., Eckardt, André, Sengstock, Klaus, and Weitenberg, Christof. 2019. Measuring Topology from Dynamics by Obtaining the Chern Number from a Linking Number. *Nat. Commun.*, **10**(1), 1728.

[173] Tarruell, Leticia, Greif, Daniel, Uehlinger, Thomas, Jotzu, Gregor, and Esslinger, Tilman. 2012. Creating, Moving and Merging Dirac Points with a Fermi Gas in a Tunable Honeycomb Lattice. *Nature*, **483**(7389), 302–305.

[174] Tonks, Lewi. 1936. The Complete Equation of State of One, Two and Three-Dimensional Gases of Hard Elastic Spheres. *Phys. Rev.*, **50**(Nov), 955–963.

[175] Trotzky, S., Pollet, L., Gerbier, F., Schnorrberger, U., Bloch, I., Prokofev, N. V., Svistunov, B., and Troyer, M. 2010. Suppression of the Critical Temperature for Superfluidity near the Mott Transition. *Nat. Phys.*, **6**(12), 998–1004.

[176] Tung, Shih-Kuang, Jiménez-García, Karina, Johansen, Jacob, Parker, Colin V., and Chin, Cheng. 2014. Geometric Scaling of Efimov States in a Li6-Cs133 Mixture. *Phys. Rev. Lett.*, **113**(24), 240402.

[177] Vincent Liu, W. 1997. Theoretical Study of the Damping of Collective Excitations in a Bose-Einstein Condensate. *Phys. Rev. Lett.*, **79**(21), 4056–4059.

[178] Vitanov, Nikolay V., Rangelov, Andon A., Shore, Bruce W., and Bergmann, Klaas. 2017. Stimulated Raman Adiabatic Passage in Physics, Chemistry, and Beyond. *Rev. Mod. Phys.*, **89**, 015006.

[179] Wang, Ce, Zhang, Pengfei, Chen, Xin, Yu, Jinlong, and Zhai, Hui. 2017. Scheme to Measure the Topological Number of a Chern Insulator from Quench Dynamics. *Phys. Rev. Lett.*, **118**(May), 185701.

[180] Wang, Chunji, Gao, Chao, Jian, Chao-Ming, and Zhai, Hui. 2010. Spin-Orbit Coupled Spinor Bose-Einstein Condensates. *Phys. Rev. Lett.*, **105**(16), 160403.

[181] Wang, Pengjun, Yu, Zeng-Qiang, Fu, Zhengkun, Miao, Jiao, Huang, Lianghui, Chai, Shijie, Zhai, Hui, and Zhang, Jing. 2012. Spin-Orbit Coupled Degenerate Fermi Gases. *Phys. Rev. Lett.*, **109**(Aug), 095301.

[182] Xiao, Di, Chang, Ming-Che, and Niu, Qian. 2010. Berry Phase Effects on Electronic Properties. *Rev. Mod. Phys.*, **82**(Jul), 1959–2007.

[183] Yang, C. N. 1962. Concept of Off-Diagonal Long-Range Order and the Quantum Phases of Liquid He and of Superconductors. *Rev. Mod. Phys.*, **34**(Oct), 694–704.

[184] Yang, C. N. 1967. Some Exact Results for the Many-Body Problem in One Dimension with Repulsive Delta-Function Interaction. *Phys. Rev. Lett.*, **19**(Dec), 1312–1315.

[185] Yang, C. N., and Zhang, S. C. 1990. SO(4) Symmetry in a Hubbard Model. *Mod. Phys. Lett. B*, **4**(11), 759–766.

[186] Zhai, Hui. 2015. Degenerate Quantum Gases with Spin-Orbit Coupling: A Review. *Rep. Prog. Phys.*, **78**(2), 026001.

[187] Zhang, Chen, and Greene, Chris H. 2013. Quasi-One-Dimensional Scattering with General Transverse Two-Dimensional Confinement. *Phys. Rev. A*, **88**(1), 012715.

[188] Zhang, Jian, and Zhai, Hui. 2005. Vortex Lattices in Planar Bose-Einstein Condensates with Dipolar Interactions. *Phys. Rev. Lett.*, **95**(20), 200403.

[189] Zhang, Ren, Cheng, Yanting, Zhai, Hui, and Zhang, Peng. 2015. Orbital Feshbach Resonance in Alkali-Earth Atoms. *Phys. Rev. Lett.*, **115**, 135301.

[190] Zhang, Ren, Zhang, Deping, Cheng, Yanting, Chen, Wei, Zhang, Peng, and Zhai, Hui. 2016. Kondo Effect in Alkaline-Earth-Metal Atomic Gases with Confinement Induced Resonances. *Phys. Rev. A*, **93**, 043601.

[191] Zhang, Ren, Cheng, Yanting, Zhang, Peng, and Zhai, Hui. 2020. Controlling the Interaction of Ultracold Alkaline-Earth Atoms. *Nat. Rev. Phys.*, **2**(4), 213–220.

[192] Zhang, Shizhong, and Leggett, Anthony J. 2009. Universal Properties of the Ultracold Fermi Gas. *Phys. Rev. A*, **79**(Feb), 023601.

[193] Zhang, X., Bishof, M., Bromley, S. L., Kraus, C. V., Safronova, M. S., Zoller, P., Rey, A. M., and Ye, J. 2014. Spectroscopic Observation of SU(N)-Symmetric Interactions in Sr Orbital Magnetism. *Science*, **345**(6203), 1467–1473.

[194] Zheng, Wei, and Zhai, Hui. 2014. Floquet Topological States in Shaking Optical Lattices. *Phys. Rev. A*, **89**, 061603.

[195] Zheng, Wei, Yu, Zeng-Qiang, Cui, Xiaoling, and Zhai, Hui. 2013. Properties of Bose Gases with the Raman-Induced Spin-Orbit Coupling. *J. Phys. B*, **46**(13), 134007.

[196] Zwierlein, M. W., Abo-Shaeer, J. R., Schirotzek, A., Schunck, C. H., and Ketterle, W. 2005. Vortices and Superfluidity in a Strongly Interacting Fermi Gas. *Nature*, **435**(7045), 1047–1051.

[197] Zürn, G., Wenz, A. N., Murmann, S., Bergschneider, A., Lompe, T., and Jochim, S. 2013. Pairing in Few-Fermion Systems with Attractive Interactions. *Phys. Rev. Lett.*, **111**(17), 175302.

Index